MW01487583

ANTHROPOLOGY
and the
Study of
HUMANITY

Scott M. Lacy, Ph.D.

THE
GREAT
COURSES®

PUBLISHED BY:

THE GREAT COURSES
Corporate Headquarters
4840 Westfields Boulevard, Suite 500
Chantilly, Virginia 20151-2299
Phone: 1-800-832-2412
Fax: 703-378-3819
www.thegreatcourses.com

Scott M. Lacy, Ph.D.

Associate Professor of Sociology and Anthropology
Fairfield University

S cott M. Lacy is an Associate Professor of Sociology and Anthropology at Fairfield University in Connecticut, where he teaches anthropology, environmental studies, and black studies courses. He is also the founder and executive director of African Sky, a nonprofit organization that serves hardworking farm families in rural Mali, West Africa. Dr. Lacy earned his Ph.D. in Anthropology at the University of California, Santa Barbara, where he started his teaching career in the Department of Black Studies as University of California President's Faculty

Fellow. Prior to arriving at his current post at Fairfield University, he taught in the Department of Anthropology at Emory University during his tenure as Marjorie Shostak Lecturer. Dr. Lacy's research interests include cross-cultural knowledge production, food systems, intellectual property rights associated with seed, and the anthropology of happiness.

Dr. Lacy has worked in Mali since 1994, when he first served in the Peace Corps. Since then, he has partnered with family farmers, teachers, community leaders, plant scientists, engineers, and a host of other knowledge specialists in Mali and throughout the world. A two-time Fulbright Scholar (in Mali from 2001–2002 and in Cameroon from 2016–2017), Dr. Lacy has presented his work as a consultant and/or keynote speaker for Engineers Without Borders, the Peace Corps, the Materials Research Society, ICRISAT Mali, the Institut d'Economie Rurale (Bamako, Mali), the Guangxi Maize Research Institute (Nanning, China), the D80 Conference, the Massachusetts Institute of Technology, and Columbia University. Dr. Lacy is working on a book manuscript that chronicles more than two decades of friendship and collaboration in southern Mali. His nonprofit and academic work has been featured in two major documentaries: *Sustaining Life* by Sprint Features (nominated for a 2009 Academy Award) and the 2017 release *Nyogonfe: Together*.

Dr. Lacy was awarded a Certificate of Congressional Recognition and Achievement from the US House of Representatives in 2011, the same year he was the inaugural awardee for Otterbein University's Global and Intercultural Engagement Award. He received a Martin Luther King, Jr. Humanitarian Award from Mothers On a Mission International and Strategic Solutions Group. Since his first years as a teacher, Dr. Lacy has received numerous teaching awards and grants from the University of California, Emory University, and Fairfield University. As an innovative instructor, he has mentored Institute for Developing Nations scholars (in Mali, Uganda, and Panama) through Emory University and The Carter Center. His applied teaching has resulted in a number of student projects, including documentary films, student research conferences, a West African drumming group, and a student-managed coffee cart that serves as an outpost for promoting sustainability.

Dr. Lacy is a coauthor of two popular textbooks, *Applying Anthropology* and *Applying Cultural Anthropology*, both published by McGraw-Hill. He has published a number of book chapters and articles that document cross-cultural knowledge production in agriculture, community development, engineering, and even nanotechnology. Dr. Lacy has appeared as a panelist on HuffPost Live and has been featured in newspapers and magazines, including *Sports Illustrated*. He is also known for his batik artwork, including one piece that toured the country from 2006 to 2009 as part of a traveling Smithsonian Folklife Festival exhibit celebrating the US National Park Service. ■

Table of Contents

Unit 2: Sole Survivors

Unit 3: Human Diversity

Unit 4: Applying Anthropology

Anthropology and the Study of Humanity

Anthropology is an interdisciplinary field that is uniquely positioned to answer some of humanity's biggest questions: Who are we? Where do we come from? And how could it be that, despite our seemingly limitless physical and cultural diversity, we are indeed a single human race? Anthropologists hold no monopoly on truth or explanations, but they do employ a wide range of methods to explore the remarkable breadth of the human condition.

This course is an introduction to academic anthropology and its 4 subfields: biological anthropology, archaeology, linguistics, and sociocultural anthropology. Over the course of 24 lectures, we will learn how anthropology and its subfields further our understanding of our world and ourselves. Specifically, we'll see how anthropologists deploy multidisciplinary methods to trace the origins of our species as well as the development of religion, agriculture, money, language, and many other pillars of the modern human experience.

Our anthropological journey is organized into 4 units. First, we begin with biological anthropology to address the question: Who are we, and where do we come from? Specifically, unit 1 focuses on biological anthropology, which studies the origins of humanity, primatology, the spread of humankind, and a re-articulation of the concept of race.

In unit 2, we move to the question of our status as the sole remaining survivors of a long line of upright walking apes. In particular, we bring in archaeology and linguistics to work out how *Homo sapiens* outlived all the other branches of our extended family tree. This exploration reveals how tools, agriculture, cities, money, and language all contributed to the survival of our species.

In unit 3, we turn to cultural anthropology to explore why people and cultures are so diverse despite our singularity as a species. We'll review the history and methods of cultural anthropology and see differences in the way people throughout the world practice and understand core pieces of humanity, including family, marriage, gender, sexuality, religion, and artistic expression.

Finally, in unit 4, we'll apply all 4 subfields of anthropology to see how this interdisciplinary approach helps us understand and work out human problems. We'll see anthropologists in action as they examine conflict, forensics, health, economic development, ecology, and even the nature and pursuit of happiness.

In sum, we will discover that anthropology digs deep into the geographic, temporal, and biological diversity of humankind to help us understand our remarkable diversity as a species. And ironically, the deeper we dig, the more we reveal the oneness of the human race.

Why Anthropology Matters

I t's helpful to start any study of anthropology with a definition of what anthropology is. Put simply, anthropology is the study of humankind over time and space. American Anthropology is categorized into 4 subfields: biological anthropology, archaeology, linguistics, and cultural anthropology. We use each of these subfields to explore the survival and diversity of humankind across time and geography. This opening lecture will launch our anthropological journey by describing what anthropology is and how it produces knowledge that matters.

Subfields

The first subfield, biological anthropology, includes everything from primatology and paleontology to evolution, biology, genetics, health, and forensic science. All of those themes will be covered in the early lectures of the course, as we answer the big questions: Who are we, and where do we come from?

The next subfield, archaeology, uncovers and interprets artifacts to reveal the histories of people who are no longer here to share their stories. Like crime scene detectives, we can look at what earlier humans left behind to work out how they lived and died.

The third subfield, linguistic anthropology, is much more than studying languages. Linguists dissect the structure of language; preserve and investigate dead languages; tell language histories; and provide a record of human migrations and cultural interaction.

Finally, cultural anthropology is an interdisciplinary subfield that explores kinship, economics, gender, development, religion, art, and just about anything else we humans do. Through cultural anthropology, we'll explore how, despite cultural and linguistic differences, we are a single human race.

While anthropologists typically specialize in 1 or more of these subfields, the lines between them are blurry. In fact, most anthropologists draw on several or all 4 subfields to investigate our world.

Applied Anthropology and Building Bridges

One technique that focuses on real-world uses for anthropology is applied anthropology. Applied anthropologists tend to differ from conventional academic researchers. In addition to traditional academic tasks, applied anthropologists may write annual reports for charities or research institutes, they may help draft legislation, or they might bring various stakeholders together to improve patient outcomes at a regional clinic.

There are countless examples of anthropologists who consider themselves applied anthropologists, and there's an even larger number with interdisciplinary training and scholarship. Some of the biggest areas of applied and/ or interdisciplinary anthropology include medical anthropology, legal anthropology, education, and international development.

Anthropologists can act as bridge builders between cultures. It's in that capacity that they're able to produce cross-cultural knowledge, that is, knowledge that draws on the experience and understanding of diverse groups who often have no common cultural ground.

Such knowledge is important because it counteracts the problem of cultural bias and blind spots in order to create benefits for a wider swath of humanity. And perhaps even more starkly, such an approach works to mitigate the power inequalities imbued in top-down development programs.

A Field Experiment

By looking at a field experiment that took place in the village of Dissan in Mali, we can see anthropology at work as a cross-cultural knowledge producer. Sorghum is a major crop in Dissan. The experiment involved 23

new varieties of sorghum collected from ICRISAT, the International Crops Research Institute for the Semi-Arid Tropics. Near Bamako, Mali, they have a research station with a well-respected sorghum-breeding program.

The participant observer had a packet of seed for all 23 varieties, plus a local variety of their choosing. Together, the participant observer apprenticed with each participating family, and used a handheld hoe to plant individual parcels of each test variety. They did 5-square-meter plots and planted 8 rows of sorghum in each plot.

They marked each parcel to identify the 24 varieties and planted all 24 packets right in the middle of the participant families' fields. The center of fields tends to be the richest, more productive parts of fields in Dissan.

The study was published in the journal *Agriculture and Human Values*, but here are some of the big takeaways:

- Unlike the official trials plots, farmers tend to test novel sorghum varieties on the perimeters of their family fields. That's because they eat what they grow, so when they test seed, they don't have the luxury of giving up their most productive spots: the centers of their family sorghum fields.

- One farmer explained that he looks for new varieties that perform despite less than perfect conditions. That's because if they work well in bad spots, they'll emerge as champions elsewhere in the field.

- Farmers insisted on intercropping the test varieties in their secondary plots. They explained that they needed to give the seed a chance. Beans and other legumes fix nitrogen in the soil, much to the delight of sorghum.

- The farmers not only loved the experiment, after the test plots, they adopted the exact opposite sorghum varieties from their stated preferences. Before planting, the households said they preferred heavy seed, but after, they adopted 5 of the 8 lightest varieties available and only 2 of the 8 heaviest varieties.

- The farmers also said they preferred varieties that

Sorghum

matured quickly in 3 months or less. However, they adopted only 1 of the 3 fastest varieties available and 4 of the 5 slowest. This observation demonstrates the importance of engaged, long-term field research. Like everyone else, these farmers don't always say what they do or do what they say.

Information like this is indispensible for the development of national and international plant-breeding programs. Scientists can't just ask family farmers like those in Dissan to make a list of desired seed traits. That's a good start, but such a list will be incomplete.

Cross-cultural projects in particular require participant observation and other anthropological approaches. Anthropology is a tool for social transformation. It's a means to creating stronger communities and a happier, healthier world.

Suggested Reading

Bernard, *Research Methods in Anthropology*.

Lacy, "Nanotechnology and Food Security."

Podolefsky, Brown, and Lacy, *Applying Anthropology*.

Questions to Consider

1. What are the 4 subfields of anthropology, and what does each subfield investigate?

2. Why is bridge building an effective metaphor for the work and relevance of anthropology?

Why Anthropology Matters

I t was my first night with my first host family in Mali, West Africa, and, sitting beside my new mother, we laughed together as waves of smoke thwarted every attempt I made to escape their path. Now, mind you, I couldn't speak or understand a thing. We gestured enthusiastically, searching for any words we had in common, words like *bonjour* and Jimi Hendrix. All the while, Rokiya, my new five-year-old sister, she sat beside me poking my leg in search of a smile. But then she jabbed me a second time, but this time with a heck of a lot more enthusiasm, and, glancing over, I found her with the kerosene lantern in her hands.

"Lampon taa!" she said. She picked up the lantern. Then, "Lampon sigi!" She placed it back on the ground. Repeating this lesson, and with a wide grin full of expectations, she placed that lantern right at my feet. So, "Lampon taa!" I responded, whisking the lantern right into the air, and everyone erupted in cheers and laughter, but went silent in a flash—they wanted more. "Lampon sigi!" I set the lantern back down. And, after an impromptu celebration of dancing and clapping, I leaped back into my own seat and announced, "Solo sigi!" Over there, I'm known as Solo. And that was my first ever original phrase in Bamanankan: "Solo sit."

Now, that may not sound so impressive now. My host family's excitement, though, it instantly affirmed that I had passed my first Bamanankan language lesson, and that's pretty much how I learned everything I know about life in rural Mali. I lived alongside my hosts. I farmed when they farmed, I ate when they ate, and I slept when they slept, because I desperately wanted to understand them, their problems, their dreams, everything. So, without formal training, and two years before I'd even take my first anthropology course in graduate school, I was intuitively using a classic anthropology research method—namely, participant observation through cultural immersion.

As a foundational tool in our anthropological toolbox, participant observation, it allows us to viscerally experience, learn, and inquire about the daily lives of the generous people and communities who, frankly, are willing to put up with us. And, significantly, participant observation, it allows us to delve much deeper than our questionnaires and interviews because it can reveal what people say as well as what they do. So, from learning how to say "Lampon sigi" to understanding the origins of agriculture and humankind, anthropology inspires us to integrate multiple perspectives to enthusiastically explore our human condition and all its history and diversity.

So, what exactly is anthropology? Well, let's start with the basic definition, and this is how I start my introduction to four-field anthropology course for undergraduates. Put simply, anthropology is the study of humankind over time and space. And bear with me, because I promise, if I were to get any more specific than that, we'd be chopping off specialized branches of our disciplinary tree. In American anthropology, we categorize our discipline into four subfields. There's biological anthropology, archaeology, linguistics, and then cultural anthropology. And we use each of these subfields to explore survival and the diversity of humankind across time and geography. Let me explain.

First, biological anthropology includes everything from primatology and paleontology to evolution, biology, genetics, and even health and forensic science. And all of those themes will be covered in the early lectures of this course, as we answer the big question: who are we, and where do we come from? Then, as we move deeper, we'll figure out how *Homo sapiens* survived, while all other upright walking apes have long gone extinct? And to do that, we'll continue to apply biological anthropology, but we'll also draw new insight from archaeology, linguistics, and cultural anthropology.

So the next subfield, archaeology, now that uncovers and interprets artifacts— that's what we call material culture. And we look at material culture to reveal the histories of people who are no longer here to share their stories. Like crime scene detectives, we can look at what earlier humans left behind to work out how they lived and died.

Next, the third subfield, linguistic anthropology, now that's a lot more than just studying language. Linguists dissect the structure of language, they preserve and investigate dead languages, they tell language histories, they even provide a record of human migrations and cultural interaction, and many of them explore that intricate chicken-and-egg relationship between language and culture. Now, linguistics help us answer our second big question about how we are the sole surviving upright walking apes, but it also helps us transition to our final major subfield, cultural anthropology.

Cultural anthropology is an interdisciplinary subfield that explores kinship, economics, gender, development, religion, art, and just about anything else we humans do. And it's cultural anthropology that helps us work out our third big question. Through cultural anthropology, we'll explore how, despite cultural and linguistic differences, we are one human race.

While anthropologists typically specialize in one or more of these subfields, the lines between them, as you'll see, are kind of blurry. In fact, most anthropologists draw on several or all four subfields to investigate our world, and that's exactly what we'll do all throughout this course. But today, to begin our adventure, I want to take you to where I acquired my passion for everything anthropological. So let's go back to Mali, so I can share with you the transformative power of this thing called anthropology.

My anthropological story begins hundreds of years ago when a cattle herder named Npan Sangare left his community in Macina, Mali, and he headed south with his cattle. He was walking them far to the south in search of early rains and grasses that could strengthen these cattle in advance of the next growing season. And, as he wandered, he came across this beautiful small lake along a seasonal river, and it is replete with birds, fish, grass for cattle, and even some shady groves. Now, his cattle drank the clean water, and Npan rested below a great palm tree, and he awoke from an inspirational dream. He stood right there in place, right beside that lake, and untying his Fulani head wrap— that's what we call a *disa*—he placed that on the ground before him, and he announced to the world, with his cattle and wife as witnesses, that he would build a village here. And today this small mud-block village is named Dissan, in honor of the *disa* and the vision of Npan Sangare.

So, fast-forward to late 20th century when some direct descendants of Npan, along with a handful of others, petitioned Peace Corps to post a volunteer there. And in 1994, I had my first visit with the people of Dissan. And, in retrospect, I must have been quite a sight, because, not yet fully acclimated to the Malian life, I arrived on the back of a beat-up, powder blue moped, and I was wearing these matching pants with a local gown-like shirt with these fish prints on it. Plus, to complete my freak flag, I also sported some red John Lennon sunglasses under a bright yellow helmet. This is not a picture I share on Facebook.

But anyway, things went well despite my wardrobe choice, and rather quickly I became an adopted son of the village, with mothers, brothers, and more. But I was working out in the fields, and my high-protein maize project was about ready to launch, when, unfortunately, I fell gravely ill and was medevaced back to the States where my convalescence required a medical separation from Peace Corps and, in a flash, Dissan was out of my life.

I went to Mail to help the poorest of the poor, but as I lay in the hospital bed, I realized the village of Dissan had done more for me than I could ever hope to do for them. So, to honor this debt, I was determined to return to Mali, but this time with more useful skills. And during my convalescence, I discovered anthropology while researching careers that might allow me to forever keep Dissan in my life.

At UC Santa Barbara, I was trained in four-field anthropology as a graduate student, and my specialization was in cultural anthropology. And, as a cultural anthropologist, I've spent the past two decades working between family farmers in West Africa and technical experts, including plant breeders, engineers, builders, and even teachers. But, that said, let's be clear about my training and role as an anthropologist.

First, I'm no plant breeder, but I've studied plant breeding and agriculture long enough to speak their language, hold a conversation, and even understand lab culture. Similarly, I'm not a Malian farmer from rural West Africa. Nonetheless, I've lived and worked alongside Bamana farming families long enough to speak their language and to be a functioning member of the community and household that was kind enough to take me in.

So, with a foot in each camp, an anthropologist like me, we can bridge the efforts and knowledge of disconnected experts with something to gain from sharing knowledge and resources. And, inspired by that original Peace Corps experience, as an anthropologist, I bridge rural farmers and research scientists, both of whom work hard toward increasing food security for farming families. And this is typical of anthropologists. As we'll see time and time again throughout this course, anthropology builds cross-cultural bridges of understanding.

Now, that said, I should probably offer a confession: I didn't really start off as a true bridge builder. I was a hard-line, farmer-first kind of thinker. I was reflecting on the past few centuries of international development, and at first I was convinced that what farmers in tough ecologies like we have in Mali, all they really needed was for scientists to just get out of the way. But, as I spent more and more time living in Dissan, and more and more time observing plant breeders in the field, I realized that nobody really has a monopoly on good or bad ideas. On the contrary, alone, both sides of plant breeding in Mali—the farmer side and the plant breeder side—they're not enough. Climate change, socioeconomics, and enduring conflict have created a really challenging food puzzle in Mali that neither scientists nor farmers alone can figure out.

To the contrary, I learned what elder teachers were telling me from my first day in the village: *bele koni kelen te se ka bele taa*—one finger cannot lift a stone. And even when a great guy name Sidi first taught me that lesson, early on I held out. He physically demonstrated how one finger simply will not lift a stone. But, as a young upstart, I licked my finger and I commenced to do exactly that, and I lifted that single stone with a single finger. But, as an American, I knew that some fingers can lift a stone. Not really true. Before my smirk could transition to a know-it-all grin, that stone, it fell right down. And Sidi, he laughed and said,

"Voilà." Sidi, you've passed away, but thank you for that early lesson. Sidi, he helped me see the path of the anthropologist as a bridge builder. So, let me give you an example of how my work as an anthropologist has helped bridge the knowledge of scientists on the one side and rural farmers on the other.

Early on in my field research, I partnered with Malian farmers to test and evaluate a couple dozen unique varieties of sorghum. Sorghum, it's this fantastic grain that grows well in drier climates. And, at one point, we cold-tested a dwarf variety that produced plants with incredibly large seed heads but on really short stalks. And we, with the farmers, we grew several plots of this seed, and these farmers were definitely impressed with this new variety's yield potential. After harvest, I asked the big question: "*Sa were, jon be na si kura dan?*—who's going to plant this variety again?" Every single farmer said the same thing: "Not me."

Now, just imagine. Say you're a scientist, a professional plant breeder, and you've developed this fecund sorghum variety. You did every single thing right. You sought out farmer input on what they want, and most farmers prioritized increased yields as their most desired quality. So, high-yielding sorghum: check. Then, you experimented until you finally developed that high-yielding variety that invests less energy on the stalk height in favor of increasing potential grain production. High-yielding sorghum; that's exactly what they wanted: check.

Yet even farmers who admired your dwarf variety chose not to adopt it for their family fields. Why? Well, the answer emerged not from the questionnaires and interviews that we did, but rather they were hidden in the plain sight of daily life. You see, there are a lot of cattle in rural Mali, and it it's not uncommon that cows and bulls break loose for an unlimited buffet in some unsuspecting farmer's field. So, with a big and tasty seed head poised right at eye level for passing cattle, this prolific sorghum seed, it's not likely to make it all the way home to the family granary. So, here in this corner of the world, the advantage of increased grain production, it simply wasn't worth the risk of a bovine catastrophe.

And no one can blame a plant scientist really for not knowing the dynamic values, constraints, and resources of every single community that might want to test new seed. What kind of researcher would have the time to conduct that kind of long-term, culturally embedded research? You're looking at one. For a cultural anthropologist like me, cultural immersion and cross-cultural research—that's my bread and butter.

With plant scientists and family farmers as partners, I deploy cross-cultural research methods to bridge our mutual resources, knowledge, and research priorities. And over the past several years, I've used this approach to assemble unlikely cross-cultural partners for mechanical engineering as well as structural engineering and even design. I've become a specialist in what I now call cross-cultural knowledge production, and I promise you, that wasn't even a thing when I was in grad school. Yet, organically, anthropology and Mali transformed me into a cross-cultural knowledge production guy.

Ultimately, I do what many folks call applied anthropology. Now, fierce advocates of applied anthropology may consider this as a fifth subdiscipline of our field, because applied anthropologists tend to differ from conventional academic researchers. In addition to peer-reviewed journal articles, university press books, and academic conferences, we applied anthropologists, we may write annual reports for charities or research institutes; we may help draft legislation, or even bring together various stakeholders to improve patient outcomes at a health clinic.

These are some of many examples of anthropologists who consider themselves applied anthropologists, and there's an even larger number of us with interdisciplinary training and scholarship. Some of the biggest areas of applied and/or interdisciplinary anthropology include medical anthropology, legal anthropology, education, and, for me, international development. And I promise, those are all topics we'll discuss in future lectures.

So, for now, let me return to the image of anthropologists as bridge builders, because it's in that capacity that we're able to produce cross-cultural knowledge, which is to say, knowledge that draws on the experience and understanding of diverse groups who often have very little common cultural ground. Such knowledge is important because it counteracts the problem of cultural bias and blind spots in order to create benefits for a wider swath of humanity. And perhaps even more starkly, such an approach, it works to mitigate the power inequalities imbued in top-down development programs—or, more plainly, it's collaborative co-learning. It's a collaborative co-learning ethos rather than that deliverer-teacher model of conventional agricultural extension and development initiatives. To illustrate this point, let's return to my favorite field experiment ever, where we can see anthropology at work as a cross-cultural knowledge producer.

After learning there were just over a half dozen unique varieties of sorghum grown in Dissan, I got a hold of 23 new varieties that no one had ever seen before. And, I promise, these weren't frankenseeds, just a wide collection of varieties collected from ICRISAT, the International Crops Research Institute for the Semi-Arid Tropics. Now, they're near Bamako, Mali, and they have a research station there with a very well-respected sorghum-breeding program. We followed their protocol. We had a packet of seed for all 23 of their varieties, plus a local variety of our choosing. And, together, I apprenticed with each participating family, and we used a handheld hoe to plant individual parcels of each test variety. Farmers loved it. We did five-square-meter plots, and planted eight rows of sorghum in each plot.

Now, by first posting, we marked each parcel to identify the now two dozen varieties we planted. That's 24 packs of seed. And we were supposed to do that right in the middle of the participating farmer's family fields. The center of fields, those tend to be the richest parts of fields in Dissan. And, thankfully, that year, we only had to plant that field once. As a participant observer, one year—that was a few years earlier—I worked with my host family to seed a five-hectare plot, by hand, no less than three times in a single season. Why? The rain never came. We planted, but the first round didn't take, so we had to repeat the process. And I'll tell you, I remember nearly crying—seriously—I remember nearly crying when I heard we had to reseed that big field yet a third time.

But in our experiment, the first planting took. So Burama Sangare, my host brother and primary farming teacher—I apprenticed with him, too—he asked me what we're supposed to do with our extra seed. To which I did one of the best things an anthropologist can ever do: I shut up. "I don't know," I said. Then I asked him, instead of the formal trials we previously planted, how does he normally test new seed that he's never seen before?

Now that's some anthropology there. We have the formal ICRISAT protocol going on, and now a parallel, farmer-led experiment. And the whole experience is teaching me how both scientific plant breeders and farmers work to secure the future of food. Now, you can read about this study in the journal *Agriculture and Human Values*, but for our purposes today, here are some of the major findings from that study.

First, unlike the official plots, farmers tend to test novel sorghum varieties on the perimeters of their family fields. Why? Well, they eat what they grow, so when they test new seed, they really don't have the luxury of giving up their most productive spots, the center of their family sorghum fields. But more importantly, Burama explained to me that when he tests new varieties, he tests them. His existing varieties already work in good conditions here. So, what Burama is looking for are new varieties that perform despite less than perfect conditions, like the periphery of the field, because if they work well in bad spots, they'll emerge as champions elsewhere in the field. And, in fact, using his alternative tests, Burama identified and adopted several varieties that year that he continues to grow and disseminate to this day.

OK, what else? Well, second, unlike official test plots, farmers like Burama insisted on intercropping the test varieties in their secondary plots. When I asked why, they basically said that they needed to give the seed a fair chance. You see, beans and other legumes fix nitrogen in the soil, much to the delight of sister sorghum. And another unpredictable conclusion was that in the secondary tests, farmers didn't even bother sectioning off or marking off unique seed packs. In fact, when we planted Burama's secondary tests, we walked together, each one of us in our separate furrows, planting two to three seeds about an inch deep after every short pace. We'd start, and Burama took out a random packet, we split up the seed, and planted and planted until we were out. Then we just repeated this process for all the test varieties.

There was a point, then, when I asked Burama, "Why aren't we marking off these different seed packets?" I remember his quite generous and surprised smile when he responded with a question: "Don't you think I'll be able to figure out if there's anything interesting in there?" He didn't need a name; he wasn't interested in this type over that type. He was looking for something that would speak for itself.

When I presented this research at the Ministry of Agriculture in Bamako, they literally giggled and even joked a bit about the clear breach of scientific protocol. But they were all ears when I showed our final, major conclusion. Amazingly, the farmers revealed the critical importance of doing fieldwork in the anthropological tradition. It seems the farmers not only loved the experiment, but—and here's the twist—after the test plots, they adopted the exact opposite sorghum varieties that their stated preferences indicated.

Seriously, it turns out sometimes people say one thing and do another. Before we planted the seed, I asked every household in the village to tell me their desired seed traits for sorghum, and they said they wanted heavy seed. But then, in this experiment, they adopted five of eight of the lightest varieties in the whole experiment, and, conversely, only two of the eight heaviest varieties. What people say is not what they do. The farmers also said they preferred varieties that matured quickly—in three months or, even better yet, less. But they only adopted one of the three fastest varieties. They did adopt four of the five slowest.

The same thing happened with farmers' stated preference for pure white seed coats, and again for their stated preference for large seed. Remarkably, farmers' selections completed contradicted every one of their stated preferences, and this observation, it demonstrates the importance of engaged, long-term field research. Like everyone else, these farmers don't always say what they do, or do what they say, and it's information like this that's indispensible for the development of national and international plant breeding programs. I mean, it's clear: we scientists can't just ask family farmers like those in Dissan to make a list of desired seed traits. That's a really good start, but as my vignette showed, that list—it's incomplete. Cross-cultural projects in particular require participant observation and other anthropological approaches, all of which begin with inquires into and engagement with local knowledge and resources.

From their methods for testing novel seed to their complex criteria for choosing the right type or types of sorghum for their own family fields, these farmers are amazing. Even before we had planted these seeds, merely by looking at these curious test varieties, they were spot-on with their predictions of weak and strong seed. We used a five-point scale. In the end, they adopted zero of the varieties they ranked unfavorably at 2.5 or less. And on the other hand, they adopted over half of the varieties they rated absolutely highest.

My Malian hosts have taught me more than I can describe in a single lecture, but that seed study remains my favorite, and it produced knowledge that served science and filled bellies at the same time. In fact, it's now the subject of a documentary project called *Nyogonfe: Together*. Think about this: without a dime exchanged between them, farmer to farmer, Malian households shared this seed with neighbors and family, and some of these varieties have spread so far we can no longer track them. And essentially some of the varieties that started out in our 5-square-meter plots spread informally at least as far as 250 kilometers away from our village.

And while all this sorghum excitement was going on, back home at UC Santa Barbara, the Engineers Without Borders team proposed a joint project with me in Mali. They wanted to work with Malians to make biofuel using jatropha nuts, which are common all over Dissan and southern Mali. So, swapping engineers for plant breeders, I applied my cross-cultural knowledge production approach to bridge science and the knowledge of blacksmiths, farmers, hunters, parents, teachers, and others in Dissan.

And after spending a year studying the jatropha projects, those engineers, that team, arrived in Dissan. But before I facilitated the start of their jatropha work, I gave them a crash course in cross-cultural knowledge production. And the long story short, by the end of the second day, that team decided to ditch their jatropha project, and instead they wanted to collaborate with the village's water pump repair team. It was during discussions with locals that the team started with jatropha questions, but it didn't take long for the conversation to shift to a topic that was far more important for Dissan. Now don't get me wrong—the folks in Dissan loved the jatropha idea, but they had a more critical priority: drinking water.

And by the time that Engineers Without Borders team left the village, the pump repair team and the volunteers had not only rebuilt the village's two water pumps, but they also had established a preventative maintenance routine that has kept those pumps running smoothly for over a decade. And, in the old days, I can remember we always had at least one broken pump on any given day in Dissan. And, given that there's almost 900 people who live in this village, keeping those two pumps running, it makes life easier and healthier for all. And best yet, long after this engineering team left Mali, their Dissan counterparts have kept those pumps working well, and they've actually become a regional pump repair team. They're called upon by communities throughout the Bougouni region of southern Mali. Once, even, they were dispatched to fix a troublesome pump as far north as a three-day bus ride.

This story and the story of our collaborative work on sorghum illustrate how and why I do anthropology. As a facilitator for cross-cultural knowledge production, for me anthropology is a tool for social transformation. It's my means to creating stronger communities and a happier and healthier world. And, as we saw with the Engineers Without Borders team, my true job is to bring disconnected knowledge experts together, to help them develop a collaborative mission, and then work myself out of the picture. And that's not to say I no longer need or work with EWB. Currently, my nonprofit is building modest schools and literacy centers throughout Mali, and we make our cement now with a manual mixer created through yet another brilliant collaboration, this time between engineers and Malian masons.

Today, we launched our anthropological journey by developing a basic understanding of what anthropology is and how it produces knowledge that matters, knowledge that connects us with the mutual potential of enriching lives. And I've focused on my personal experience today in hopes that it gives you a concrete example of the cross-cultural and interdisciplinary bridge building that anthropology promotes.

But bridge building, it's not the unique domain of professional anthropologists; it's something that all of us can do. And a great first step in becoming an effective bridge builder is to learn what the four subfields of anthropology have to teach us about human diversity and our unity as a single race. And that's what we'll be doing for the rest of this course, and I'm grateful and delighted to be with you on this anthropological journey.

Science, Darwin, and Anthropology

As humans, it's in our nature to ask big questions. And as a meaning-seeking species, perhaps the biggest question we can ask is: Who are we, and where do we come from? In this lecture, we'll consider how people have answered this question over the past 3,000 years or so. First, we'll look at prescientific and nonscientific ideas. Then, we'll see how the emergence of science, including anthropology, completely changed the way we understand ourselves and our origins.

Prescientific Answers

Long before the emergence of science, humanity had woven a rich tapestry of colorful origin stories and folktales to give meaning and order to the world.

An example comes from the Channel Islands just off the coast of Santa Barbara. There, the ancestors of the Chumash people gave an answer to the question, "Who are we, and where do we come from?"

Before the age of microscopes and Bunsen burners, the ancestors gave an answer with certainty, but they really didn't concern themselves with empirical proof. In this case, the answer was that a female deity named Hutash planted seeds on Santa Cruz Island, and those seeds grew into the first people.

Spectacularly, Hutash created a rainbow bridge for people to cross over the Pacific to the mainland of California. And those who made it over are the ancestors of contemporary Chumash people. But those who fell into the ocean became dolphins.

Positivism and a History of Science

The philosopher Auguste Comte examined how people have answered the big human questions over time. Ultimately, he spelled out an evolutionary schema to explain how humanity developed into the scientific knowledge producers we are today.

Comte's schema goes under the name of positivism, which identifies 3 main stages in the human quest for truth: the theological phase, the metaphysical phase, and the positive phase.

In the theological phase of human development, humans associated the unknown exclusively with unpredictable supernatural forces like the gods of ancient Greece. Blaming Zeus or someone else in the pantheon mollified humans because they had an explanation for what's going on, and the human mind craves explanations.

Eventually, however, humankind transitioned into another phase on the road to scientific knowledge: the metaphysical phase. In this phase, thinkers like Aristotle used reason and observation to sharpen their understanding of the world. Aristotle theorized that the universe consisted of 3 types of substances: earthly matter that can be seen and felt; celestial matter that can be seen but not felt; and spirit matter that can't be seen or felt.

This explanation of the universe traveled widely, from Thomas Aquinas in the Christian tradition, to Averroes and Maimonides of the Muslim and Jewish traditions, respectively. All of these influential thinkers incorporated Aristotle's explanation of the cosmos. Simply put, they did this because Aristotle's explanation placed the earth at the center of the cosmos, surrounded by the heavens. It required a prime mover or deity.

Aristotle's ideas about the universe held strong well into the Middle Ages, largely because they aligned so well with existing religious explanations of the universe. But eventually, people like Galileo picked up instruments like telescopes and determined that the earth couldn't possibly be at the center of the universe.

A critical invention that inspired people like Galileo to test and correct our knowledge of the universe came from mid-1500s Europe. That's when and where Johannes Gutenberg built a printing press that, unlike the wood and ceramic print blocks that preceded it, made viable the mass production of books.

Amazingly, without cars, trains, or factories, the Gutenberg press spread across Europe in half a century. A faster and cheaper way of making books meant that humans had progressively more books, and therefore more access to information.

Charles Darwin

The anthropological elder Charles Darwin noticed that humans are related to every living organism. He wasn't the first person to come up with that idea, though. His grandfather, Erasmus Darwin, wrote *Zoonomia* some 60 years before Charles Darwin's *On the Origin of Species*. And the ideas in that text will sound quite familiar to readers of *Origin of Species*, namely, that all living organisms share a common ancestor.

Just before the elder Darwin started writing, a naturalist

Charles Darwin

named Carl Linnaeus began publishing *Systema Naturae* in 1735. In this multivolume classic, Linnaeus classified any living thing he could set his eyes on. In essence, he gave us our very first comprehensive family tree of all living organisms.

One question that challenged Erasmus Darwin and Linnaeus alike was the question of how species emerge. They could see and classify the planet's biological diversity, but they weren't clear on the actual mechanisms of evolution.

In the testable and correctable tradition of science, Charles Darwin picked up the mantel and continued the efforts of his grandfather. Darwin explained that the mechanism of evolution is natural selection. Specifically, organisms change gradually over time. As a result sometimes new species emerge, and sometimes they die out.

He broke this process down into 4 pieces:

1. Within a single species, there is variation in traits. Imagine, for example, starfish in the Caribbean. Some are red, but some are pinkish and barely visible because they blend so well into the ocean floor.
2. Not all of these starfish will be equally successful at reproducing. Darwin called this differential reproduction. Sea creatures like manta rays and a few types of sharks love eating up starfish. And because they can easily spot the red ones, it's the red ones that tend to get eaten unlike their camouflaged counterpart. As a result, fewer red starfish survive to reproduce.
3. Additionally, there is heredity. The pink starfish transmit their pink color to their offspring, as do the red ones provided that some diver or shark does not snatch them up.
4. In our ocean water example, the pink starfish color is an advantageous trait because it reduces the likelihood of being discovered by a predator. As time rolls on, the pink starfish become the most common

type of these starfish, until eventually it is the sole survivor.

Testing and Correcting

Unlike the Chumash origin story, Darwin's answers were produced through science. That means his answers were contingent truths that scientists were supposed to test and correct. And that's exactly what people did.

One of the earliest revisions of Darwin's ideas came from Gregor Mendel, a monk living in what is now the Czech Republic. Darwin theorized that heredity operates through blending: An organism is a blend of its 2 parents.

Mendel knew about cross-breeding plants to produce desirable traits like high yields. Mendel planted huge numbers of pea plants at his monastery to test his ideas about how traits pass from parent to offspring.

Mendel observed 7 key traits of his plants. He watched how traits like pod color were expressed in 1 of only 2 ways: yellow or green. Mendel noticed that only 1 of 4 plants tended to have yellow pods.

He noted the same ratio for other traits like seed shape and plant height. What he discovered were dominant and recessive genes. A yellow pod is rare because it only emerges when both parents contribute a recessive gene for pod color; any other combination produces green.

Mendel wasn't the only person testing, correcting, and improving our understanding of evolution. Hugo de Vries helped us begin to imagine and test ideas about mutations, Francis Crick and James Watson revealed to us the structure of DNA, and now we've mapped out the entire human genome.

Social Darwinism

It didn't take long for people to metaphorically apply the idea of evolution outside the biological world. Influenced by Darwin's ideas, scholars like Herbert Spencer and William H. McGee ushered in social Darwinism, which asserted that social systems and cultures travel the same trajectory as evolving plants and animals do: They go from simple to complex.

In the late 19th century, social Darwinism ruled the day because it was considered a "scientific" way to explain and justify some serious inconsistencies and inequalities on the planet. Simply enough, people were poor because they were less evolved.

The arrival of social-Darwinist thinking emerged with the arrival of the earliest anthropologists. While other biologists sorted out the rest of the animal kingdom, this new field called anthropology tried to make sense of human diversity, biological and cultural.

Harvard's own Frederick Ward Putnam revealed the new discipline to the American public at the Chicago World's Fair in 1893, which included an anthropology exhibit. The fair was a celebration of technology in the new scientific age.

Among other sights, there was a pair of marble statues to represent the "ideal" man and woman. The sculptures were created based on measurements taken from Harvard and Radcliffe students. Putnam and others were imposing the biological model of evolution upon culture and cultural achievements.

Today, we consider this idea to be the very definition of ethnocentrism and pseudoscience, but back then, people were essentially arguing that some people are further along on the evolutionary path, biologically and culturally.

Such thinking was embraced by those who could then justify disastrous inequalities like colonization and manifest destiny as moral burdens. Rather than understanding colonization and the Atlantic slave trade as evil-rooted violence, whites of European extraction believed they had a moral duty to extend their colonial reach, no matter the human cost.

More Testing and Correcting

Biological determinism and social-Darwinist thinking were so pervasive at the dawn of the 20th century that their influence even reached into the Supreme Court of the United States.

In 1896, just as anthropology was emerging as an academic discipline, the Supreme Court justices consulted the science of the day and upheld racial segregation as constitutional in the *Plessy v. Ferguson* case.

In line with the day's anthropological theories, Justice Henry Brown wrote the majority decision and said that so long as one race was inferior to the other, the Constitution is powerless to put them on the same plane.

Justice John Marshall Harlan's dissent saw the upholding of segregation as planting seeds of racial hate, but he was clearly in the minority. So, just like their anthropology contemporaries, most of the Supreme Court justices used science and prevailing ideas about evolution to defend racial inequalities as "natural" and segregation as inevitable.

There's a silver lining that has been at work throughout this lecture: the testable and correctable nature of science. We're always learning more.

Suggested Reading

Bolotin and Laing, *The World's Columbian Exposition*.

Darwin, *The Origin of Species*.

Newkirk, *Spectacle*.

Stocking, *Victorian Anthropology*.

Questions to Consider

1. Beyond survival of the fittest, what is natural selection and how does it work?

2. What is social Darwinism and why did it become so influential on the edge of the 20th century?

Science, Darwin, and Anthropology

As humans, it's in our nature to ask big questions. And, as a meaning-seeking species, maybe the biggest question that we could ask is this one: who are we and where do we come from? Well, today we'll consider how people have answered this question over the past 3,000 years or so. First, we'll look at prescientific and nonscientific ideas, but then we'll see how the emergence of science, including anthropology, completely changed the way that we understand ourselves and our origins.

Now, long before the emergence of science, humanity had woven a rich tapestry of colorful stories and folktales to give us meaning and order to the world that we live in. And, as a result, when we start to think about who we are and where we come from, there's no shortage of answers. But we're going to start with a trip to the Channel Islands just off the coast of Santa Barbara, where we'll take our big question—who are we and where do we come from—to the ancestors of the Chumash Nation.

Before the age of microscopes and Bunsen burners, the ancestors gave us answers, and those were answers with certainty, but they didn't really concern themselves with empirical proof. In this case, they'd tell us that a female deity named Hutash she planted seeds on Santa Cruz Island, and those seeds grew into the first people. And, spectacularly, Hutash created a rainbow bridge for people to cross over the Pacific into mainland California. And those who made it over, they're the ancestors of contemporary Chumash people. But those who fell into the ocean became dolphins. And you know what? That might explain why dolphins like to swim alongside boats—we're long-lost cousins.

Anyway, imagine you were born to a Chumash family; it was a thousand years ago. You're a brilliant hook and harpoon maker for fishing, you live in a beautiful coastal settlement, and undoubtedly you'd be a major fan of this Hutash. But, despite your certainty about her role in your origins, you're going to be hard-pressed to find archaeological evidence of that rainbow bridge—of migration, yeah, but a bridge and people seed? Not going to happen.

But you know what? That's OK. If you're still in character, you have certainty—you do not need proof. You do not need archaeological evidence. Certainty requires faith, not evidence. You have faith, and that faith helps you understand the world and your place in it. So, whether you're a Chumash elder living over a thousand years ago, or maybe you're a 21st-century substitute teacher from Cuyahoga Falls, Ohio, all humans seek answers to some of the same questions, big and small, every day. And if we don't get them, if we don't understand what's going on, we tend to—and pardon my technical term here—we freak out. And the bigger the question, the bigger the potential freak-out.

Just think of a young child—that'll be a metaphor for humanity's age-old quest to understand the overwhelming mystery of our origins. So, zap, lightning shoots across the sky, then thunder booms, sending that child right under the table in fright. Now that's scary, scary stuff for a kid who's having a hard enough time trying to figure out how to walk and talk. That kid just doesn't have the mental apparatus to understand why electricity is ripping through the air. So how are we going to calm him down? What are we going to say to calm his young, shaken nerves? If you're like my parents, you're going to use the bowling thing. "Don't be afraid, Jimmy. That's just your grandfather bowling with the angels. Remember, he loved bowling? I think that was a strike. Hopefully he'll get another one."

And right after that, suddenly, now that he understands what's going on, this kid can get out from under the table and function again. In fact, now that thunder and lightning means that grandpa's getting a strike, and so Jimmy can barely wait for the next one. So here he went, from a fearful breakdown to curious fascination. This is what good answers do, they ground us.

So, with that being the case, we're going to go ahead and check out a new kind of good answer. How about some scientific answers? Scientists are part of a long tradition of humans seeking answers about our world and our origins, and let's jump into that line right now. Let's get some perspective from the philosopher Auguste Comte, because he already did exactly what we're doing today. He examined how people have answered the big human questions over time. And, helpful to us, he spelled out an evolutionary schema to explain how humanity went from being frightened children under the table to the scientific knowledge producers that we are today.

Comte's schema goes under the name of positivism, and positivism identifies three main stages in the human quest for truth. There's the theological phase; the metaphysical phase; and of course, last, the positive phase. So, in the frightening early phase of human development, when we're under the table, we associate the unknown exclusively with unpredictable supernatural forces like the gods of ancient Greece. Why are we suffering from drought? Did we offend Zeus? Blaming him or someone else in the pantheon mollifies us because we have an explanation for what's going on. And the human mind, as we've talked about already, it needs these explanations; it craves them. So, eventually, humankind transitions into another phase on the road to scientific knowledge, that middle phase, and it's Aristotle who's going to provide a great example of this one.

Thinkers like Aristotle used reason and observation to sharpen their understanding of the world. When he looked at the natural world we live in, he saw a dynamic planet of constant change, but it was nestled in this perfect universe, right in the center of a perfect universe. And, like a simplified periodic table, he theorized that the universe consists of three different types of substances. One, first, there's this earthly matter, and that can be both seen and felt, like you and me. But then there's celestial matter. Now that can be seen, like when we look into the stars, but you're not going to feel that. Last, there's the spirit matter, and that can't be seen and it can't be felt.

It was this explanation of the universe that Aristotle used, and this explanation traveled widely. It went from Thomas Aquinas in the Christian tradition to Averroes and Maimonides of the Muslim and Jewish traditions, respectively. And all of these influential thinkers incorporated Aristotle's explanation of the cosmos. Why? Well, quite simply, it placed the Earth at the center, surrounded by the heavens, and you know what? That Earth required a prime mover or a deity. So, Aristotle's ideas about the universe held strong well into the Middle Ages, and the reason is because they aligned so well with existing religious explanations of the universe. So why would humanity move on from the comfort of their Aristotelian model of the universe? Well, the more people thought, and the more they sort of considered this idea, the more they started testing and correcting what we were thinking. And it was people like Galileo who picked up instruments like telescopes, and they eventually determined that our Earth couldn't possibly be at the center of our universe.

But there was another critical invention that inspired people like Galileo to test and correct our knowledge of the universe. Maybe you can you guess what it was that ushered us into Comte's third phase, that positive phase? And if you need a hint, think about mid-1500s in Europe, because that's when Johannes Gutenberg built a printing press that, unlike the wood and ceramic print blocks that preceded it, made viable the mass production of books—of knowledge. And so think about this: amazingly, without cars, trains, or factories, this Gutenberg press spread across Europe in half a century. And a faster and cheaper way of making books meant that humans had progressively more books, and more access to information, both religious and otherwise. And these books and our collective human library, eventually that will seed a revolutionary string of scientific pioneers, from Copernicus, and Herschel, and Newton to Darwin, Mendel, and Crick.

And so the collective work of the early methodologists like Newton established a new way to explore our universe and ourselves. So now, through the scientific method, we can test and correct each other's ideas to produce knowledge that, by design, evolves. Now, take Darwin for example. Sure, he was on to something major with natural selection, but as we'll discuss in a moment, he was completely off with his explanation for heredity. Other scientists tested and corrected Darwin's ideas to improve and refine our collective understanding about the scientific answer to the foundational anthropological question: who are we and where do we come from? And instead of a story about humans coming to North America on a Pacific rainbow, scientists tell a new story to explain our origins. And, in a word, their story is the story of evolution.

And as we'll soon see, anthropology, as an interdisciplinary science, is uniquely anchored to this concept of evolution. Now, it took millennia for humans to collectively work out science and the idea of biological evolution, but for most of our existence, the majority of us sided with Aristotle. And the answer was essentialism. We, like all living things, we've always existed the way we exist today. The exact opposite of evolution is called stasis. Now, we can imagine why early religious figures like Aquinas, why they accepted Aristotle's essentialism, and even promoted it. It's because it aligned with literal interpretations of the biblical creation. So, it was thinkers like Galileo, they certainly understood prevailing theories about our universe, but when ideas in their heads didn't match up with what they saw with their own eyes, that cognitive dissonance, that's what's going to precipitate the rise of science.

No matter how many times he read that the Earth was at the center of the universe, Galileo just couldn't help recognize the evidence: we're spinning around the Sun, not vice versa. Galileo didn't understand everything about our place in the universe, but he challenged us to continue looking, and to continue refining what we know: test and correct. So, rather than looking up to the stars with Galileo, it's our anthropological elder, Charles Darwin, who turned his sights from planets to people. And what he noticed is that we're related to every living organism.

Think about that—every living organism. Actually, he's not the first one to come up with this idea. It was, oddly enough, his grandfather, Erasmus Darwin, who wrote this book *Zoonomia*. It was about 60 years before Darwin's *On the Origin of Species*. And the ideas in this text are going to sound quite familiar to those of you who've read *Origin of Species*—namely; all living organisms share a common ancestor. And, as a protoanthropologist, Erasmus Darwin relied on comparative anatomy, paleontology, and even embryology, because he was testing theories on how species, including our own, emerge. But, remarkably,

Erasmus Darwin even wrote poetry to express this idea. Bear with me for a second and let's hear a few lines from his 1802 poem *The Temple of Nature*:

Organic life beneath the shoreless waves
Was born and nurs'd in ocean's pearly caves;
First forms minute, unseen by spheric glass,
Move on the mud, or pierce the watery mass;
These, as successive generations bloom,
New powers acquire and larger limbs assume;
Whence countless groups of vegetation spring,
And breathing realms of fin and feet and wing.

Just before the elder Darwin started writing these poems and his books, there was another guy, a naturalist named Carl Linnaeus, and he began publishing *Systema Naturae* in 1735. And it was in this multi-volume classic from the history of science where Linnaeus classified any living thing he could set his eyes on. In essence, what he's doing is he's giving us our very, very first comprehensive family tree of all living organisms. But wait. There's one question that challenged Erasmus Darwin and Linnaeus alike, and that was: how do these species really emerge? What's going on here?

Well, they could see and classify the planet's biological diversity, but they really weren't clear on the actual mechanisms of evolution. So, in the testable and correctable tradition of science, it's Charles Darwin who picks up the mantle, and he continues the efforts of his grandfather. And, as the famous story goes, he worked out the answer to his grandfather's question while sailing around the world on the HMS *Beagle*. Darwin explained the mechanism of evolution, that thing that drives evolution, is natural selection. Specifically, what did he mean? Well, he says organisms change gradually over time, and as a result sometimes new species emerge, and sometimes they die out. So, Darwin said, it's natural selection that going to make this possible, and in order for us to sort of grapple with what he's thinking about, let's break it down into four pieces.

First, as we can plainly see with our own eyes, there's variation in traits. Imagine, for example, we're staring deep into the blue Caribbean water, and there on the sandy ocean floor we can see hundreds of starfish. There's a lot of red ones, there's some pinkish ones, and the pinkish ones you can barely see because they're blending so well into that beautiful sand. That's step one. These starfish, all of them: one species. But, as you see, there's two different traits—there's the pink, and there's the red.

Let's move on to part two. Not all of these starfish will be equally successful at reproducing; some of the starfish are not even going to reproduce at all, and it's this that he calls differential reproduction. So, building on our Caribbean metaphor, think about this: sea creatures like manta rays and a few types of sharks absolutely love eating up starfish. And because they can easily spot the red ones, it's the red ones that tend to get eaten, unlike their camouflaged counterpart. So, as a result, fewer of these red starfish are going to survive to reproduce—differential reproduction.

All right, let's move to the third piece of natural selection. This is called heredity. Now, go back to the pink starfish. They're going to transmit their pink color to their offspring, as do the red ones, provided that some diver or shark doesn't snatch them up. And that brings us to the fourth and final piece of natural selection, and that is, in our ocean water example, the pink starfish color. That's what we're going to call an advantageous trait, because it's going to reduce the likelihood of being discovered by a predator. So, in the large picture, as time rolls on, the pink starfish is going to become the most common type of these starfish, until eventually it's the sole survivor.

So, if you have the first three—that would be a variable trait, which is hereditary, and if it leads to differential reproduction—bingo, that's it. That is the process of evolution through natural selection according to Darwin. Now, Darwin's going to publish this answer and even more in his book *On the Origin of Species* in 1859, and it's this book that really revolutionized how scientists answer our central question: who are we and where do we come from?

But let's not forget, unlike the Chumash origin story, Darwin's answer was produced through science, and that means his answers are contingent truths that we're supposed to test and correct. And that's exactly what people did. And one of the earliest revisions of Darwin's ideas came from a monk, Gregor Mendel, living in what is now the Czech Republic. And, as part of his evolutionary ideas, Darwin theorized that heredity operates through blending. It's like mixing paint. You pour some white paint with some red paint, what are you going to get: it's pink paint. Darwin said that's what happens with us. We're born as a blended version of mom and dad.

However, what Mendel discovered was that when parents produce offspring, maybe a better heredity metaphor might be M&M candies. Think about this: when your mom contributes a bunch of yellow M&M's and your dad contributes a bunch of blue ones, you mix it all up in a bowl; you don't get green M&M's. And so, like most farmers, Mendel and a lot of others knew about crossbreeding plants to produce highly desirable traits like, maybe, high yield. So, Mendel planted tens of thousands of these pea plants at his monastery, because he was going to test how traits pass from parent to offspring, and he observed seven key traits of his plants. He watched how traits like pod color were expressed in only one of two ways—they're either going to be yellow or they're going to be green, nothing in between.

So, like a sports fan keeping stats on a team, Mendel notices that, hey, only one of four of these plants are having yellow pods. And then he noted the same ratio as happening with other traits like, say, seed shape, and height. What he's discovering here is dominant and recessive genes. A yellow pod is going to rare. Why? It was because it only can emerge when both parents contribute a recessive gene for pod color; any other combination, that's going to produce green.

Mendel wasn't the only person testing, and correcting, and improving our understanding of evolution. Just as Mendel helped tighten the screws with regards to heredity, fellow scientists continued to refine our knowledge of evolution and of our biological origins. Hugo de Vries helped us begin to imagine and test ideas about mutations. Meanwhile, Francis Crick and James Watson, eventually they're going to reveal to us the structure of DNA, and with that we've mapped out the entire human genome. So, by remaining open to new and progressively better explanations about our origins, scientists completely built a much more comprehensive explanation for how evolution works.

From Darwin's days on, the theory of evolution was an efficient and compelling way for us to understand who we are and where we come from. In fact, it didn't take long for people to metaphorically apply this idea of evolution outside of the biological world. Influenced by Darwin's ideas, it was scholars like Herbert Spencer and William H. McGee who ushered in social Darwinism, which asserts that just as biological organisms evolve from simple, single-cell organisms to complex things like humans, giraffes, penguins, social systems, they do the same thing. Social systems and cultures travel the trajectory from simple to complex societies—sociocultural evolution.

So, in the late 19th century, social Darwinism ruled the day because it was considered a scientific way to explain and justify some serious inconsistencies and inequalities on the planet. Simply enough, people were poor because they were less evolved, and people lived in houses versus huts because they're just further along as humans evolve. The arrival of social Darwinist thinking emerged with the arrival of the earliest anthropologists. And while other biologists sorted out the rest of the animal kingdom, this new field called anthropology tried to make sense of human diversity, both biological and cultural. And it was Harvard's own Frederick Ward Putnam who revealed the new discipline to the American public at the Chicago World's Fair in 1893.

Now, this fair was branded the Columbian Exposition and it included an anthropology exhibit. Now, the fair was a celebration of technology—it was technology in a new scientific age, some 400 years after Christopher Columbus. And visitors were stunned when they entered this amazing, magnificent building that was aptly titled the White City. Now, this White City, it was imposingly beautiful, and it was a massive constellation of exhibition halls displaying the day's newest machines and inventions. It was supposed to be a celebration of modernity—what have we achieved these 500 years since Columbus sailed the ocean blue? There was even a pair of marble statues. They represented the ideal man and the ideal woman, and the sculptures were created based on measurements taken from Harvard and Radcliffe students. You see, Putnam and others were imposing that biological model of evolution upon culture and cultural achievements.

Today, we consider this idea to be the very definition of ethnocentrism—well, ethnocentrism and pseudoscience—but back then, these folks were essentially arguing that some people are further along on the evolutionary path, both biologically and culturally, than others. Biology and culture? One and the same. And from a distinctly Western perspective, they categorized people on a continuum from savage to barbarian to civilized and even enlightened. And once visitors got their fill of the enlightened White City, it was time to be presented with a satellite exhibit, a satellite exhibit called the Midway. Well, that's going to be where you and I can see how the rest of the world lives.

I tell you, man, the contrast is going to be stark. Rather than an imposing hall of new technology, the anthropological exhibit was a veritable Epcot Center from another era. Putnam, he organized this historic global exhibit as a living museum of what he called primitive people, primitive people who collectively helped visitors see and measure for themselves the progress of modern humanity on display in the White City. He actually got this idea from an exhibition that was earlier in Paris, but the early French anthropologists and others brought in so-called natives to populate ethnographic village displays of the people who lived in the French colonies of Africa and Asia. So, by the time the fair finally opened, the Midway—I'll tell you what—it was a hit. If you came in through the entrance just off of Washington Park in Chicago, you were instantly transported to another world—there was sights, sounds, smells, tastes.

There's the Brazilian music hall pumping out South African rhythms, and an ostrich pen, and a Chinese theater, and the exotic souvenirs in the Indian bazaar. You'll have a blast visiting those people in the model villages tucked in every single corner. You've got the Javanese, you've got the Dahomey, the Irish, the Pacific Islanders—there's even a brilliant recreation of a street in

Cairo. Now, you can imagine that, in a world that would have to wait another century for the Internet and YouTube, visitors were absolutely smitten with this remarkable display of humanity. And I tell you, they were. Visitors flocked to gaze at these exotic peoples and lives from across the world, buying up new foods, souvenirs, and enjoying amusements like a Ferris wheel all along the way. And that really pleased the fair organizers, because this whole Midway thing was designed as a moneymaker. People and artifacts from around the world gave visitors a window into the broader world from which to measure the modernity of the White City. So, from Turkey and colonial India to Native America and Africa, there was a stark difference and an intentional contrast between the White City and everything else.

So here are some of the earliest anthropologists teaching people, the general public, that much of the world's population is slightly less civilized and ultimately less human. Now, this type of thinking was embraced by those who then could justify disastrous inequalities like colonization, manifest destiny— these were moral burdens. So, rather than understanding colonization and the Atlantic slave trade as evil-rooted violence, whites of European extraction had a moral duty to extend their colonial reach, no matter what the human cost. It was part of a mission to "civilize savage peoples." They're lagging too far behind on the evolutionary scale, and that's at least what we were thinking. So consider, for example, a young Congolese man named Ota Benga. Not only was Ota Benga exhibited at a similar exhibit at the St. Louis World's Fair in 1904, but he was also on display at the Bronx Zoo in 1906. Think of that. People's views on evolution at the time made it possible for the Bronx Zoo to actually house and display a fellow human in the Monkey House.

Biological determinism and social Darwinist thinking were so pervasive at the dawn of the 20th century that its influence even reached into the Supreme Court of the United States, and the Supreme Court justices had something to say about the matter. It was 1896, just as anthropology was really, truly emerging as an academic discipline. The justices consulted the science of the day, including anthropology, and, frankly, they upheld racial segregation as constitutional. It was the Plessy v. Ferguson case, and that science, that's what we saw afoot at the World's Fair: technology with a major side order of social Darwinism.

So, in line with the day's anthropological theories, Justice Henry Brown wrote the majority decision, and he said that so long as one race was inferior to the other, our Constitution is powerless to put them on the same plane. Now you need to note, Justice John Marshall Harlan, he gave the dissenting opinion, and he actually said that he saw the upholding of segregation as planting seeds of race hate, but he was clearly in the minority. So, just like the anthropology, most of the Supreme Court justices used science and prevailing ideas about evolution to defend racial inequalities as natural—segregation, it's inevitable.

So this, my friends, is the nefarious beginnings of anthropology in America. Yikes! But wait, don't you give up the ship just yet, because there's a silver lining that we've seen at work throughout this entire lecture, and that's the testable and correctable nature of science. The production of truth, yes, but remember, in science it's always contingent truth. We're always ready to learn more.

So, 20th-century anthropologists continued to test theories about racial difference and human diversity. And it won't be a spoiler to jump ahead a half a century to 1954 when ideas about human evolution and diversity matured to the point that they're going to influence the Supreme Court again, and this time they'll completely reverse their 1896 decision. And this decision in '54 was Brown v. Board of Education. So, in future lectures, we'll learn more about the people whose research dismantles social Darwinism, but for now we've got to return to biological evolution to continue our quest to understand our origins. Next, that quest will take us to the primate family and primatology. We'll see you there.

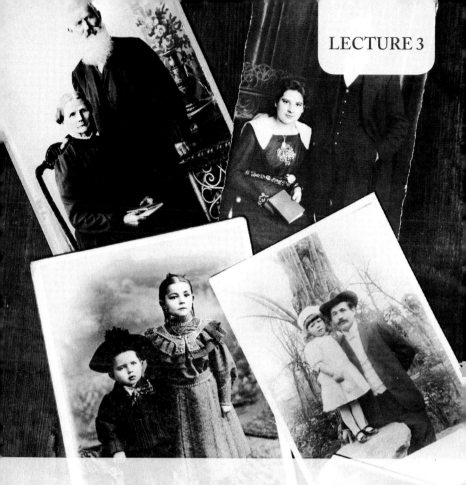

Our Primate Family Tree

This lecture travels back in time over 63 million years to introduce an ancestor shared by every living primate on Earth today. In this course, we've looked at how, over the past few thousand years, people have answered the question: Who are we, and where do we come from? This lecture shows what the primate order has to say about that big question. It starts by considering how and why anthropologists integrate primatology into our study of humankind, and then takes a closer look at our primate family tree.

We Are Primates

Humans have a long and amazing line of primate ancestors. The more we interact with and study our primate cousins, the more we see the depths of our shared genetic heritage.

A remarkable study from the Yerkes National Primate Research Center at Emory University shows the human-like depths of our primate cousins. The researchers did a simple game with capuchin monkeys. Two monkeys side by side in separate cages took turns trading marbles for food treats. They were trained to give a marble to the research assistant in order to receive a food treat, starting with grapes.

As the experiment went on, one of the monkeys began receiving inferior cucumber bites instead of grapes, while the other began receiving more grapes. The one receiving cucumber bites began to react angrily, showing a human like reaction: discontent at unfair treatment.

From our behavioral similarities down to our shared genetic heritage, nonhuman primates can teach us so much about being human. In some cases, these differences begin to disappear the more we probe them. That's exciting because we're getting more and more precise about what it means to be human.

Susan Savage-Rumbaugh

Primatologists collect data and test theories to help us understand our humanity, including our primate roots. Generally, primatologists tend to specialize in one or more areas, from primate genetics and anatomy, to cognition, behavior, and social organization.

To get a better idea of what primatologists think about, this lecture will visit 2 exceptional primatologists to see them in action. First up is Susan Savage-Rumbaugh, an exceptional primatologist known for cognitive and behavioral research.

Her star collaborator in this research is a bonobo ape named Kanzi. As a bonobo, Kanzi shares almost 99% of our genome, and together, Susan Savage-Rumbaugh and Kanzi definitely teach us that our human-bonobo differences could be more cultural than they are biological.

Kanzi lives at a primate research center in Iowa, and he's helped us test language as one of the boundaries between humans and the other apes. Susan Savage-Rumbaugh and Kanzi speak through lexigrams or symbols representing words, but Kanzi clearly understands Savage-Rumbaugh's spoken English.

Videos exist of Kanzi and Savage-Rumbaugh making a campfire. As Kanzi methodically completes one task at a time, Savage-Rumbaugh talks to him with helpful banter. She reminds him that there's a lighter in her pocket, after which Kanzi digs in and grabs it.

Kanzi not only builds, lights, and tends a true campfire: He also makes a s'more with a toasted marshmallow and puts the fire out with a bucket of water. He demonstrated a curious ability to remember and manage complex tasks.

Capuchin monkey

Laurie Santos

This lecture's second primatologist is Laurie Santos at Yale. After the economic collapse of 2008, Santos wondered if other primates share human inclinations toward risky economic behavior. She found that they definitely do.

Santos and her students taught primates how to use money to buy snacks. Then, in a series of experiments, she tested their financial strategies by setting up a monkey market.

First, the researchers taught capuchin monkeys to use money. They gave the monkeys a wallet with a dozen aluminum coins, which they used to trade for food. At the monkey market, the capuchins eventually got to make a choice between 1 of 2 snacks.

Then Santos kicked it up a notch. Prices at the monkey market changed, and some choices were cheaper than others. So, are capuchin monkeys bargain hunters like us? Santos and her research tell us yes.

Additionally, Santos never saw a monkey save coins for the future.

They love shopping and eating, and they seem as human as us when it comes to blowing the budget.

Shared DNA

Now that we've seen primatologists in action, let's get back to the primate family tree through the lens of genetics and biology. When we explore our origins, genetics can dramatically change the way we understand our place in the world. Remarkably, we share a surprising number of genes with all living things.

Let's take a tour of the primate family to see how and where we stack up to our remarkable cousins. In essence, we need to figure out how a common ancestor could eventually evolve into such a wide variety of primates, including humans.

There are 3 major types of primates: prosimians, monkeys, and apes. Technically, humans fit into the ape category. Nonetheless, we all share a common ancestor. We've all been on the same evolutionary freeway, and the only reason why chimpanzees are different from orangutans and humans is that we each took different exits to continue our evolutionary journey.

One important note: This is not to say that humans were once chimpanzees. We share a common relative with modern chimpanzees.

Prosimians and the MRCA

Now, it's time to meet the common ancestor shared by primates. Her name is MRCA, an acronym for most recent common ancestor. This MRCA kind of looked like a squirrel with a long tail.

In the world inhabited by the MRCA more than 60 million years ago, there were none of the current primates because they hadn't evolved just yet. Instead, proto-primates were living an arboreal lifestyle in the trees.

Ultimately, some proto-primates splintered off to become the prosimian branch of our family tree. Just over 60 million years ago the first prosiminans—the lemurs and lorises—diverged from the rest of the primates, and then the tarsiers broke off a few million years later.

Lemurs, lorises, and tarsiers are all prosimians, but the tarsiers are closer cousins to humans because they stayed on the evolutionary freeway for those extra few million years. Prosimians are small, often nocturnal primates with large eyes. Some of them leap through the trees while others are skilled climbers.

Lemurs, like all prosimians, used to be found in many places, but now the only lemur populations are in Madagascar. The smallest lemurs, like the dwarf lemur, would easily fit in the palm of a human hand. But larger lemurs, like sifakas, can weigh up to about 15 pounds.

Outside of Madagascar, the other prosimians—lorises and tarsiers—are found in Africa, India, and Southeast Asia. They too are mainly nocturnal tree-dwellers who are usually strong climber or jumpers.

New World and Old World Monkeys

As the prosimians continued to evolve on their own, the rest of the primates carried on down the evolutionary freeway until the next

Lemur

group broke off on its own. First the New World monkeys broke off some 40 million years ago, and then the Old World monkeys followed suit around 15 million years later.

Let's take a look at the Platyrrhini, another term for the New World monkeys. New World monkeys can be found in Central and South America. They're tree-dwellers and they like to eat leaves, fruit, and occasionally insects.

On the small end of the Platyrrhini spectrum are the marmoset and tamarins. They can't change their facial expressions, they don't have opposable thumbs, and they frequently produce twins. Once those twins are born, males tend to carry infants on their back when the children aren't nursing.

The bigger side of the New World Monkey classification is the Cebidae family. These monkeys are also limited to Central and South America, but they are much larger than the marmosets. The howler monkey, for example, can grow to over 20 pounds.

With the New World monkeys and the prosimians departed, the rest of the early primates moved on together, further down the primate evolution freeway. They were becoming more and more human all along the way.

Take the Old World monkeys. Humans have much more in common with Old World monkeys than we do with prosimians and new world monkeys. Take teeth, for example. Humans have 8 teeth on each side of each jaw.

Old World monkeys also have 8 teeth per side. This is different from the New World monkeys, who have 9. Humans also have the same types of teeth as the Old World monkeys: 2 incisors first, then a canine tooth, then 2 premolars, and then 3 molars. The New World monkeys have an extra premolar that Old World monkeys and humans got rid of.

Old World monkeys can be found in Africa, the Middle East, and parts of Asia. Baboons and macaques are examples of Old World monkeys. Driving home the depths of our shared genetic path through evolution, the rhesus macaque is the most frequently used primate in human medical research.

Apes

With the prosimians, Old World monkeys, and New World monkeys branched off, all that is left is the apes. Things picked up quickly for the earliest apes. Around 17–18 million years ago, the gibbons took off on their own path. The orangutan and gorilla did the same a few million years after that.

Around 7 million years ago, humans left the common evolutionary path. That left the bonobo and chimpanzee to split up on their own. This explains why we share some 99% of our genome with the bonobo and chimpanzee, but only 80% or so with the lemur. Humans, chimps, and bonobos were together the longest on the primate evolution freeway.

Suggested Reading

De Waal, *The Bonobo and the Atheist.*

Goodall, *In the Shadow of Man.*

Sterling, Bynum, and Blair, *Primate Ecology and Conservation.*

Questions to Consider

1. How does learning about the primate order help us understand what it means to be human?

2. What are some fascinating similarities and stark differences between us humans and the rest of the primate family?

3. What are the origins of the primate order, and where exactly do humans fit among the other primates?

Our Primate Family Tree

Today, we're going to travel back in time over 63 million years to meet an ancestor that you and I actually share with every living primate on Earth. And this is an important journey because already in this course we've looked at how, over the past few thousand years, people have answered the question: who are we and where do we come from? So we're going to see what the primate order has to say about that very big question. And while the details may be innumerous, primatologists actually have a pretty simple and straightforward answer. Who are we? We're primates—apes, actually. And where do we come from? Well, we come from a long line, an amazing line, of primate ancestors. And it's because we share this primate family tree, that it's not surprising that the more that we interact with and study our primate cousins, the more we see the depths of our shared genetic heritage.

Now, don't get me wrong, we share much more than our genetics, but today we're going to see how our shared evolutionary heritage with the primates is also evident in the way we think and act. Not all primatologists are anthropologists, but anthropologists do rely on the work of primatology to discern what it is that makes us human. So, let's start by considering how and why how anthropologists integrate primatology into our study of humankind, and then we'll take a closer look at our primate family tree.

Imagine the last time you saw a monkey or an ape. Maybe you were at the zoo, or maybe watching a movie with some simian costar. One of my childhood favorites was the short-lived series called *Lancelot Link, Secret Chimp*. This cast of chimpanzees, they chased each other, they double-crossed each other; they were international spies. They wore these human clothes, they skied, they drove flashy sports cars; they even talked, albeit through dubbed voices. But I loved that show, but why? What is it that makes apes wearing sunglasses and driving convertibles so funny?

Well, from *Lancelot Link* to that "move it-move it" lemur in the animated film *Madagascar*, we human primates really enjoy seeing our primate cousins doing human things. It's as if we might be hardwired to enjoy blurring those lines between them and us. Every time we see a chimpanzee wearing a tuxedo or playing a video game, it tickles our funny bone, because they're amazingly similar yet entirely different from us. And it's peering into that divide that primatologists test the boundaries of what makes us human. And as we'll see, the deeper we look into the lives of our primate cousins, the more we're going to see ourselves.

Take, for example, a remarkable study from the Yerkes National Primate Research Center at Emory University, and I'll tell you, if this experiment doesn't show you the human-like depths of our primate cousins, nothing will. So, briefly, noted primatologists Sarah Brosnan and Frans de Waal published a 2003 study titled "Monkeys Reject Unequal Pay." Now, the researchers did a simple game with capuchin monkeys. Two monkeys side by side in separate cages took turns trading marbles for food treats, like cucumbers or grapes.

The first monkey has been trained to basically grab a marble and put it in the hand of a research assistant the second that research assistant reaches out. And that's exactly what monkey one does—grabs a marble, puts it in the hand of the research assistant. The research assistant then reaches in and gives that monkey a grape—brilliant. Monkey number two sees that and is getting ready for his grape. So now we have monkey number two. Monkey two see that the game is on, he grabs his marble, gives it to the research assistant, and in return, a grape. They're happy; they're eating their grapes together, maybe even smiling back and forth.

But we get to round two. Let's go back to monkey number one again. So here's monkey number one. He does the game right, he sees the hand, the research assistant's putting that hand out, so he grabs the marble, puts it right in the hand, and, again, a grape: eats it—brilliant. Monkey number two sees that, he's ready for his turn. Let's go over to that one. Monkey number two grabs a marble and, with some anticipation for this grape, pops it into the research assistant's hand, reaches out, and guess what he gets? He gets a cucumber. Now, he's not really too upset, he eats that cucumber, but you can see he's ruminating, his eyes, he's like, "Wait a minute, that guy got a grape over there. What's this cucumber thing?" But he's cool, he eats it, you know, things happen; you and me, we do the same thing, we just roll.

But here it is; we get to number three. Back to monkey, first monkey. Monkey holds out the hand. After giving the marble to the research assistant, and this time it's really even more exciting, because for one marble this monkey gets two grapes. So, the monkey's psyched, man—one marble, two grapes, he's eating it, he's just strolling, he's doing great. Number two's sort of looking over here like, "Okay, I see what's going on here. I'm going to give my marble." So he gives the marble to the research assistant, and you know what? When he gets his treat, it's another cucumber. It's not a grape, it's a cucumber. So he gets kind of upset. He starts munching on that thing, but I tell you, he's not happy about it. He's thinking, "That guy got two grapes, I get one cucumber, what's going on with this place? I don't think I like my job."

So, it gets worse, and let me tell you, we get to the last round, and, bottom line, we go over here to monkey number one, and monkey number one goes and puts the marble in the hand of the research assistant. With the hand still out from the monkey, the research assistant puts those two grapes right back into the hand of the primate, monkey number one. Monkey one eats it just like before and is still just having the best time of his life. Man, this is a great, great day of the game; I'm getting two grapes for one.

But monkey number two, he's kind of getting suspicious here. He's over here ready to give that marble, but you see his eyes. So he grabs that marble and he puts it in the hand of the research assistant, and he's ready, but then when he sees that he's going to get just another piece of cucumber, he gets mad. He reaches his hand out of that cage, he picks up the cucumber, he doesn't even eat it, he actually whips it at the primatologist; he's mad. And once that hits the primatologist, he's running around like a little kid, just banging his hand against the wall. You've got to feel bad for this little guy. He thinks like me, he acts like me—he's so human. He gets upset when he's not treated fairly, and you know what? I totally relate.

My dear friends, this is exactly why, from our behavioral similarities down to our shared genetic heritage, that nonhuman primates can teach us so much about being human. As an anthropologist, I'm not directly involved in primate research, but I have so much fun with primatology in the classroom, because, after all, to understand our humanity is to understand our unique place amongst all the primates. And so, over the years, we've defined ourselves by what separates us from the rest of the primate order and our shared evolutionary history, while our differences reveal our uniqueness as humans. We walk upright, we use tools, we have language, and we make fire—we're *Homo sapiens*.

But one of the reasons it's so fun to teach and keep up with this primate research is that the more and more we probe these differences, they actually begin to disappear, and that's very exciting because we're getting more and more precise about what it means to be human. I mean, think about this: if other apes can learn English, if they can learn language, that might not be the distinguishing factor that we presume it to be. Maybe our humanness is somewhere else—it's something else.

For example, what about fire and tool use? That's definitely a human thing. Can other apes, our closest primate relatives, learn to make a campfire? Can they roast marshmallows? The answer's going to surprise you. But how do we go about gathering that information? What methods can scientists use to study nonhuman primates? Besides trading grapes and cucumbers for marbles, what is it that primatologists do to produce this knowledge? Well, actually, there are a range of methods, from behavioral observations to genetics, biology, and even fossil analysis. And using all of these tools, primatologists collect data and they test theories to help us understand our humanity, including our primate roots.

Now, generally, primatologists tend to specialize in one or more areas, from primate genetics and anatomy, to cognition, behavior, and even social organization. So, to give you a better idea of what primatologists actually think about, let's visit two primatology labs and let's see some primatologists in action. Then we can review the evolutionary timeline of our primate order so we can actually mark the moment that we became human.

Let's go ahead and take a look at two exceptional primatologists who do cognitive and behavioral research, and we're going to see how some shared genetics are going to yield some really compelling behavioral similarities. First, Susan Savage-Rumbaugh. Now, she's an exceptional primatologist doing amazing cognitive and behavioral research. Her star collaborator in this research is a bonobo ape named Kanzi. And, as a bonobo, Kanzi shares almost 99 percent of our genome, and together Susan Savage-Rumbaugh and Kanzi definitely teach us that our human-bonobo differences could be perhaps a lot more cultural than they are biological. Let me explain.

Kanzi lives at a primate research center in Iowa, and he's helped us test language as one of the boundaries between humans and the other apes. Susan Savage-Rumbaugh and Kanzi speak through lexigrams, or symbols that represent words, but he cleverly understands Susan's spoken English as well. And I've seen videos of Kanzi and Savage-Rumbaugh making a campfire, and Kanzi methodically completes one task at a time. And all along, Savage-Rumbaugh talks to him, not like a child but with regular language just like you and me, and she reminds him, "Kanzi, there's a lighter in my pocket," after which Kanzi digs in and grabs it. Kanzi not only builds, lights, and tends a true campfire, but get this; he also makes a s'more, toasts a marshmallow, and puts the fire out afterward with a bucket of water. Kanzi not only communicated with Savage-Rumbaugh, but he demonstrated a curious ability to remember and manage complex tasks.

An even funnier example of Kanzi's ability to learn, remember, and manage complex tasks comes from an old TV show named *Champions of the Wild*. And it's on that show that Kanzi played the classic video game Pac-Man. Think about that. On the one level, it's fun to see a non-human ape enjoying an arcade game, but there's a lot going on there. Pac-Man. To last more than 10 seconds in this game, a player's going to have to keep track of a lot of things. You have to move Pac-Man all around this maze and eat these pellets, and that's easy enough. But wait, you have to avoid those ghosts as they move about the maze as well independently, and they're ready to catch you—they're trying to get you. And, to add an extra cognitive layer, when you eat one of those four power pellets in all the corners, you can actually reverse the game and you can start to chase and eat the ghosts. But hold on a second here, it's only when the ghosts are blinking. Now, this may be simple brainwork for a human like you and me, but what about a bonobo? Apparently, it's not at all out of the question.

So, we'll return to Kanzi and primate language studies in a future lecture, but for now let's visit our second primatologist, Laurie Santos at Yale. Now, Santos explores the cognitive dimensions of primate life, and one of her most intriguing research areas: monkeynomics. You heard that right, monkeynomics. So, basically, after the economic collapse of 2008, Santos wondered, do other primates share our human inclinations toward really risky economic behavior? And what she found was they do. Santos and her students taught primates how to use money to buy snacks. Then, in a series of experiments, she tested their financial strategies by setting up a monkey market.

First, the researchers taught capuchin monkeys to use money, kind of like the previous experiment we heard about. They gave monkeys a wallet and, instead of marbles, this time they had a dozen aluminum coins, and they used those coins to trade for food. But at the monkey market, the capuchins eventually got to make a choice between one of two snacks, generally like an apple, maybe a cube of wiggly Jell-O. But right away the researchers were thrilled. They discovered that the monkeys, when confronted with a choice, they stop and really think about it. Just like you and me, they think about, "Well, which daily special should I order today?"

But wait, that's when Santos kicks it up a notch, because then she actually plays around with the prices. Prices at the monkey market change and some choices become cheaper than others. So, let's think about that: are capuchin monkeys bargain hunters like us? Santos and her research tell us most definitely yes. When presented with an apple slice that is half the cost of the Jell-O, the monkeys went bananas for that bargain. They wanted more bang for each buck. And what's even more remarkable is that these monkeys also were just like us— when they had money to spend, they spent it, and Santos never saw a monkey save coins for the future. They love shopping and eating, and they seem, just like us humans, to just blow the budget whenever we get a dime or two.

Santos has done all kinds of monkeynomics research, including a foray into themes like risk aversion and loss prevention. From Laurie Santos and her cognitive research to Susan Savage-Rumbaugh's work with Kanzi and other bonobos, primatologists have a unique point of view when it comes to exploring our big anthropological question: who are we; where do we come from? And at the nexus of biology and culture, it's their research that helps us explore and refine our definition of humanity.

When we realize that apes can use English to communicate with us, when we see monkeys exasperated when they expected equal pay—albeit cucumbers and grapes—or like Santos where we saw monkeys making the risky economic decisions just like the ones that led to the 2008 collapse of Wall Street. When we see primates learn to do things that we consider to be human, our entire idea of being human has to change. So, surprisingly, to truly understand what it means to be human actually is going to require some insight and some comparative research with all primates, because we define what it means to be human by identifying both how we differ from and what we share with the entire primate order.

OK, now that we've seen primatologists in action, let's get back to the primate family tree through the lens of both genetics and biology. And when we explore our origins, it's genetics that can dramatically change the way we understand our place in the world. Remarkably, we share a surprising number of genes with all living things. You name it, and we share part of our genome with it. Sounds strange, but daffodils, for example, well over 20 percent. Can you imagine that? Next spring, when you see those daffodils emerge, you remember, from a strictly genetic perspective, that's your cousin—kind of. How about this? How much do you think we share with, say, dogs or mice? Which is our closer relative? Dogs, they might be our best friends, but genetically speaking, we're much closer to mice, and that's going to explain why mice play such a significant role in human medical research.

But, when we move from flowers and mice to strictly the primate family, that's where it's going to get fun, so let's take a tour. Let's see the primate family and understand how we stack up with our remarkable cousins. In essence, we need to figure out how a common ancestor could eventually evolve into such a wide variety of primates, including us talking apes. So, to start, let's put it simple. There are three major types of primates: we've got prosimians, we've got monkeys, and we've got apes. And, technically, humans fit into the ape category. Nonetheless, if you go back far enough, we actually all share a common ancestor. We've all been on the same evolutionary freeway, and the only reason why chimpanzees are different from orangutans and humans is that we each took different exits to continue our evolutionary journey.

But one important note: we're not saying that we were once chimpanzees. Let's be clear about this. We share a common relative with modern chimpanzees, and even Darwin a century and more before, he warned us—

and I'm going to quote him right here—he says: "We must not fall into the error of supposing that the early progenitor of the whole Simian stock, including man, was identical with, even closely resembled, any existing ape or monkey." So, from lemurs and baboons to bonobos and humans, we are all primates, and if we go back far enough, we're going to see that we share a common ancestor. So now, let me introduce you. Her name is MRCA—Most Recent Common Ancestor. Now, this MRCA—brace yourself—she kind of looked like a squirrel with a long tail.

So travel back with me some 60 plus million years to the world inhabited by our MRCA. There are no humans, there are no baboons, no gorillas; none of the current primates exist yet. We had not yet evolved into those forms of primates. Instead, we are all living the arboreal MRCA life. And I tell you, life in the trees was groovy. But wait—suddenly, for one reason or another, some of our fellow proto-primates exited the freeway. They left the rest of us to pursue their own unique evolutionary path. And, ultimately, they splintered off to become the prosimian branch of our family tree.

Just over 60 million years ago the first prosimians—lemurs and lorises—diverged from the rest of us early primates, and then the tarsiers broke off a few million years later. So here we are: lemurs, lorises, and tarsiers are all prosimians, but the tarsiers are closer cousins to us humans because they stayed with us longer on that evolutionary freeway. It was a few extra million years. So, prosimians are small, often nocturnal primates with very large eyes. Some of them are leapers while others actually are great climbers. And lemurs, like all prosimians, used to be found all over in many places, but now the only lemur populations are in Madagascar. The smallest lemurs, like the dwarf lemur, would fit in the palm of your hand. But bigger lemurs like the sifakas can weigh up to about 15 pounds. And, man, those sifakas are amazing. With their powerful legs, they can jump over 30 feet into the air as they leap from tree to tree.

Outside of Madagascar, the other prosimians—lorises and tarsiers—are found in Africa, India, Southeast Asia, and they too are mainly nocturnal tree-dwellers, and they're usually really good climbers or really strong jumpers. And, in fact, one of our oldest primate fossil remains that we've discovered goes back 55 million years, just after the prosimians broke off on their own evolutionary path. Remarkably, this fossil gives us a snapshot of the divergence of the tarsier from the rest of the early primates, and, I promise, it doesn't disappoint.

Found by a farmer in the Hubei Province in China, this fossil is an early primate with hands, feet, small eyes—not the big ones—like monkeys. They had small eyes like monkeys, not the big eyes like prosimians. But it also has other characteristics more commonly associated with tarsiers. For example, it was super small, only about 25 grams—that's about an ounce—and it shared similar skull and limb proportions with tarsiers. Scientists named this creature *Archicebus achilles*. *Archicebus*, that's the genus name, refers to the guy's super long tail, which he used for balance while living high up in the trees. Now, the species name, *achilles*, points to its more monkey-like heel, so it's definitely a cross between the two. So, as the prosimians continued to evolve on their own, the rest of us early primates carried on down the evolutionary freeway until the next group breaks off on its own. First the New World monkeys, and they're going to break off some 40 million years ago, then the Old World monkeys followed suit around 15 million years later.

So, let's take a look at the Platyrrhini, often referred to as New World monkeys. New World monkeys can be found in Central and South America. They're tree-dwellers and they like to eat leaves, fruit, and maybe a few insects here and there. And on the small end of the Platyrrhini spectrum, we have the marmoset and tamarins. Now, these guys are remarkable primates. They can't change their facial expressions, they don't have opposable thumbs, and they frequently produce twins. Then, once those twins are born—get this—the males are the ones that tend to carry their infants on their back, as long as the children aren't busy being nursed.

So, let's look at the other side of these New World monkeys. This is called the Cebidae family. These monkeys are also limited to Central and South America—the New World—but they're much larger than marmosets, like the howler monkey, for example. He can grow to over 20 pounds. Another example of the Cebidae family are the capuchin monkeys. Remember the cucumber thrower? That's the other side of this family.

So let's move on to the Old World monkeys. With the New World monkeys and the prosimians departed, the rest of us early primates moved on together, further down the primate evolutionary freeway, and we were becoming more and more human all along the way. So, ultimately, another group is going to depart, and that's going to be the Old World monkeys. We have a lot more in common with Old World monkeys than we do with prosimians and even the New World monkeys, even down to our teeth.

Take a moment, use your tongue, and check out your own dentition. If you start at the front on one side and you count back to your molars, you're going to find that most of us have about eight teeth on each side. Our Old World monkey cousins also have eight teeth per side, and that's different from the New World monkeys that got off the freeway earlier, they have nine. But it goes deeper than that. If you look closely, you'll see that we have the same types of teeth as the Old World monkeys. So, there are two incisors first, then there's that canine, then there's two premolars, and then three full-fledged molars. Now, the New World monkeys have an extra premolar, a third one that Old World monkeys and humans got rid of. So, Old World monkeys can be found in Africa, the Middle East, and Southeastern Asia, the so-called Old World. Baboons and macaques are two examples of Old World monkeys. And driving home the depths of our shared genetic path through evolution, the rhesus macaque is the most frequently used primate in human medical research.

Apes. So here we are, all alone, finally. With the prosimians off on their exit, as well as the New and Old World monkeys taking off on their own, all that's left is us apes. That's us. And I tell you what; things are going to pick up quickly for us early apes. It was around 17 or 18 million years ago that the gibbons took off on their own path, then the orangutan and the gorilla did the same. The final phase of this primate and ape story comes after the gorilla, and that's about 7 million years ago when we humans were the ones that left our common evolutionary path. And that actually left the bonobo and the chimpanzee to go on and split off on their own, which explains why we share some 99 percent of our genome with both the bonobo and the chimpanzee. Yet, if we go all the way back to the prosimians, we're only at about 80 percent or so. So humans, chimps, bonobos were all together the longest on this primate evolutionary freeway.

And, my friends, this is the primate family tree. From the Most Recent Common Ancestor through prosimians, monkeys, and then the apes, we can trace our early human origins through primatology. In fact, the fossil record and genetics both allow primatologists and others to pinpoint the very moment in our primate history when we became human, when we diverged from the ancestors of the modern bonobo and chimpanzee.

So, to recap, let's remember how primatology fits within anthropology. Essentially, it helps us answer our big question: who are we and where do we come from? First, the primates we see in zoos and in the wild are as modern as you and I; they're not less evolved humans. Second, we—all of us primates— evolved from a common ancestor some 63 million years ago. And last, we're humans, and therefore primates—one of the apes, more specifically—and we're most closely related to the bonobo and the chimpanzee.

Anthropology is the study of humankind over time and space, and that most certainly includes our primate family and history. Why? Well, as we explore the depths our similarities, both biological and behavioral, with our primate cousins, our shared evolutionary history is magnificently illuminated. Today, we saw how our primate cousins emerged, each in their own way, over the past 60 million years, including us upright-walking apes, but we've yet to consider what is unique about our human branch of this primate tree. If we've got nearly identical genetics to a bonobo, what is it exactly that makes us human? What defines us, not as primates, but as humans? These are the questions that lead us from primatology to paleoanthropology, the study of our earliest human ancestors, and that's the discipline we'll turn to in our next lecture.

Paleoanthropology and the Hominin Family

This lecture introduces more knowledge about the primate family tree. Already, we've gained insight into human origins by looking at the field of primatology. In this lecture, we'll turn to paleoanthropology to trace our earliest human ancestors through to modern *Homo sapiens*. Like primatology, paleoanthropology will give us some great perspective as we explore answers to our foundational anthropology question: Who are we, and where do we come from?

Recent Arrivals

Humans are relatively recent arrivals. The Earth is 4.5 billion years old. Humans began to diverge from our closest primate cousins around 7 million years ago.

From a paleoanthropologic point of view, the moment our ancestors started walking on 2 feet demarcates our human origins as primates. Paleoanthropologists are the human origins specialists in the anthropology family. They search for paleoecological evidence, namely fossil evidence, to reveal details about the lives and biology of the earliest humans.

Paleoanthropologists teach us that the earliest apes who started walking upright lived some 7 million years ago. Slowly, over time, hominins (upright walking apes who are direct human ancestors) evolved into modern humans.

Sahelanthropus tchadensis

In Chad in central Africa, the researcher Michel Brunet and his team made a game-changing paleoanthropological discovery. They unearthed a famous fossil

that is certainly one of the earliest hominins on the human family tree.

In 2001, Brunet's team discovered the remains of a primate skull that dates back some 6–7 million years. The skull ended up being identified as *Sahelanthropus tchadensis*.

Sahelanthropus tchadensis is remarkable because:

- It dates back to the same timeframe humans separated from the chimpanzees and bonobos.
- This early primate was likely a facultative bipedal. In other words, it walked on 2 feet.
- It's our oldest known primate ancestor whose fossil remains indicate that it was bipedal. And that is where we started our human story: with these earliest upright-walking apes.

An important question: How can someone look at a skull and determine if a primate ancestor walked on 2 feet?

- When all we have is just a skull, a major indicator for bidpedalism can be seen in the foramen magnum—the

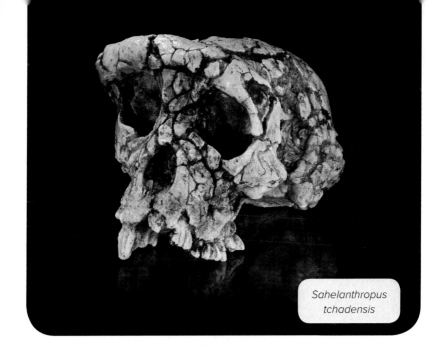

Sahelanthropus
tchadensis

place where the skull meets the spine. In a gorilla or chimpanzee skull, the foramen magnum is toward the rear of the skull, not the center, as it is with humans.

- Bipeds have a more centrally located foramen magnum than nonhuman primate relatives. So does *Sahelanthropus*.

Ultimately, Brunet and his colleagues argue that this fossil is one of our earliest bipedal ancestors, close to the time we split from chimpanzees and bonobos. Some think *Sahelanthropus* was more of a peripheral human ancestor, and others challenge both of these arguments because of a dearth of any other evidence.

Ardipithecus

East of Chad toward Ethiopia, other paleoanthropologists, Timothy White and Berhane Asfaw, discovered one of the earliest known fossils of our human ancestors. Beginning in the early 1990s, Asfaw and White's team began uncovering some extremely fragile fossil remains. They were so delicate that it took 15 years to preserve and analyze what they found. They also had to do CT scans to make it easier to piece together the fragmented remains.

Their results were groundbreaking. Their major find was the skeletal remains of a female who lived 4.4 million years ago. She was bipedal, but she could get around in the trees as well. Scientists dubbed this new discovery *Ardipithecus ramidus*.

In the late 1990s and into the 2000s, Yohannes Haile-Selassie discovered more fossil evidence of *Ardipithecus* in Ethiopia. The remains he unearthed were confirmed as part of the *Ardipithecus* classification, but were an older type that lived over 5.5 million years ago. They named this early human ancestor *Ardipithecus kadabba* because *kadabba* means "oldest ancestor" in the local Afar language.

Raymond Dart and *Australopithecus africanus*

From time to time, humans have found fossils that beg further inquiry. Take Raymond Dart, for example. Dart was a professor of anatomy living in South Africa in the early 20th century.

One day Dart came across a captivating fossil. It was a skull with jawbone. It was relatively complete, and Dart knew something was odd about his find. It clearly wasn't a modern human, but it certainly didn't resemble any other modern ape. Its skull, for example, was smaller than a human skull, but larger than a chimpanzee's.

The skull was named the Taung child: "Taung" because of the town in South Africa where it was found by quarry workers in 1924, and "child" because it turns out that this specimen died as a child. In fact, we're pretty sure that the Taung child died as the result of an eagle attack.

Dart proposed we classify this skull as *Australopithecus africanus* (meaning "southern ape from Africa"). Distinct from humans and other primates, this represents another major era on our evolutionary timeline and our evolutionary family tree.

In the case of the Taung child discovery, and Dart's theory that it represents a major moment in our evolutionary path, anthropologists challenged and tested these ideas rigorously.

At first, many were skeptical of the idea that *Australopithecus* was a direct human ancestor, but over the decades, more and more fossils emerged to support the theory that at least some australopithecines were indeed a direct ancestor of *Homo sapiens*.

Ultimately we've learned that these australopithecines lived some 4 million years ago in Africa, largely eastern Africa.

There were many types. *Australopithecus afarensis*, for example, is definitely a direct human ancestor. Perhaps the most famous of all the australopithecines is Lucy, a female *Australopithecus afarensis*. Her discovery in 1974 was celebrated widely because it is rare to find such an old yet fairly complete and well-preserved skeleton.

Homo habilis

Just as Raymond Dart analyzed the Taung child fossil and concluded it didn't fit any existing classification, the same thing happened to the famous paleontologists Mary and Louis Leakey.

In the early 1960s, the Leakeys found the remains of an early human that caught their attention. It wasn't a modern human, but it wasn't an *Australopithecus* either. It was somewhere in between. Its brain and body was a little larger, and it lived more recently than our *Australopithecus* ancestors.

The Leakeys created a new intermediary classification that clarifies our path from *Sahelanthropus* to *Homo sapiens*. Additional discoveries throughout East Africa affirmed the status of this newly revealed ancestor: *Homo habilis*, or "handy human."

Homo habilis lived in Africa about 2.5 million years ago. Their brain size had grown. Researchers have found *Homo habilis* remains with very crude stone tools, something they had yet to find with earlier ancestors like *Ardipithecus*.

Homo erectus

The next ancestor on the timeline is *Homo erectus*. The first known *Homo erectus* discovery was found in 1891 by a Dutch surgeon, Eugène Dubois, who clearly saw it was not human. Like Dart and the Leakeys,

Dubois created a new classification for his ape-like skull discovery. He called it *Pithecanthropus erectus* ("ape who stands erect"). Later fossils filled in the gap of our knowledge, and eventually paleoanthropologists modified the name of this ancestor to *Homo erectus*.

The first *Homo erectus* populations appeared just under 2 million years ago. Their cranial capacity could exceed 1000 cubic centimeters, clearly a major leap beyond the 750 cubic centimeters of *Homo habilis*. These ancestors had quite an amazing toolkit compared with everyone who came before them. And *Homo erectus* can be found outside of Africa.

Some scientists separate the *Homo erectus* populations into three distinct groups based on their geographic distribution: *Homo ergaster* (Africa); *Homo erectus* (Asia); and *Homo heidelbergensis* (Europe).

Archaic Homo sapiens

Homo erectus appeared 1.9 million years ago and went extinct as recently as 150,000 years ago.

Somewhere out of that *Homo erectus* population emerged archaic *Homo sapiens*.

The archaic form of *Homo sapiens* first came into the fossil record when the Leakeys found some remains dated over 100,000 years ago. That was 1967. Contemporary teams have revisited the fossils and the site; they dated these early *Homo sapiens* to around 190,000 years ago. The most celebrated fossils are called the Omo skulls, and they're our oldest known definitively *Homo sapiens* fossils.

As of yet, we've only found early *Homo sapiens* fossils in eastern Africa, leading us to believe that modern humans are likely rooted to a population of *Homo sapiens* who lived in Africa before spreading out to the other continents.

Testing and Correcting: The Hobbit

Anthropologists are never satisfied with stagnant theories. They test and retest their theories and ideas. One recent "test-and-correct" challenge they've dealt with relates to the "hobbit," a small human that

lived in Indonesia within the past 20,000 years or so.

After the bones of this small, recent human were discovered in 2003, the popular press labeled it a hobbit. The scientific name was *H. floresiensis* because their remains were found on the island of Flores.

Specifically, this was a population of 1-meter-tall hominins who not only had a complex tool kit, but lived long after we thought every hominin species but *Homo sapiens* had gone extinct. And they had relatively small brains for being such sharp toolmakers.

The hobbits mess up the clean storyline of better tools through bigger brains as the road to survival as well as the old idea about humans being the sole remaining hominins for the past 40,000 years.

For now, most folks tend to keep our hominin-evolution storylines in tact by subscribing to 1 of 2 theories:

1. Some people argue that the small humans are smaller simply because of a genetic condition that emerged in the region and remained isolated.

2. Another theory states that on a small island with limited resources, larger animals like humans would likely evolve to become smaller to better compete for limited resources.

Regardless, the fossil record is clear and it's deep. Walking upright appears to have been a game changer for us, and in many ways, we can consider bipedalism as the foundation of our humanity, biological and otherwise.

Suggested Reading

Dawkins and Yan, *The Ancestor's Tale*.

Tattersall, *Masters of the Planet*.

Questions to Consider

1. When did the first upright walking apes migrate beyond Africa?

2. How were our pre-*Homo sapiens* ancestors different from modern humans?

3. What are some examples of groundbreaking fossil discoveries that help us construct our early history as upright walking apes?

Paleoanthropology and the Hominin Family

Today, we're going to build on our knowledge of the primate family tree. Already, we've gained insight into human origins by looking at the field of primatology, but in this lecture we'll turn to paleoanthropology to trace our earliest human ancestors through to modern *Homo sapiens*. And, like primatology, paleoanthropology will give us some great perspective as we explore answers to our foundational anthropology question: who are where; where do we come from? We're going to be talking about a story that is about 7 million years old, so let's first grab a little perspective, because it's kind of hard to grasp timelines that include millions of years. So, let's use a straightforward metaphor to get some perspective on the origins of humanity.

Think of a clock. Not a 12-hour clock, but a 24-hour clock. And if we take the entire past 4.5 billion years since the Earth was created, and if we jammed all those years into a single 24-hour day, what time do you think humans would emerge? When should our human history start? Some might say 4:00 am, with the origins of life, but then it's really not until late in the afternoon, almost dinnertime—about 5:30-ish—when we get the first multicellular organisms. It's not till 11:00 pm till we even see dinosaurs. So, if triceratops emerge around 11:00 pm, when do we humans show up? Well, you'll be surprised to hear it's at 11:58 and 43 seconds, and again that's pm, just about a minute and change before midnight.

So, our 24-hour clock shows us where the story of our origins fits within the meta-story of our planet Earth, and it's clear we are recent arrivals on a vastly ancient planet. And when I reflect on the magnitude of the ecological, the cultural, and the biological inheritance that each and every one of us is born into, the human story and our place in the universe become all the more magnificent, and perhaps, depending on your worldview, even a little magical.

So let's recap our timing here. As humans, we are Earthlings, and the Earth is 4.5 billion years old. So, in some ways, our story could begin there. Or we're mammals, so we could start about 200 million years ago with the earliest of those. But in our previous lecture we went back 63 million years ago to the Most Recent Common Ancestor that we share with all modern primates, so today let's get right to that human branch of our tree. That's going to diverge from our closest primate cousins right around 7 million years ago.

From a paleoanthropology point of view, the moment our ancestors started walking on two feet, that moment demarcates our human origins as primates. Paleoanthropologists are the human origins specialists in the anthropology family. They search for paleoecological evidence—namely, fossil evidence— to reveal details about the lives and biology of the earliest humans. And it's from this fossil evidence, and without spoiling the story of our ancestors, paleoanthropologists teach us that the earliest apes who started walking upright lived around 7 million years ago, and slowly, over time, hominins— upright walking apes who are direct ancestors to us humans—evolved into modern humanity.

Now, along the way, we're going to see that the way we walk transformed our bodies, from the size of our brains to the shape of our pelvis. And when we finally start making stone tools some 2.5 million years ago, things are really going to start to change. Now think about that, just that—in 2.5 million years, we went from stone tools to Wi-Fi and hoverboards. But wait; let's slow this down just a minute here. Let's get right to our origins with *Sahelanthropus*.

So come with me now to the Djurab Desert in Chad, and it's here in central Africa where we're going to find Michel Brunet and his team with a game-changing paleoanthropological discovery. After decades of work in places like Afghanistan and Cameroon, Brunet and his team went to Chad and they unearthed a famous fossil that is certainly one of the earliest hominins on the human family tree. This takes us back to 2001, when Brunet's team discovered the remains of a primate skull that dates back approximately 6–7 million years. Now, to make some sense of that time frame, we could count back in time, year by year, one second at a time. If we counted down those years one second at a time, it would take us about 11.5 weeks to get there. No sleep, no pauses: one second equals one year—11.5 weeks.

Now, Brunet's skull is commonly referred to as Toumaï, or its scientific name TM266-01-06—bottom line, let's just say TM266 for short. TM266 ends up being identified as *Sahelanthropus tchadensis*, or to translate that, "Sahelian human from Chad." Now, this nearly complete skull's fractured surface still plainly reveals a prominent, wide, raised brow, and its canine teeth were smaller than other apes, who have longer canines for making those scowling threats and for fighting.

Now, this weathered skull is aged to the point it almost looks like driftwood in the shape of a jawless skull. But, despite its age, TM266 and *Sahelanthropus tchadensis* are remarkable because they date back to the same time frame that we separated from the chimpanzees and bonobos. It fits the timeline. And they were distinct from all other primates except our human line. This early primate was likely a facultative biped. Or, in other words, to one degree or another, it walked on two feet. Ultimately, it's our oldest known primate ancestor whose fossil remains indicate that it was bipedal. And that's where we started our human story, with these earliest walking, upright apes.

But hold on just a minute here. How in the world can someone look at a skull like the Toumaï skull and determine if it walked on two feet? Well, that's classic paleoanthropology and primatology. When all we have is just a skull, one major indicator for bipedalism can be seen in a thing called the foramen magnum. Now, that's the place where the skull meets the spine. So, if you take a look at a modern gorilla or a chimpanzee, its foramen magnum is actually toward the rear of the skull, not the center as with humans. So it's kind of like a lollipop with the stick toward the side of the candy instead of right in the middle. And one biological outcome of our transition to bipedalism is that our spine is more centrally located where it connects to the skull. Us bipeds have a more centrally located foramen magnum that our nonhuman primate relatives do not. Ultimately, Brunet and his colleagues argue that this fossil is actually one of our earliest bipedal ancestors close to the time we split from chimpanzees and bonobos. We haven't found anything, nothing bipedal anywhere, that is older than this discovery, so you can imagine there are some people who might challenge Brunet's conclusions.

Some actually think that *Sahelanthropus* was more of a peripheral human ancestor, and others challenge both of these arguments because of, frankly, a dearth of any other evidence. We have plenty of fossils out there to tell the story of our human origins, just not so many that date back 7 million years. So, remember, anthropology is testable and correctable, so we're going to continue to challenge these hypotheses as we discover more of the fossil record. But, in the meantime, let's go to a starting point that most everyone can agree on, and that's going to be *Ardipithecus*.

So, let's stay in Africa, and this time we're going to go east from Chad toward Ethiopia, and we can actually meet other paleoanthropologists who discovered one of the earliest known fossils of our human ancestors. My friends, meet Dr. Timothy White and Berhane Asfaw. Beginning in the early 1990s, Asfaw and White's team began uncovering some extremely fragile fossil remains. They were so delicate that it took 15 years to preserve and analyze what they found. They even had to use CT scans to make it easier to piece all these things together. The wait was worth it; their results were groundbreaking. They had discovered the remains of early human ancestors who lived over 4 million years ago. Their major find was the skeletal remains of a female who lived 4.4 million years ago. She was bipedal, but she could get around in the trees as well. Scientists dubbed this new discovery *Ardipithecus ramidus*, which in the local Afar language, *ardi* is "ground," and *ramid* means "root." So really, what we're talking about is "ground ape root." Let's stick with the name Ardi, just for easy sake.

How does Ardi compare with modern humans? Well, she was barely 100 pounds. She stood at about four feet tall. She even had a reduced canine tooth, which is one of the ways our human ancestors started to differentiate themselves from other apes on our unique evolutionary path. But remember, anthropology is a science—that means testable and correctable. So, as such, anthropologists have continued to test and correct our understanding of *Ardipithecus* and all our human ancestors.

It was in the 1990s and into the new century that Yohannes Haile-Selassie discovered more fossil evidence of *Ardipithecus* in Ethiopia. In fact, the remains that he unearthed were confirmed as part of the *Ardipithecus* classification, however they were older. They lived around 5 1/2 million years ago, so they named this early human ancestor *Ardipithecus kadabba*, because *kadabba* means "oldest ancestor" in the local Afar language. So, if we're looking to make a family portrait wall of our major human ancestors, then we'll certainly include both *Ardipithecus ramidus* and *Ardipithecus kadabba* because they are great representatives of our earliest bipedal ancestors.

But wait a second. How do we know all this to be true? How do humans go about making sense of bones and other remains from millions of years ago? Well, from time to time, humans have found fossils that beg further inquiry. Take Raymond Dart, for example. Dart was a professor of anatomy living in South Africa in the early 20th century, and one day Dart came across a captivating fossil. It was a skull with a jawbone, and it was a remarkable skull, because it was relatively complete, and Dart knew something was a little off with this find. He looked at it, and he said, this is clearly not a human, at least not a modern human, but it sure wasn't any other modern ape he knew of. So, this skull was smaller than a human skull, but larger than a chimpanzee. What was it?

Well, the skull was named the Taung Child—Taung because of the town in South Africa where it was found by quarry workers in 1924, and child because, as it turns out, this poor chap died as a kid, and we're pretty sure that he died as the result of an eagle attack. Dart proposed that we classify this skull as *Australopithecus africanus*, translated as "southern ape from Africa," and distinct from humans and the other primates. Southern apes, or Australopithecine, represent a whole other major era on our evolutionary timeline and our evolutionary tree.

But remember what we anthropologists do: we test and correct our ideas. And that means that once Dart comes up with this intermediate ancestor theory about *Australopithecus*, we've got to test it. And that we did, over and over again. And in the case of the Taung Child discovery, and Dart's theory that it represents a major moment in our evolutionary path, anthropologists challenged and tested these ideas rigorously, because at first many were skeptical of the idea that *Australopithecus* was a direct human ancestor. But over the decades, more and more fossils emerged to support the theory that at least some Australopithecines were indeed our direct *Homo sapiens* ancestor.

Ultimately, we learned that these Australopithecines lived some 4 million years ago, and they're only found in Africa, mainly eastern Africa, and there were many, many types. Remember our friend the Taung Child? We found out that he was definitely an upright walking ape, but he wasn't the type of *Australopithecus* that's directly related to us *Homo sapiens*. But don't worry, we found other versions of *Australopithecus—Australopithecus afarensis*, for example. Now, they are definitely our direct ancestors. And, while quite similar, and members of the same genus, *Australopithecus africanus*—that's the Taung Child—split from *Australopithecus afarensis* because they started down their own evolutionary path. And our direct *Homo sapiens* ancestors did the same.

Now, I'm sure you've heard of this next celebrity hominin before. She's perhaps the most famous of all the Australopithecines—*afarensis*, *africanus*, all of them. Now, here's a hint. Think the Beatles. Something about a sky with diamonds? Lucy. She was an *Australopithecus afarensis*, and they are our direct human ancestors. Her discovery in 1974 was celebrated widely because it is so rare to find such an old yet relatively complete preserved skeleton. And rumor has it that the team named the skeleton Lucy as the Beatles and "Lucy in the Sky with Diamonds" played through the team's celebration of their discovery.

So, let's recap. Where are we on the evolutionary path toward modern humanity? We started some 7 million years ago when our ancestors became those strange, upright walking members of the primate order, and our earliest fossil records of bipedal ancestors, from after the point we diverged from chimpanzees and bonobos, is the skull in Chad of *Sahelanthropus tchadensis*. Next, we discovered another key ancestor, *Ardipithecus*, who lived 4.5 million years ago. And last, we met *Australopithecus*, and we figured out that, as a genus, they come in different forms, multiple species, some of which are direct human ancestors, while others found their own evolutionary paths separate from the hominin line that leads directly to you and me and all modern humanity.

Before we fill in the gap between *Australopithecus* and *Homo sapiens*, let's just take a moment to observe a few basic changes that, over the course of millions of years, transformed who we are as humans. In particular, let's take note of a couple key trends as we move closer and closer toward *Homo sapiens*. First, *Sahelanthropus* through to *Homo sapiens*, we're going to see remarkable growth of our brain. *Sahelanthropus* and *Ardipithecus* had brains that were 400 cubic centimeters or smaller, and that's going to contrast with the 1,350-cubic centimeter brains that you and I are using to make sense of this lecture today. Second, we eventually distinguish ourselves as unique primates by walking upright, as well as our ability to make tools. So, wait—spoiler number one: as we continue this story, our brains are going to get bigger and bigger, from around 400 cubic centimeters to even as high as 1,450 cubic centimeters as recent as 20,000–30,000 years ago. And spoiler alert two: our tools and our tool making are just going to get better and better the closer we get to *Homo sapiens*.

Just as Raymond Dart analyzed that Taung Child fossil and concluded it did not fit any existing classification, the same thing happened to the famous paleoanthropologists Mary and Louis Leakey. In the early 1960s, before we found Lucy, but well after we had established an extensive fossil record of *Australopithecus*, the Leakeys found the remains of an early human that caught their attention. It wasn't a modern human, but it wasn't an *Australopithecus* either. It was somewhere in between the two. Its brain and body were a little larger, and it lived a little more recently than our *Australopithecus* ancestors. For these and other reasons, the Leakeys did what Dart did almost a half a century earlier. They created a new intermediary classification, one that clarifies our path from *Sahelanthropus* to *Homo sapiens*.

Additional discoveries throughout east Africa affirmed the status of this newly revealed ancestor: *Homo habilis*, translated "handy human." So, as we've done at each phase of our human origins story, let's add *Homo habilis* to our hominin family tree. And since the Leakeys first theorized about *Homo habilis*, many more fossils have been found, but they are all found in Africa. So that means it's apparent that our early ancestors were all still exclusively living in Africa.

Our knowledge of *Homo habilis* is clear. We know they lived in Africa about 2 1/2 million years ago; that their brain size—note the number—had grown to around 750 cubic centimeters. Now, remarkably, we find *Homo habilis* remains with very crude stone tools, something we had yet to find with any other ancestor, including *Ardipithecus*. More recently, a new generation of Leakey paleoanthropology has taken root. And, in 2012, Maeve Leakey, the daughter-in-law of Louis and Mary Leakey, discovered fossil remains in eastern Africa, and she argues that the fossils "confirm the presence of two contemporary species of early humans." Now, what she's arguing here is that we may need to break down the *Homo habilis* category into two subcategories.

Others, like Timothy White, who discovered *Ardipithecus ramidus* if you remember, disagree because he believes that the fossil record does not yet warrant the creation of a completely new branch of the human family tree. For the rest of us, we can review the evidence as new fossils emerge to help us finish this story. But for now, I want you to see that despite our neat and tidy review of these six major human ancestors, the actual path from one to the next is far more complicated than our abbreviated history may make it seem.

The next ancestor that we'll add to our timeline and family tree is *Homo erectus*. The first known *Homo erectus* discovery was actually found in 1891 by a Dutch surgeon, Eugène Dubois, who clearly saw it wasn't human like you and me. So, like Dart and the Leakeys, Dubois created a whole new classification for his ape-like skull discovery. He called it *Pithecanthropus erectus*—"ape who stands erect." And later fossils filled in the gap of our knowledge, and eventually paleoanthropologists modified the name of this ancestor to *Homo erectus*.

Now, to get to the first *Homo erectus* populations, we're going just under 2 million years ago, not too long since the emergence of *Homo habilis*. And, nonetheless, we can see why paleoanthropologists established yet another classification for our human family ancestors. These *Homo erectus* folks were different. First, their cranial capacity could exceed 1,000 cubic centimeters, and remember, we started around 400-some. This is clearly a major leap beyond *Homo habilis*. Second, the timeline for *erectus* is much closer yet to our *Homo sapiens* origins, less than 2 million years ago.

Third, these ancestors had quite an amazing tool kit compared with everyone who came before them. So, rather than basic Oldowan choppers that *Homo erectus* were making, more sophisticated tools were getting sharper and more specialized for the hunter-gatherer lifestyle. The evolution of tools, however, is actually getting us into archaeology, so we'll return to these tools in a future lecture. And quite significantly, back to *Homo erectus*, for the first time since we started this history, we can find *Homo erectus*, or hominins, outside of Africa.

Some scientists separate the *Homo erectus* populations into three distinct groups based on their geographic distribution. There's *Homo ergaster* from Africa, *Homo erectus* in Asia, *Homo heidelbergensis* from Europe, and as we move forward, let's just remember two major lessons from *Homo erectus* and their phase of the human story. First, until *Homo erectus*, we do not see any human relatives outside of Africa, period. We don't have any compelling tool kits, either. But *Homo erectus* changes all of that. Essentially, they're the ones that pave the way for our next and final group in our lineage, and it might sound familiar: *Homo sapiens*.

So, my dear friends, here we are. After millions of years of evolution, we find ourselves on the verge of *Homo sapiens*. The timeline sure has sped up, if you've noticed. Think about where we've been. Sixty-three million years ago, we're sharing all our ancestors with modern primates. Starting around 7 million years ago, *Sahelanthropus* emerged, but later died out by the time Ardi comes around. Then Ardi shows up 5.5 million years ago, but they also faded away after just a couple million years. It was approximately 4 million years ago that *Australopithecus* roamed the Earth, but again, like all previous ancestors, they only lived in Africa. But around 2.5 million years ago, *Homo habilis* came forth, and they worked out how to make some basic tools. And they too faded out, but some 2 million years ago, finally, living alongside *Homo habilis* for a while, *Homo erectus* rears his head around 1.9 million years ago, and then goes extinct as recently as 150,000 years ago.

So that's our pre-story. Somewhere out of that *Homo erectus* population, we emerge as *Homo sapiens* 1.0 or *Homo sapiens* archaic. We understand our *Homo sapiens* roots by dividing that phase of our history into two parts: the archaic and the modern periods. Now, the archaic form of *Homo sapiens* first came into our fossil record when our friends the Leakeys found some *Homo sapiens* remains dated to about 100,000 years ago. But that was 1967, and more contemporary teams have revisited those same exact fossils and the site, and with their more modern technologies, they dated these early *Homo sapiens* to actually more like 190,000 years ago. So, the most celebrated fossils from this *Homo sapiens* line, the archaic version, are what we call the Omo skulls, and they definitely are our oldest known *Homo sapiens* fossils, at least for now.

As of yet, we've only found early *Homo sapiens* fossils in eastern Africa, leading us to believe that modern humans are likely rooted to a population of *Homo sapiens* who lived in Africa before spreading out to other continents, and we'll look at this migration closely in our next lecture. But for now, let's wrap our overview of the story of human evolution by looking at modern *Homo sapiens* and how we differ from our early ancestors.

There are four major takeaways from the transition of our ancestors from early hominin to *Homo sapiens*. First, brain capacity. It expanded to 1,450 cubic centimeters, though we started at less than 400. And in the past 30,000 years our brains have actually gotten a little smaller. These days, the average adult male cranial capacity is around 1,350 cubic centimeters. Second, very sophisticated pre-industrial tools and weapons—we make those, no one else did. Third, regional adaptations. Those are going to lead to differences like skin color. And fourth, we are racially one, and we're the only remaining hominin species left on Earth: *Homo sapiens*.

Now, those are all themes for our future lectures, but to solidify everything we've learned today, let's see one final example of our evolutionary history and the value of anthropology's test-and-correct ethos, and I promise it's a quick one. Now, as we saw today, the storyline was fairly simple. After evolving right along with some of the other apes, we hominins eventually take our own distinct evolutionary path as upright walking humans. Now, the fossil records, primatology, and genetics all concur that this threshold to humanity opened some 7 million years ago.

And in the 7 million years since we started this journey, our story has been relatively simple. Our brains and bodies got bigger as time moved on, our tools improved rapidly once we developed them, and they apparently helped us avoid the fate of all other hominins: extinction. So we stood upright, we got bigger brains, got cooler tools, and eventually took over the world. That's true-ish, but don't forget, this is anthropology; we're never satisfied with stagnant theories. We test and retest our ideas to improve our knowledge, and as that new evidence emerges, we gleefully reconcile it with the existing theories, revising our story of human evolution.

Now, one recent test-and-correct challenge we've dealt with recently relates to J. R. R. Tolkien's hobbit. Not really the hobbit, but it's a small human that lived in Indonesia within the past 20,000 years or so. After the bones of this small yet very recent human were discovered in 2003, the popular press labeled it a hobbit. Now, the scientific name was *Homo floresiensis* because their remains were found on the island of Flores, Indonesia. Specifically, what we found was a population of very small—one meter tall—humans who not only had a complex tool kit, but lived long after we thought every hominin species but *Homo sapiens* had gone extinct. And they had relatively small brains for being such amazing toolmakers.

The hobbit messes up our clean storyline of better tools through bigger brains as the road to survival, as well as that old idea about us being the sole remaining hominins for the past 40,000 years. So what do we do with such a find? Well, we deal with it. We make sense of it. So, for now, most folks tend to keep our hominin evolution storyline intact by subscribing to one of two theories. Some people argue that the small humans were smaller simply because of a genetic condition that emerged in the region and remained isolated, while a second theory states that on a small island with limited resources, larger animals like humans would eventually evolve to become smaller to better compete for limited resources.

Regardless, the fossil record is clear and it's deep. Walking upright appears to have been a game changer for us, and in many ways, we can consider bipedalism as the foundation of our humanity, biological and otherwise. We'll continue our anthropological journey with a closer look at our early human tool kit to retrace humanity's expansion out of Africa.

Tracing the Spread of Humankind

The previous lecture retraced our origins to the first bipedal apes some 7 million years ago. Remarkably, we *Homo sapiens* are the sole survivors of what used to be a vastly diverse hominin family tree. This lecture adds to our human story by bringing in some geography to retrace the spread of humankind. Specifically, we'll see how humans migrated into just about every corner of the world, including the Americas. Starting in East Africa, we'll watch *Homo sapiens* spread into the Middle East, Asia, Europe, and eventually the Americas.

Naia

In 2007, a software engineer and his avid scuba diver friends accidently discovered one of the oldest human skeletons ever found in the Americas.

They named her Naia, after the Naiads, who were mythical Greek water nymphs. Naia may have been alone the past 12,000 years, but she was one of the very first Americans.

Naia's presence on the Yucatan Peninsula tells us that humans have been in the Americas for at least 12,000 to 13,000 years. Her remains rank among the oldest, most complete human skeletons ever discovered in the Americas.

Homo erectus

After over a century of revolutionary fossil discoveries, we now have a clear idea of when the first hominins migrated out of Africa. Appearing nearly 2 million ago, *Homo erectus* fossils line the trail of their collective migrations into the Middle East, Europe, and Asia. They showed up in Java about 1.8 million years ago, and in Europe and Asia not long after that.

The Zhoukoudian site near Beijing, for example, has produced fossilized remains from some 40 different *Homo erectus* individuals. Those fossils and others show us that by 800k years ago, Homo erectus was living as far as Southeast Asia, the Middle East, and southern Europe.

Homo erectus, however, has been extinct for 400,000 years, and unlike *Homo sapiens*, they never made it to northern Asia, northern Europe, or the Americas.

Three Theories

After *Homo erectus* went extinct, it was up to *Homo sapiens* to settle the rest of the globe. But why did *Homo erectus* go extinct after spreading across Africa and into southern Asia and Europe? How did *Homo sapiens* come to populate the world? Three major theories seek to answer these questions:

1. In the regional continuity model, modern Europeans have a direct genetic link to the *Homo erectus* populations that preceded them in Europe. Proponents of this model typically point to continuity and

differences between *Homo erectus* and modern humans in Asia and Europe.

2. The complete replacement theory states that an intrepid line of *Homo sapiens* spread out from Africa and populated the globe starting some 60,000 years ago. In the process, they ultimately replaced *Homo erectus* and all other hominin populations.

3. The partial replacement model adds the element of interbreeding. According to the partial replacement model, modern *Homo sapiens* spread rapidly well beyond the African continent some 60,000 years ago. Along the route they interbred with and eventually replaced all other hominins, including the Neanderthal.

As scientists, anthropologists love new evidence, and that's exactly what they received in 2015 when Beijing paleoanthropologists discovered 47 human teeth in a cave in southern China. The teeth dated to around 100,000 years ago. This discovery puts modern humans in eastern Asia early enough to lend support to the partial replacement model for the peopling of the world.

The Genetic Trail

In addition to fossils, paleoanthropologists also rely on molecular genetics to retrace the global dispersal of humankind. Just as all humans share a common ancestor with all living primates, we also share a much more recent earliest common ancestor with all living *Homo sapiens*.

Geneticists have traced our mtDNA (mitochondrial DNA) and our Y-chromosome DNA back to Africa.

Y-chromosome DNA is genetic material males inherit from their father's line. Y-chromosome Adam is how we refer to the *Homo sapiens* male ancestor that all living humans share. By dating the frequency of human mutations, geneticists estimate that this man lived in Africa, perhaps as early as 100,000 years ago.

Similarly, we've studied mtDNA from across the world in order to determine the woman from whom all modern humans descend. MtDNA is how we genetically explore our maternal inheritance.

● When a human egg cell is fertilized, chromosomes from

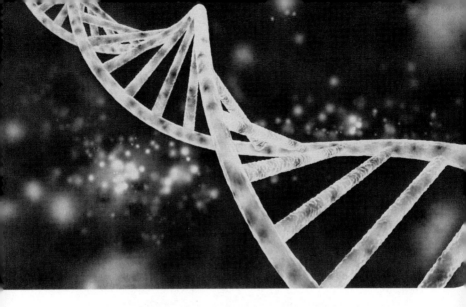

a sperm cell enter the egg and combine with the DNA in the egg's nucleus. The sperm cell's mtDNA, however, never enters the egg, so the fertilized egg retains the mother's exact copy of her mtDNA.

- That's why mtDNA is such an amazing window into our evolutionary past. Based on their analysis of mtDNA from people around the world, scientists estimate that our maternal line as humans dates back to mtEve some 200,000 years ago. Like Y-chromosome Adam, she too would have lived in Africa.

It's important to note the mtEve was not the first *Homo sapiens* woman to walk the Earth. There would have been human women before her and there would have been human women contemporary with her. But it was this particular woman who passed her mitochondria down, through an unbroken female line, to all humans living today.

In any case, as the *Homo sapiens* population grew, they migrated out of Africa and populated the entire planet.

Entering the Americas

For the final segment of this lecture, we'll look at the fossil record, genetics, and climate history to map out the peopling of the Americas.

The Last Glacial Maximum covered continents with ice sheets as recently as 30,000 years ago, and these ice sheets effectively closed the doors of human migration into Northern Europe and North America. But then, around 20,000 years ago, the ice started melting.

Glacial melting increased sea levels and revealed new opportunities for human expansion. It opened up new routes into North America through Siberia. The glacial melt created coastal and interior corridors abundant in the kind of flora and fauna that enticed human hunters and fishers.

Archaeological evidence supports this. In the northern Yukon, the Bluefish Caves are 3 small caves where early humans butchered animals, perhaps as early as 25,000 years ago. That's major: 25,000 years ago, this site and any population who lived there would have been cut off from the rest of North America by those massive ice sheets. The people who spent time in the Bluefish Caves lived on the edge of these ice sheets, hunting large game like mammoth and caribou.

Similarly, further to the west, on the Siberian side of things, there is a site on the Yana River with hundreds of tools made by mammoth hunting humans who lived there around 27,000 years ago.

There's nothing conclusive that shows human presence in the Americas south of the Yukon before 15,000 to 16,000 years ago, but that changed. It was at Fort Rock Cave in Oregon where archaeology unearthed dozens of prehistoric shoes. Some of these dated to 10,000 years ago. With sea levels down, a new land bridge connected Siberia with Alaska, opening the migration into all of the Americas.

Near Murray Springs, Arizona, researchers uncovered a site where nomadic hunters chased big game like mammoths and bison. The stone tools and fossilized bones date the site to over 12,000 years ago, and it's one the earliest human sites in the Americans below Beringia and the Yukon.

Another early site exists at Monte Verde, Chile, where archaeologists found foraged plants, plenty of tools, a hearth, and even some simple animal skin tents with poles. This site dates to 12,500 years ago.

The Clovis Point

Remarkably, most of the people who first settled in the Americas shared a similar tool kit featuring a fluted stone tool called a Clovis point. It's about 2 fingers wide and shorter than a pencil.

In 1932, the archaeologist Edgar Howard came to Clovis, New Mexico. He'd caught word that a local road crew had uncovered unknown quantities of ancient bones. He excavated the site and discovered layers and layers of mammoth bones. Between the bones were seemingly countless specialized spearheads that early Americans had skillfully manufactured for big-game hunting.

Clovis points had amazingly sharp edges that were fluted at the end. Knappers shaped these distinctive lance-like points and would insert them into spears and shafts. People used Clovis points to settle areas from the Pacific Northwest, the American South, and into the Great Lakes. Since that first discovery back in New Mexico, we've now found thousands of Clovis points all across North America.

Genetics

An exciting 21st-century discovery was a comprehensive study of the prehistoric Native American genome. A group of UC Berkeley researchers sequenced the genome of 23 ancient humans found in North and South America, including the DNA of a 12,600 year old boy who was buried in Montana with over 125 artifacts.

Scientists compared these ancient genomes with the genomes of 31 living Native Americans, Pacific Islanders, and Siberians. The results are astounding. They show us that these early Americans share DNA markers with living 21st-century Native Americans, as well as with people from the Pacific Islands and Siberia.

Moreover, this consequential study clears up 2 enduring questions about the first Americans:

1. The first Americans arrived through Siberia after the last ice age was drawing to a close around 20,000 years ago.
2. This population remained in the north for several thousand

Clovis point

years before splitting into northern and southern branches.

The waves of humans coming into the Americas had to wait until the ice barrier melted down, just about 15,000 years ago. That is the same time frame as the documented entry of humans beyond the Yukon.

Anthropologists always remain open to new discoveries, and there is certainly a chance that, in the future, they may discover earlier, pre-Clovis sites in the Americas. Anthropologists may very well find evidence for human migration outside of the Beringian model.

Suggested Reading

Graf, Ketron, and Waters, eds., *Paleoamerican Odyssey*.

Harari, *Sapiens*. New York: Harper, 2015.

Stringer, *Lone Survivors*.

Questions to Consider

1. How does archaeology help us determine when humans first arrived in the Americas? How and when did they arrive?

2. What role did climate play in the peopling of the Americas?

Tracing the Spread of Humankind

I n our previous discussion, we retraced our origins to the first bipedal apes some 7 million years ago, and, remarkably, we learned that we *Homo sapiens* are the sole survivors of what used to be a vastly diverse hominin family tree. So today, let's add to our human story. Let's bring in some geography to retrace the spread of humankind, and, specifically, we'll see how humans migrated into just about every corner of the world, including the Americas.

Our journey today begins with actually a thrilling scuba adventure. Now, just imagine, you and your buddies are on a sunny Cancún vacation, and with your tanks filled and fins on your feet, it's time to jump right in that water and explore a long, mysterious tunnel. So, splash, the water's warm and super clear. And with just a little encouragement from your buddies, you just keep going deeper and deeper until, suddenly, this narrow tunnel opens up into this dark and monstrous cave. Flash, you turn on your scuba high beams, and they reveal a sensational, subaqueous world, and it's packed with hauntingly intact skeletons. There's pumas, saber-toothed tigers, and even something that kind of looks like an elephant. But wait, what's that over there? Yeah, over there.

Alone in this skeletal zoo, you also discover the well-preserved bones of a teenage girl. But this isn't just any girl. When researchers examine your find, this turns out to be a teenage skeleton from someone who lived some 12,000 years ago. Literally, you've made the discovery of a lifetime; or, to be honest, thousands of lifetimes for that matter. And get this: that scuba story actually happened. In 2007, a software engineer and his avid scuba diver friends accidently discovered one of the oldest human skeletons ever found in the Americas, and they named her Naia, after the Naiads, who were mythical Greek water nymphs. Naia may have been alone the past 12,000 years, but now we can all celebrate her as one of the very first Americans.

We'll start our conversation today with Naia because her remains are exactly what we need to retrace the spread of humankind across the globe. And it's as simple as that. When we find human skeletons and artifacts in a particular region of the world, they enable us to confirm or document our great human migration, both in timeline and in trajectory. Naia's presence on the Yucatan

Peninsula, for example, clearly tells us that humans have been in the Americas for at least 12,000–13,000 years.

But, there are plenty of other discoveries that can help us document when our *Homo sapiens* ancestors migrated and settled on new continents, so let's take a look. Essentially, we'll be Paleolithic detectives, traveling the world over. We'll map out the story of how humans populated the Earth. Starting first in east Africa, we'll watch *Homo sapiens* spread into the Middle East, Asia, Europe, and eventually the Americas.

Now, the 2014 Yucatan discovery was a game changer, and in a way, Naia could be dubbed the Lucy of Paleo-American studies. And that means that her remains rank amongst the oldest, most complete skeletons ever discovered in the Americas. And even more exciting to us anthro-folks is the fact that her DNA actually documents a direct link to the present Native American lineages. Ultimately, this young woman, Naia, affirms quite clearly that humans have been in the Americas for over 12,000 years. Now, just think about that for a moment. Then, place this archaeological fact on the shelf, so we can continue searching for other discoveries that will reveal even more details about the spread of humankind into the Americas.

For context, let's review how and when humans first migrated from Africa. Let's look at the big picture. After over a century of revolutionary fossil discoveries, we've now got a clear idea of when the first hominins migrated out of Africa. Appearing nearly 2 million years ago was *Homo erectus* and his fossils that line the trail of their collective migrations into the Middle East, Europe, and Asia. They show up in Java about 1.8 million years ago and in Europe and Asia not too long after that. The Zhoukoudian site near Beijing, for example, has produced fossilized remains from some 40 different *Homo erectus* individuals—males, females, adults, children. And those fossils and others show us that by 800,000 years ago, *Homo erectus* was living as far as Southeast Asia, the Middle East, and southern Europe.

Now, *Homo erectus*, they got around, but *erectus* has been extinct for 400,000 years, and unlike *Homo sapiens*, they never made it to north Asia or northern Europe, let alone the Americas. So how did *Homo sapiens* pull this off? Well, the short of it is that *Homo erectus* went extinct, and then it was up to us *Homo sapiens* to settle the rest of the globe. But why did *Homo erectus* go extinct after spreading across Africa and into south Asia and Europe? How did *Homo sapiens* come to populate the world? Let's review the three major theories out there.

First, let's check out the Regional Continuity model. In this theory, as *Homo erectus* spread into Asia and Europe, these regional populations then took their own evolutionary paths to become modern Europeans and Asians. In other words, modern Europeans have a direct genetic link to the *Homo erectus* populations that preceded them in Europe. Now, proponents of this Regional Continuity model typically point to continuity and differences between *Homo erectus* and modern humans in Asia and Europe.

Now, in complete contrast to the Regional Continuity model, our second theory, the Complete Replacement theory, states that an intrepid line of *Homo sapiens* spread out from Africa and started populating the globe some 60,000 years ago. And, actually, this complete replacement begins well before that, with a relatively small population of *Homo sapiens* in Africa. Now, this theory states that these distinct populations eventually spread out of Africa and eventually populated the modern world. And, in the process, they replaced *Homo erectus* and all other hominin populations. And if this Complete Replacement theory is true, one consequence is that modern Asians and Europeans would share direct, more recent genetic ties to each other and early *Homo sapiens* in Africa.

Now, the third and final theory of how *Homo sapiens* outlived *erectus* and populated the globe is actually a compromise. It combines the Regional Continuity and the Complete Replacement models, and as a result, this last model, the Partial Replacement model, adds the rather salacious element of interbreeding. According to the Partial Replacement model, modern *Homo sapiens* spread rapidly well beyond the African continent some 60,000 years ago, and along the route they interbred with and eventually replaced all other hominins, including the often-maligned Neanderthal.

And, in defense of our dear Neanderthal cousins, despite the fact that they're not our direct descendants, they really weren't different from us. They were a bit shorter and stocky, and they had kind of big noses. They were well adapted to colder climates, but they were amazing big game hunters. And archaeologists tell us that, like early *Homo sapiens*, they also buried their dead. Researchers recovered DNA from Neanderthal fossils and, because of the Neanderthal Genome Project, we now have great evidence to better discern the relationship between *Homo sapiens* and Neanderthal.

Now, this may come as a shock to some of us, but proponents of all three of our human geography theories accept that there is significant evidence for interbreeding between modern humans and more archaic members of the hominin line like Neanderthal. In fact, this interbreeding may have contributed to their decline. But despite the fact that we outbred and outlived Neanderthal, believe it or not, they're still among us to this day. Have you ever seen one? Well, would you believe you're looking and listening to one right now? Yeah, that's right, it's me. And I tell you what, I'm not alone.

You see, most non-African populations still actually carry Neanderthal genes in their modern genome. When I recently had my DNA analyzed, I discovered that 2.3 percent of my genome matches that Neanderthal genome. Imagine that. It's right in line with the general average, which is closer to about three percent. But I digress. Let's get back to the question of how and when humans populated the globe.

As scientists, anthropologists love new evidence, and that's exactly what we received in 2015 when Beijing anthropologists discovered 47 human teeth in a cave in southern China. These teeth dated to around 100,000 years ago. Actually, this discovery puts modern humans in east Asia early enough to lend support to that Partial Replacement model for the peopling of the world. Testable and correctable—that's one of the reasons I love this anthropology.

So what have we discovered thus far? First, we learned that modern humans who populated the Earth arose from an east African population who date back some 200,000 years ago. And, second, we learned that the geographic distribution of *Homo erectus*, Neanderthal, and other late hominins also extended beyond Africa, but unlike *Homo sapiens*, they've all gone extinct, leaving us to explore and inhabit the planet.

Putting the fossil record aside for just a moment, let's also consider our genetic trail, because, in addition to fossils, paleoanthropologists also rely on molecular genetics to retrace the global dispersal of humankind. And, just as all humans share a common ancestor with all living primates, we also share a much more recent earliest ancestor with all living *Homo sapiens*. Now, geneticists have traced our mitochondrial DNA and our Y-chromosome DNA back to Africa.

You see, we may share a common ancestor, but our DNA isn't identical. When a new mutation occurs and endures, that genetic marker is passed on to every subsequent generation. Our individual genomes carry these genetic markers, and they help us group you and me with people who share many of the same markers. So, thanks to molecular genetics, it's time we meet the ancestors that you and I share with every living human on the planet. It's time we meet mitochondrial Eve and Y-chromosome Adam.

as our *Homo sapiens* family grew, we migrated out of Africa and populated the entire planet, but it was our entry into the Americas that was our most recent game-changing mass migration. So, for the final segment of this lecture, we'll look at the fossil record, genetics, and even climate history to map out the peopling of the Americas.

It may sound like a strange place to start, but we need to understand a little about the world's climate over the past 30,000 years or so. Specifically, when we think about glaciers, we can see why *Homo sapiens* were the first hominin to make it to the Americas, and the short of it was this. We needed to get through the most recent ice age before we could migrate from Asia into the Americas.

The Last Glacial Maximum covered continents with ice sheets as recently as 30,000 years ago, and these ice sheets effectively closed the doors of human migration into Northern Europe and North America. But then, around 20,000 years ago, the ice started melting, and glacial melting actually increased sea levels and revealed new opportunities for human expansion into the New World. Literally, it opened up new routes into North America through Siberia. This glacial melt created coastal and interior corridors abundant in the kind of flora and fauna that would entice *Homo sapiens* hunters and fishers. So, fellow paleo-detectives, let's assemble the evidence.

Remember Naia from our scuba adventure? Well, her presence in the Yucatan Peninsula tells us that humans have been in the Americas for at least 12,000 years, and that timing fits well with our knowledge of the receding ice sheets since the Last Glacial Maximum. If it was 20,000 years ago that the ice sheets began to recede and opened new paths into the Americas, then we should expect to find fossils lining our route from Siberia. Now, let's do some archaeology to dig deeper and find the trail that led Naia and her ancestors into the Americas.

First, let's head to the northern Yukon, to the Bluefish Caves. It's basically three small caves where early humans butchered animals probably around 25,000 years ago. And this is major. Twenty-five thousand years ago, this site, and any population who lived there, would have been cut off from the rest of North America by those massive ice sheets. The people who hung out in Bluefish Caves lived on the edge of the ice sheets, hunting large game like mammoth and caribou. Similarly, further to the west, on the Siberian side of things, we discovered a site on the Yana River with hundreds of tools made by mammoth-hunting humans, and they lived there around 27,000 years ago.

So our stage is set. Here in Beringia, on the northern side of the Last Glacial Maximum, right here in the Yukon and Siberia, we see evidence of human populations on the frontier. But below that, we haven't found any evidence of human presence in the Americas. But what happens once the glaciers melt and the ice sheets open new coastal and interior corridors to the south?

Well, there's nothing conclusive that shows us human presence in the Americas south of the Yukon before 15,000–16,000 years ago, but that changes fast a couple thousand years later. What's the evidence? Well, maybe it was destiny that the company Nike emerged in Oregon, because it was Fort Rock Cave in Oregon where archaeology unearthed dozens of prehistoric shoes, some which dated to 10,000 years ago. So, with sea levels down and a new land bridge connecting Siberia with Alaska, migration is open to all of the Americas. But it wasn't just the Pacific Northwest that attracted early human inhabitants.

Near Murray Springs, Arizona, for example, we uncovered a site where nomadic hunters chased big game like mammoths and bison. The stone tools and fossilized bones date to over 12,000 years ago, and it's one of the earliest human sites in the Americas below Beringia and the Yukon. So, like those who lived near Fort Rock Cave, the Murray Springs mammoth hunters arrived only after the glacial melt made that possible. And these hunters left us one of the most prolific archaeological sites in Paleo-American studies. And finally, there's one other early site at Monte Verde, in Chile, where archaeologists found foraged plants, plenty of tools, a hearth, and even some simple animal skin tents with poles. And this site dates to 12,500 years ago.

So, the archaeological record shows us that once the Last Glacial Maximum began to fade, humans migrated south into the Americas. We have Naia's skeleton in Central America, we have artifacts in Monte Verde in South America, then there's those shoes back in Oregon, and they all tell us that humans have been in the Americas for thousands of years. So, set the clock back 13,000–14,000 years ago, and we have definite human evidence in North and South America. Then, within a few thousand years, *Homo sapiens* were everywhere across North and South America. But were they all related? Are they all rooted in those mammoth-hunting traditions coming through Beringia? The quick and easy answer is yes—their tool kits and genetics tell us so.

Remarkably, most of the people who first settled in the Americas shared a similar tool kit featuring a fluted stone tool called a Clovis point. About two fingers wide and shorter than a pencil, the Clovis point is more American than even jazz and apple pie. It's one of the first technological revolutions this continent has ever seen. We figured this out when archaeologist Edgar Howard came running to Clovis, New Mexico, in 1932 because he caught word that there was this local road crew and that they had uncovered all sorts of quantities of unknown, ancient bones. So he ran there, he excavated the site, and he discovered layers and layers of mammoth bones. But there between the bones were seemingly countless specialized spearheads that early Americans had skillfully manufactured for big game hunting.

What makes the Clovis point so unique is its amazing sharp edges that were fluted at the end. Now, knappers shaped these distinctive lance-like points and they'd insert them into spears and shafts. It's as if they were available at some prehistoric Ace Hardware store. All across North America, people used Clovis points to settle areas from the Pacific Northwest, the American South, and even into the Great Lakes. And since that first discovery back in New Mexico, we've now found thousands of Clovis points all across North America. And much in the same way smartphones bind contemporary humans into bands of iPhone and Android tribes, the Clovis point was a uniquely American invention. And that's what the Clovis point teaches us about the relationship between early Americans and the people who came out of Siberia. But what about genetics?

Well, one of the most exciting 21st-century discoveries in Paleo-American studies was a comprehensive study of the prehistoric Native American genome. It was a group of UC Berkeley researchers who sequenced the genome of 23 ancient humans found in North and South America, including the DNA of a 12,600-year-old boy who was buried in Montana with over 125 artifacts. These scientists compared ancient genomes with the genomes of 31 living Native Americans, Pacific Islanders, and Siberians, and the results are astounding. First of all, they showed us that these early Americans shared DNA markers with living 21st-century Native Americans, as well as with people from the Pacific Islands and Siberia. Moreover, this consequential study clears up two enduring questions about the first Americans.

One, the first Americans arrived through Siberia after the last ice age was drawing to a close starting around some 20,000 years ago. Second, this population remained in the north for several thousand years before splitting into northern and southern branches. And what's the evidence? Well, we've discovered there's plenty of evidence, and it's not just a bunch of Clovis points. We consulted the archaeology of Siberia, the Bering Strait, Alaska, and all the Americas, and, if we map out the earliest sites across North and South America, we see a revealing timeline. Before 15,000 years ago, we don't see human sites anywhere except for right up there in Beringia and the Yukon. And climate scientists told us there was very good reason for the late entry into the rest of North and South America, and that was the ice age. The waves of humans coming into the Americas had to wait until that ice barrier melted down just around 15,000 years ago. But hold it, isn't that the same time frame as the documented entry of the humans beyond the Yukon? You bet it was.

Now, we anthropologists always remain open to new discoveries, and there is certainly a chance that, into the future, we may discover earlier, pre-Clovis sites in the Americas, and we may very well find the evidence for human migration outside of the Beringian model we discussed today. A great example is archaeologists who have investigated the possible presence of pre-Clovis humans along the Savannah River in South Carolina at a place called the Topper site. Nevertheless, we've got to keep on digging. And, in the meantime, let's stick with our testable and correctable story of humankind.

So there we have it—Paleo-American studies, the final chapter of how early humans spread throughout the world: the arrival of *Homo sapiens* in the Americas. Twelve thousand six hundred-year-old DNA from a single prehistoric Montana boy profoundly changed the way we understand the history of humankind in the New World. And just like Naia's DNA from the Yucatan, his DNA is directly related to contemporary Native American genome.

In fact, this lineage, his lineage, accounts for approximately 80 percent of the contemporary Native American genome, from Native Alaskans in the north, down to the Aymara people of South America. And even more fascinating is the fact that the Montana boy's DNA also indicates that he's a descendant from a line of humans with roots tracing back to Beringia and Siberia, and ultimately, of course, to Africa.

We *Homo sapiens* have long been a wandering species. Our forbearers never stopped moving, and as a result, for better or worse, all parts of the globe bear the mark of modern humanity. And yet, no matter how far and wide we've traveled, there remains the compelling fact that we can trace all of our roots to a shared geographical homeland, and to the ancestors common to us all. And the big picture couldn't be clearer. Through anthropology, we discover not only the contours of the spread of humankind, but, with genetics and geography, we reveal new ways to understand the oneness of the human race.

Anthropology and the Question of Race

S o far in this course, we've seen some of the remarkable ways anthropologists explore the origins and diversity of humankind. And if there's one thing we've learned, it's that our genetics, our biology, and even the archaeological record all converge on one anthropological truth: Despite our physical and cultural differences, we are a single human race. Yet enduring ambiguities remain between race, skin color, and biology. This lecture aims to unpack those while acknowledging that anthropologists, archaeologists included, have had much trouble coming to terms with the concept of race.

Social Darwinism

In the early days of anthropology, Darwin's ideas about the biology of evolution proved so convincing that many leading thinkers of the time started applying this idea far beyond biology. Folks like Herbert Spencer argued that just as biological organisms evolve from simple, single-cell organisms to more complex ones, human cultures evolve from simple societies to highly complex ones.

Their assertion was clear: humans evolve culturally from savage to civilized. This idea of social evolution, or social Darwinism as Spencer described it, became a popular explanation for the diversity of humankind across the globe.

Darwin's book came out in the mid-1800s. As the close of the trans-Atlantic slave trade transitioned into the dawn of European empires, people in Darwin's age were exposed to new and diverse world cultures.

Inspired by Darwin's ideas on biological evolution, social Darwinists used evolution as a way to explain human diversity and to classify people and cultures across

the globe. Humans who lived in small, simple societies (like a single-cell organism) were deemed "less evolved" than those who lived in large, complex ones.

As the scientific community ushered in the idea of social Darwinism, the idea spread because it helped make sense of the seemingly limitless human diversity on our planet. Suddenly it became easy to explain why some people were entering the so-called modern world, with industrial factories, powerful trains, and upper-class values, while others were stuck in the past.

The idea of race emerged as an all-encompassing way to describe these vast human differences. The differences Europeans saw in humans from Africa or Asia were attributed to the fact that we all come from distinct lineages which we called races.

Upon reflection, many thinkers further developed this idea of human races, to the extreme that proponents of this way of thinking coined and used the term "white man's burden" to describe the idea that white men had the moral obligation to use their advanced state to uplift non-whites the world over. The concept of the white man's burden was used to justify the colonization of the non-white world.

In 1896, the US Supreme Court heard the case *Plessy v. Ferguson*, in which the justices considered whether the Constitution allowed for discrimination based on white versus non-white races. In a 7-to-1 ruling, the court held that if nature created people and races that were not equal, the Constitution could do nothing to change that.

Eugenics

With millions of dollars of support from wealthy families and foundations, special research centers were built to find solutions to the white man's burden. In 1904, for example, the Station for Experimental Evolution (SEE) was established at Cold Spring Harbor on Long Island.

Researchers at this institute were charged with identifying ways to prevent defective germ plasm, supposedly the biological and primitive remnants of our savage past, from spreading any further.

The researchers at SEE not only developed a series of recommendations on how to proceed with this project, they also started to identify family lines in the area that were tainted by criminality, mental disorders, and physical deformities. The idea was simple: Find a way to limit the reproduction of what they deemed to be defective human germ plasm.

Shockingly, their official recommendations included lethal chambers as the most effective solution, but they advised policy makers and others that the US population was not ready to accept this approach. Instead, other strategies were unleashed, namely marriage and sterilization laws. In 1907, for example, the state of Indiana enacted sterilization laws to make sure criminals and other undesirables could not pass on their defective humanity to another generation.

In 1927, the Supreme Court upheld a Virginia statute that promoted compulsory sterilization of the so-called unfit.

The idea of applying eugenics to improve humankind spread far beyond US borders, and was embraced by many, including Adolf Hitler and his Nazi party. In fact, when Nazi war criminals were tried in Nuremburg, they actually cited the words of US eugenics leaders and Chief Justice Oliver Wendell Holmes to defend their genocidal actions.

Chief Justice Holmes's words from the majority opinion in the aforementioned 1927 case demonstrate how pervasive and public these ideas were. Holmes wrote: "It is better for all the world, if instead of waiting to execute degenerate offspring for crime, or to let them starve for their imbecility, society can prevent those who are manifestly unfit from continuing their kind."

Fortunately, Holmes's perspective and the eugenics project in general began to lose ground in the US in the 1930s and '40s. But this was precisely the time when Nazi Germany was slaughtering millions of Jews, homosexuals, and anyone else they viewed as undesirable.

Franz Boas

Apart from eugenics scholars, there were other science-minded people who took another approach towards exploring and understanding human diversity. The father of American anthropology, a Polish immigrant named Franz Boas, is a great example.

Boas didn't see culture as the product of our biology. Instead, he looked to see how one's environment shapes unique cultural traditions. Because he looked beyond biology to explain human diversity, he argued that there wasn't really a scientific basis for ranking people and cultures on the spectrum of primitive to civilized.

For example, rather than ethnocentrically judging Arctic igloo dwellers as primitive hunter-gatherers, Boas saw them as uniquely adapted to a challenging environment. Ultimately, Boas learned and taught that cultures are neither able to be ranked nor biologically based. They're relative, and as such people from different cultural traditions are equally and fully human. This idea is basically what we call cultural relativity. Some cultures aren't better or worse than others; they're just different.

Franz Boas saw Arctic igloo dwellers as uniquely adapted to a challenging environment.

Theories

When biological determinists saw people who looked different and lived differently from themselves, they ascribed the difference to biology. And as such, they divided humanity into biological races. For example, 18th- and early 19th-century anthropologists who advocated this position commonly described our human origins and races in 1 of 2 ways:

1. Some argued that Africans, Asians, and Europeans (for example) each had distinct origins: multiple Edens, so to speak.

2. Others theorized that despite the fact that humanity shares common origins, humankind split into multiple, biological races like the ones we see listed in the US Census.

The human genome and our knowledge of our hominin ancestors definitively refute both of these theories.

The Myth of Race

If anthropologic evidence tells us that different skin colors and cultures are manifestations of a single biological race, why does the myth of multiple human races persist?

One reason is clumsy use of the term *race*. The myth of multiple races endures in universities, for example: Financial aid and admission applications still use multiple racial categories.

Toward the end of the 20th century, for example, members of the America Anthropological Association gasped with incredulity at the argument put forth by Richard Herrnstein (a psychologist) and his coauthor Charles Murray (a political scientist) in their book *The Bell Curve*.

The controversial authors looked at both genetic and environmental influences on human diversity. The controversial authors ultimately argued that, compared with black Americans, both East Asians and white Americans historically performed better on IQ tests because of their biology—their genetic heritage.

Critics and supporters of this argument fiercely debated these findings. Mel Konner from Emory University, who is a Harvard-trained M.D. as well as an anthropology Ph.D., effectively described why anthropologists and others like Noam Chomsky and Stephen Jay Gould fervently refuted *The Bell Curve's* conclusions. Specifically, he said that Murray and Herrnstein presented strong evidence that genes play a role in intelligence, but their evidence in no way supported their theory that genetics explain the documented differences between the IQs of black and white Americans.

Similarly, in 1994, the members of the AAA put forth an official statement on race to clear up the pervasive myth of multiple biological races. Simply put, they said that all humans are members of a single species, *Homo sapiens*. In their words, "Differentiating species into biologically defined races has proven meaningless and unscientific as a way of explaining human variation."

Health Questions

If racial categories are social constructions, not natural divisions in the human species, then how do anthropologists explain the existence of pervasive health inequalities that affect some groups more than others? For example, why do 40% or more of African Americans have hypertension, when only roughly 30% of non-Hispanic whites are afflicted with this condition?

The anthropologist Clarence Gravlee has an answer. In his well-cited article titled "How Race Becomes Biology: Embodiment of Social Inequality," Gravlee shows that race, despite being a social construction, is very real in terms socioeconomic and health outcomes. Those health outcomes, good or bad, are passed on through the generations.

Gravlee shows that an individual's exposure to racism over a lifetime actually increases one's risk of infant mortality, diabetes, stroke, and high blood pressure, not to mention lower life expectancy. All of these individual health outcomes can impact future generations because any one of them can adversely impact both prenatal and postnatal development and health. Keep in mind that this is the biology of racism rather than the biology of race.

Suggested Reading

Baker, *From Savage to Negro*.

Black, *War against the Weak*.

Jablonski, *Living Color*.

Questions to Consider

1. How has anthropology revised its understanding of race since the late 19th century?

2. To what extent, if any, can socially constructed races become biological?

Anthropology and the Question of Race

S o far in this course we've seen some of the remarkable ways anthropologists explore the origins and diversity of humankind, and if there's one thing we've learned, it's that our genetics, our biology, and even the archaeological record all converge on one anthropological truth: despite our physical and cultural differences, we are one human race.

But wait just a minute. If that's true, why did the last US Census form that I filled out ask me to identify my race? And they gave you and me multiple choices. There are colors, like white and black; geographic regions like Asian and Pacific Islander; and, heck, if you look at the ways the US Census Bureau has categorized races over time, there used to be religion races like Hindu; linguistic ones, too, like Spanish-speaking. They list just about everything under the sun except for one thing: the human race. So what is it? Are there multiple races, or are our biological anthropology friends correct when they say there's only one race, the human race? Today, as our capstone discussion for part one of our course, we're going to use anthropology to unpack enduring ambiguities between race, skin color, and biology.

First, we'll see how anthropology itself has been rather confused about the race concept, and then we'll see how anthropologists eventually came to understand that, biologically speaking, race is nothing but a social construct—we literally made it up. It has nothing to do with skin color and geography. But the story is a little more complicated than that, so let's get right to it.

From the very beginning, anthropologists have been thinking about this thing we call race, and we've struggled with it from day one. Through biological anthropology, we've come a long way from the days when we thought race, biology, and skin color were all one and the same. But, that said, this project is far from complete. Anthropologists have been working on the question of race for well over a century, and we've still got a lot to learn.

Archaeologist Charles Orser, for example, looked back at how archaeologists have understood race over the decades, and he found that many of them tend to confuse race with ethnicity and socioeconomic status. And his groundbreaking research, no pun intended, helps us separate biological versus cultural constructions of race. And we'll return to that question at the end of our lecture, but first, let's look back to the roots of this issue. Let's see how and why anthropologists, archaeologists included, have had so much trouble coming to terms with this concept of race. Let's start at the beginning, anthropology's earliest version of this thing called race.

As we discussed in previous lectures, Darwin's theory of natural selection was a seductive idea that formalized a testable and correctable answer to the question of our human origins. And his scientific approach can't tell us why there's life on this planet, but it did provide a compelling answer to the question of how—once there was life on this planet—we became human. Actually, his book *On the Origin of Species* didn't exclusively focus on human origins. His theory explained how all species emerged.

Anyway, in the early days of anthropology, Darwin's ideas about the biology of evolution proved so convincing that a lot of leading thinkers of the time started to apply this idea far beyond the scope of biology. Folks like Herbert Spencer argued that just as biological organisms evolve from simple, single-cell organisms to more complex ones like you and me, human cultures also evolve from simple societies to highly complex ones. And their assertion was clear: humans evolve culturally from savage to civilized. And this idea of social evolution—or social Darwinism, as Spencer described it—became a popular explanation for the diversity of humankind across the globe.

Think about this. Darwin's book comes out in the mid-1800s, just as the world he's living in grows smaller and smaller. As the close of the transatlantic slave trade transitioned into the dawn of European empires, people in Darwin's age were exposed to new and diverse world cultures. And, inspired by Darwin's ideas on biological evolution, social Darwinists used evolution as a way to explain human diversity, and to classify people and cultures across the globe. Humans who lived in small, simple societies—like a single-cell organism—were deemed less evolved than those who lived in complex ones with White Houses, Supreme Courts, and microbreweries. But this was the dominant idea of scientists of the era. They literally split the world into culture grades from savage to barbarian and civilized. And when you compare living in a mud hut to life in a two-story suburban home with satellite internet, this idea almost feels intuitive.

So, as the scientific community ushered in this idea of social Darwinism, the idea spread widely because it helped make sense of the seemingly limitless human diversity on our planet. Suddenly it became easy to explain why some people were entering the so-called modern world, with industrial factories, powerful trains, and upper-class values, while others were stuck in the past. Those who were further along on the evolutionary path not only had superior biology, but that advanced biology was the foundation for advanced civilization.

Essentially, biology and culture were an interconnected and inseparable . entity, and the idea of race emerged as an all-encompassing term to describe these vast human differences. So, from our physical appearance, our geography, and our religious traditions, they were all functions of our biology, and to track and make sense of all this human diversity, we used the idea of multiple races. So the differences that Europeans saw in humans from Africa or Asia were attributed to the fact that we all come from distinct lineages, which we called races.

Now, upon reflection, many thinkers further developed this idea of human races, and they did so to the extreme that proponents of this way of thinking coined and used the term White Man's Burden to describe the idea that white men had the moral obligation to use their advanced state to uplift nonwhites the world over. Essentially, the concept of the white man's burden was used to justify the colonization of the nonwhite world.

Consider this. It's an excerpt from Rudyard Kipling's 1899 poem "White Man's Burden." It clearly embraces the idea of moral uplift through imperialism.

> Take up the White Man's burden—
> Send forth the best ye breed—
> Go bind your sons to exile
> To serve your captives' need
> To wait in heavy harness
> On fluttered folk and wild—
> Your new-caught, sullen peoples,
> Half devil, half child.
> Take up the White Man's burden

And it continues. These so-called half-devil half-child people of the world were seen as slightly less human, and the concept of race helped early anthropologists classify people and their unique biological heritage. Some races were ahead of others, and you know what? Even the US Supreme Court agreed. In 1896, the Court heard the case Plessy v. Ferguson, in which the justices considered whether or not the Constitution allowed for discrimination based on white versus nonwhite races. And what did the justices determine? Well, in a 7:1 ruling, the Court held that if nature created people and races that were not equal, the Constitution could do nothing to change that. And as a result, Plessy v. Ferguson cemented race-based segregation as fully legal in a country where all men are created equal. They didn't question the idea that there are multiple, unequal, and biological races; they accepted inequality as a biological outcome, and no piece of paper, not even the Constitution, could change that.

This idea gained favor as a neat and clean explanation for human diversity, and the world order, with whites on top, seemed to confirm this idea as fact. The idea was so compelling that the US government and corporate sectors joined forces to make use of this idea in hopes of purifying the human race. After all, if humans could breed hogs and horses to new levels of fitness and strength, why wouldn't we apply these same methodologies to improve human stocks? Believe it or not, they tried.

With millions of dollars of support from the wealthiest families and foundations like Rockefeller, Carnegie, and the Harrimans, and more, special research centers were built to find solutions to this white man's burden. In 1904, for example, the Station for Experimental Evolution was established at Cold Spring Harbor on Long Island. Researchers at this institute were charged with identifying ways to prevent defective germ plasm, the biological and primitive remnants of our savage past, from spreading any further, and it's the eugenics creed that helps clarify this mission, and I'll quote:

> I believe that I am the trustee of the germ plasm that I carry; that this has passed on to me through thousands of generations before me; and that I betray the trust if, that germ plasm being good, I so act as to jeopardize it, with its excellent possibilities, or, from motives of personal convenience, to unduly limit offspring.

The researchers at the SEE not only developed a series of recommendations on how to proceed with this project, they also started to identify family lines in the New York, New Jersey, and New England area that were tainted by criminality, mental disorders, and physical deformities. The idea was simple: to find a way to limit the reproduction of what they deemed to be defective human germ plasm.

Shockingly, their official recommendations included lethal chambers—and that's their words, not mine—as the most effective solution, but they advised policymakers and others that the US population was not really ready to accept this approach. So, instead, less revolting strategies were unleashed, namely marriage and sterilization laws. In 1907, for example, the state of Indiana enacted sterilization laws to make sure criminals and other undesirables could not pass on their defective humanity to other generations. And, in 1927, the Supreme Court actually upheld a Virginia statute that promoted compulsory sterilization of the so-called unfit. And in the majority opinion for that case,

Chief Justice Oliver Wendell Holmes declared: "Three generations of imbeciles is enough."

Actually, early on, this social Darwinist movement was championed by Darwin's cousin, Francis Galton, who labeled this race-improvement project eugenics. And with eugenics research centers at major US universities all the way from Yale to Stanford, it was actually California that most whole-heartedly embraced the sterilization idea as the best bet for improving the human stock. And believe it or not, in the first 10 years of sterilization laws, California alone sterilized nearly 10,000 people, nearly half of whom were women deemed passionate or bad girls.

The idea of applying eugenics to improve humankind spread far beyond US borders and was embraced by many, including Adolf Hitler and his Nazi Party. In fact, when Nazi war criminals were tried in Nuremburg, they actually cited the words of US eugenics leaders and Chief Justice Oliver Wendell Holmes to defend their genocidal actions, because, in their eyes, they were simply applying science to eliminate what they saw as defective human germ plasm.

Let me return to Chief Justice Holmes's majority opinion from the 1927 case I mentioned a moment ago, because I think his words demonstrate just how pervasive and public these ideas were. Holmes wrote:

> It is better for all the world, if instead of waiting to execute degenerate offspring for crime, or to let them starve for their imbecility, society can prevent those who are manifestly unfit from continuing their kind.

Fortunately, Holmes's perspective and the eugenics project in general began to lose ground in the US in the 1930s and '40s. But, of course, this was precisely the time when Nazi Germany was slaughtering millions of Jews, homosexuals, and anyone else with the potential for polluting humanity's biological future.

So how did we get past this obscene application of biology and human difference? Well, it took decades to unmask the pseudoscientific foundations of this movement, but all along there were researchers and others searching for more valid explanations for human diversity, and their work culminated in ideas that now allow us to see the difference between so-called race or ethnicity with things like criminality and poverty. And what saved anthropology as a knowledge production system that's there to explain human diversity is the fact that it was and remains to this day a science. Its ultimate aim is to produce contingent truths; explanations that are testable and correctable; explanations that are tested and improved as we increase our understanding of our origins, our biology, and our behavior.

So, while eugenics scholars were hard at work at Cold Spring Harbor, Yale, Harvard, Stanford, and elsewhere, there were other science-minded scholars who took another approach toward exploring human diversity. The father of American anthropology, a Polish immigrant named Franz Boas is a great example of these scholars who looked beyond biology and eugenics to understand why some folks live in igloos while others enjoy indoor plumbing. And we'll come back to Boas later in our course, but for now we can get a brief preview of the idea he offers as an alternative to the biological determinism of eugenics advocates. While social Darwinists would argue that one's biology determines whether or not they live as primitive or civilized people, Boas didn't succumb to two foundational ideas of strict biological determinism.

First, he did not see culture as the product of biology. Instead, he looked to see how one's environment shapes unique cultural traditions. And second, because he looked beyond biology to explain human diversity, he argued that there really wasn't a specific scientific basis for ranking people and cultures on the spectrum of primitive to civilized. People who live in igloos and huts aren't less human or civilized than cosmopolitan New Yorkers, they're just different. Equally modern, just different. Not rankable, just different.

In short, rather than ethnocentrically judging the Arctic igloo dwellers as primitive hunter-gatherers, Boas saw them as uniquely adapted to a very challenging environment. And just as an ice-fishing Inuit man might have trouble figuring out how to get by in New York City, an urbanite like Boas, if left to his own devices, would have similar problems figuring out how to feed and shelter himself in the Arctic Circle. Think about it. When he did his field research among the Inuit on Baffin Island, his questionable hunting skills and vulnerability weren't a function of Boas's biology. He wasn't primitive or biologically less evolved than his Arctic hosts; he was just far from home—far, far away.

Ultimately, Boas learned and taught that cultures are neither rankable nor biologically based; they're relative. And as such, people from different cultural traditions are equally and fully human. This idea is basically what we call cultural relativity. Some cultures aren't better or worse than others, they're just different. So, when biological determinists saw people who looked different and lived differently from themselves, they ascribed the difference to biology. And as such, they divided humanity into biological races. For example, 18th- and early 19th-century anthropologists who advocated this position commonly described our human origins and races in one of two ways.

Some argued that Africans, Asians, and Europeans, for example, each had distinct origins—multiple Edens, so to speak—and they resulted in multiple, distinct, and biologically defined races, kind of like you and your next-door neighbors. You're both human families, but you come from two distinct family lines. Others, however, theorized that despite the fact that humanity shares common origins, humankind split into multiple biological races like the ones we still see listed in the US Census. The human genome and our knowledge of our hominin ancestors definitively refute both of these theories. Different skin colors, yes, but there's only one race: the human race.

So, if all of our anthropological evidence tells us that different skin colors and cultures are manifestations of one biological race, why does the myth of multiple human races persist? My first response is that it still surprises me to see how clumsily we all use this term race. The myth of multiple races endures everywhere I look. Even universities, institutions of higher education including the one that I work for, still use multiple racial categories on financial aid and admissions applications. I mean, think about this. If the first document my students complete on their path to graduation is that document that reinforces the idea of multiple races, well, if that's the case, anthropologists like me are going to have our work cut out for us.

So, toward the end of the 20th century, for example, the members of the America Anthropological Association gasped at the argument put forth by Richard Herrnstein—a psychologist—and his co-author Charles Murray, a political scientist, in their book *The Bell Curve*. The authors explored the very question we've been discussing today. They looked at both genetic and environmental influences on human diversity, and when they analyzed the connection between intelligence and what they conceived as race, the controversial authors ultimately argued that, compared with black Americans, both east Asians and white Americans historically performed better on IQ tests because of their biology, their distinct genetic heritage.

Critics and supporters of this argument fiercely debated these findings, and it wasn't just anthropologists who sharply challenged Herrnstein and Murray's methodology and conclusions. My former colleague, Mel Konner from Emory University, who's a Harvard-trained MD as well as an anthropology PhD, effectively described why anthropologists and others like Noam Chomsky and Stephen Jay Gould fervently refuted *The Bell Curve*'s conclusions. Specifically, he said that Murray and Herrnstein presented strong evidence that genes do play a role in intelligence, but their evidence in no way supported their theory that genetics explains the difference between IQs of black and white Americans.

Similarly, in 1994, the members of the America Anthropological Association put forth an official statement on race to clear up the pervasive myth of multiple biological races. Simply put, they said that: one, all humans are members of a single species, *Homo sapiens*; and two, and I'll use their words here: "Differentiating species into biologically defined races has proven meaningless and unscientific as a way of explaining human variation and, in this case, intelligence."

Put another way, the experts all agree that genetics and heredity play important roles in determining human variation—that's a given. But they differ on the question of whether genetic variation correlates with conventional, racialized constructions like white versus black. And anthropology clearly tells us there is no such correlation. We're genetically one race, the human race, and traits like skin color and intelligence fail to actually map onto the racialized categories most people use to classify human diversity today.

But hold on just a minute. Before we can fully embrace the idea that race, as we've known it, is indeed a social construct, one urgent question remains. If racial categories like Hispanic and Asian are social constructs, not natural divisions in the human species, if they're not truly biological categories, then how in the world do anthropologists explain the existence of pervasive health inequalities that affect some groups more than others? Why do around 40 percent or more of African-Americans have hypertension, when only 30 percent of non-Hispanic whites are afflicted with this condition? And why are Native American populations more than twice as likely to be diagnosed with diabetes compared with non-Hispanic whites?

Well, anthropologist Clarence Gravlee has our own answer. In his well-cited article titled "How Race Becomes Biology: Embodiment of Social Inequality," Gravlee shows us that race, despite being a social construction, is very real in terms of socioeconomic and health outcomes. And those health outcomes, good or bad, are passed on through the generations. So, first, Gravlee shows us that an individual's exposure to racism over a lifetime actually increases one's risk of infant mortality, diabetes, stroke, and high blood pressure, not to mention lower life expectancy. And all of these health outcomes can definitely impact future generations, mainly because any one of them can adversely impact both prenatal and postnatal development and health.

For example, in the six months following the events of September 11, 2001, the likelihood that a mother would deliver a low birth weight baby rose by 34 percent. Actually, let me clarify. A 2006 study by Diane Lauderdale of the University of Chicago, reviewed Arab-American birth records and found that the incidence of low birth weight deliveries rose 34 percent for only women with Arabic names. What Gravlee and Lauderdale want us to think about here is the cyclical nature of social inequality and racism.

If you're an expecting mother, it's clear that, from the womb on, your child will feel the biological impact of racism, not race. And apply this generational cycle over and over, and we can begin to explain high rates of diabetes and hypertension among black and Native American populations. This is the biology of racism, not the biology of race. Gravlee's big idea here is a game changer, and it has implications for how scientists and health workers classify human diversity. And here's the take-home message: racism, racial inequality, and xenophobia become embodied in biological well-beings of racialized groups—not races, groups that have been racialized. Remember, biologically we're one race: *Homo sapiens*.

So, in retrospect, we now see how conventional racial categories are indeed social constructions. And, biologically speaking, they're inconsistent with actual patterns of human genetic diversity—that is, until the biological impacts of racism and inequality take root and spread into future generations. But remember, that's the biology of racism, not the biology of race. Heck, even the US Census Bureau, which has misused the term race for centuries, now even they make it clear that when they use the term race, they're not talking biology. Here is their clarification as posted on the official website. These are their words, not mine:

> The racial categories included in the census questionnaire generally reflect a social definition of race recognized in this country and not an attempt to define race biologically, anthropologically, or genetically.

So let's wrap this up and clear up the enduring ambiguity associated with the way we understand the relationship between race and human variation. Already in this course we've identified evidence that reveals our human origins. Now let's think about how our origins and our evolutionary history critically inform our understanding of race and human variation.

First, we know that, technically, we're primates—upright-walking apes to be precise. And we know that that the earliest humans split off from our common ancestors with contemporary chimpanzees and bonobo apes. And that happened some 7 million years ago. After that, early humans found creative solutions to filling bellies and producing future generations. In the lectures from unit one, we examined the Paleolithic record, including the fossil remains of ancestors like *Ardipithecus* and *Homo habilis*. And this fossil record clearly shows us how early humans have been migrating in and out of Africa, populating the planet, for some 2 million years.

When we added genetics to our investigation, we found that anthropologists trace modern humanity back to an original east African population of early *Homo sapiens*. With shared origins, we are one: one species, one race—the human race. Now, those early *Homo sapiens* migrated near and far from the equator, and eventually, through evolutionary processes, variations in skin emerged, as did hair types, and skull shapes, and more. But those variations were, as they say, skin deep—surface level. Different skin colors, yes, but different genetic origins? Absolutely not. We are one: one human race.

Archaeology and Human Tools

I n the deep past, the hominin family tree had many branches. This lecture looks at the factors that helped *Homo sapiens* remain the sole remaining survivor on that family tree. Specifically, the lecture turns to archaeology to dig up some answers. First, we'll clear up what archaeologists actually do, and then we'll visit a few archaeological sites to see how the creation of tools is one of the reasons humans are the sole survivors.

What Archaeology Does

Archaeology can not only tell us a lot about when and where humans have lived, but it can also tell us wonderfully curious details about how humans lived. Archaeologists try and figure out what humans were up to, long after they're gone. An archaeologist tracks down, classifies, and analyzes the material remains of people who are no longer present to speak for themselves.

By analyzing these material remains, we learn not only about the daily lives of our ancestors, but also about how humans adapted over the years on the path to our sole-survivor status.

Experimental Archaeology

There are many specializations in archaeology, but 3 are especially relevant to this lecture: experimental archaeology, underwater archaeology, and cultural resource management.

Experimental archaeologists not only explore and interpret the archaeological record, but they recreate the technology and tools of their research populations to probe their lives and minds.

Polynesia provides an excellent example of experimental archaeology. While living on a remote Polynesian island with his wife, the explorer Thor Heyerdahl researched regional cultural traditions as well as the local flora and fauna. He eventually turned his gaze toward human migration and became engrossed in exploring the possibility of human contact between pre-Columbian Polynesian and American populations.

Heyerdahl set out to test his theory that the earliest Americans were in contact with Polynesian groups over 1000 years ago. In 1947, he built a balsa raft and set sail from Peru with Polynesia as his destination.

The journey started in late April. Some 3.5 months and over 4000 miles later, their raft, the *Kon-Tiki*, reached the Tuamotu Islands in French Polynesia. This remarkable journey doesn't prove definitively that ancient Peruvians visited the Polynesian Islands, but it does show us that pre-Columbian contact between South Americans and Polynesians is entirely plausible. Humans could have, technically speaking, made it from Peru to Polynesia without modern equipment or materials.

Underwater Archaeology

Over 70% of the earth's surface is covered in water. If archaeologists are going to search for traces of humanity in every nook and corner of the world, some of them better look beneath the surface of the sea. That's where underwater archaeology comes in.

One exciting example of underwater archaeology research is the recovery and preservation of shipwrecks. Just off the coast of Beaufort, NC for example, is the site where researchers discovered the pirate Blackbeard's flagship, *Queen Anne's Revenge*.

Besides navigation equipment, cannons, guns, and gold, archaeologists also unearthed kitchen utensils and personal possessions like game pieces and smoking pipes, all of which help us reconstruct what life was like for sailors in the early 1700s.

Interestingly, Blackbeard's ship doctors dealt with everything

from bullet wounds and dysentery to scurvy and burns. The ship's medical gear included measuring devices as well as a mortar and pestle for mixing up medicine. There were even a clyster pump for delivering enemas and a mercury syringe for treating syphilis.

Cultural Resource Management

This lecture's 3rd example of archaeological work is cultural resource management. The idea of cultural resource management is an intentional parallel to natural resource management. Just as scientists preserve and study natural resources, some archaeologists apply their talents to preserving, studying, and sharing a wide variety of cultural resources.

For the Tennessee Valley Authority, as a large-scale example, cultural resource management included the management of over 9,000 archaeological sites, the relocation of cemeteries, and collaborative research and management of cultural resources on Native American lands.

Another example of cultural resource management work takes us to the Nashville Sounds baseball stadium on the banks of the Cumberland River. Because the stadium was built in an area known for artifacts from early Native American settlements, cultural resource management archaeologists were on the scene to facilitate excavations and ensure that important archaeological artifacts were not destroyed or disturbed by the construction process.

Tool Making

Next, this lecture will visit some archaeological sites to dig up the early history of human tool making. Its first destination is a dried-up riverbed in Kenya. That's where, in 2011 and 2012, anthropologists discovered the oldest hominin tools ever found. They found over 100 artifacts that show us that upright apes, or hominins, were making tools over 3 million years ago.

The next oldest tool unearthed dates to 2.6 million years ago, and was found in neighboring Ethiopia. This tool style, which we call the Oldowan toolkit, is a very early type of a manufactured stone tool that was widely adopted by our hominin ancestors. Archaeologists and paleoanthropologists have found Oldowan tools throughout east Africa and all the way down into southern Africa.

And they also found, in these Oldowan sites, clear use of fire. Our *Homo habilis* ancestors were quick to develop and adopt Oldowan tool production as part of their survival strategy. Oldowan tools opened up new ways of preparing meals. In general, the Oldowan tool style is known as a chopper. It had a single intentionally crafted blade, which was perfect for butchering.

Acheulean Handaxe

This lecture's next destination is a site named Saint-Acheul in northern France. That is where, in 1847, researchers discovered fossil evidence of the next great leap forward in tool making. This new tool type, now referred to as the Acheulean style, has been found in southern Africa, across east Africa, into the Mediterranean world, southern Europe, and Asia.

With its revolutionary design, the Achulean tradition produced an all-in-one household gadget that slices,

Stone axe head

dices, chops, grinds, cracks, digs, and more. These handaxes were slightly larger than the Oldowan tools they came to replace in the archaeological record. Uniquely, they had a signature teardrop shape, and they were symmetrically crafted on both sides.

Acheulean tool making essentially emerged with the arrival of *Homo erectus*. Confirming the *Homo erectus* connection, archaeologists discovered Achuelan tools in Africa that date back to well over 1.5 million years ago, and then slightly after that in Asia and into Europe.

The Microlith

As knappers (or tool makers) became more specialized, they became experts at selecting and shaping the best materials into a remarkable toolkit. In the beginning of the Acheulean tradition, knappers started with what is called a core stone.

Then, they used a hammer stone to chip away at the stone tool they pictured in their mind. Over time, knappers found more and more uses for the sharp flakes that splintered off from the core

stone during the crafting of a large handaxe. The archaeological record shows that knappers were clearly in the business of making core stone tools AND flake tools just as *Homo sapiens* emerged in the fossil record some 200,000 years ago.

One last major technique in stone tools burst into the archaeological record around 12,000 years ago. As glaciers melted and sea levels rose, humans were releasing their newest stone technology: the microlith.

The Northern Hemisphere was covered in ice for about 1.5 million years. Our ancestors created tools for cooperatively hunting and processing the huge mammoth and other big game that roamed the earth during the Paleolithic era.

But, after 1.5 million years, the ice age drew to a close. And melting ice sheets and rising sea levels weren't the only changes humans were going to have to get used to. As landscapes dramatically transformed from tundra to forests, savannahs, and mountains, the animal population changed too.

The big game followed the path into extinction. Humans needed a new toolkit to chase the more plentiful, yet smaller and flighty, prey like deer, boar, and rabbits.

Microliths answered this need. These small stone tools were made of flint or chert, another sedimentary rock. Highly skilled knappers manufactured points that were less than a centimeter wide.

By making an array of very small stone tools that could be affixed to arrows and spears, Mesolithic humans effectively transitioned out of the ice age and into a radically new world. In other words, we see once more how tool making allowed us to adapt to our environment, embolden our food system, and sustain our ongoing survival.

Suggested Reading

Fagan, *Ancient Lives*.

Henrich, *The Secret of Our Success*.

Waters and Jennings, *The Hogeye Clovis Cache*.

Questions to Consider

1. What is experimental archaeology, and how is it different from other specializations in archaeology?

2. What are some of the oldest tools on Earth, and how does the history of stone tools help us retell the geographic and temporal history of humankind?

Archaeology and Human Tools

In the deep past, the hominin family tree had many branches, and we've talked about *Australopithecus* and *Homo habilis* and *Homo erectus*, among others, but of all these species, only *Homo sapiens* has survived. Think about that. Imagine a healthy and robust tree, but reimagine that tree now, and this time every single branch is dead, all except for one, just one living branch with one leaf. My friends, that's us. It's kind of like one of those terrible, horror movie classics, where one by one everyone starts disappearing, and here we are, the last ones left. So maybe we'd be wise to consider how we've stayed around all these years if we want to avoid the fate of *Homo erectus* and *Australopithecus*.

So, today, we're going to begin looking at the factors that helped *Homo sapiens* remain the sole remaining survivor of the family tree. Specifically, we'll turn to archaeology to dig up some answers. First, we'll clear up what archaeologists actually do, and then we'll visit a few archaeological sites to see how the creation of tools is one of the reasons we're sole survivors.

As we discussed in our previous lectures, archaeology is one of the four major subfields of anthropology. We've seen how biological anthropology, our first subfield, teaches us about our origins, our evolutionary history, and even the very nature of human variation. So what does archaeology, as the second major subfield of anthropology, bring to the table? Well, archaeology can not only tell us a lot about when and where humans have lived, but it can also tell us wonderfully curious details about how humans lived. From the mundane details of daily food production to the development of agriculture, urban centers, and iPads, archaeology covers it all.

But how? What do archaeologists do that we don't typically get from historians, social scientists, and others who explore the complex epic of our humanity? Well, put simply, they tell the history of humankind by revealing the history of things. Now think of it this way. If I were knocking at your door this very moment, if you greeted me and let me in, what would I see? What would all those things in your home reveal to me about your life and the things that you value?

Well, as I enter your foyer, I see three pairs of worn jogging shoes and a rolled up yoga mat, and that tells me you're likely an avid runner who at least dabbles in some yoga or some other fitness routine. But then I walk through your amazing kitchen and I notice that your trash is filled with take-out dinner boxes like the ones I see over there on your coffee table. Fifty percent of them are sushi containers, 30 percent look like they're from a Mexican restaurant, and over there 20 percent are pizza boxes. So, you know what? If you ever come to eat at my house, I now know three kinds of meals you're sure to like. So, just by looking at your things, I can learn about you. And that's what archaeologists do. They investigate, retrieve, and interpret the material remains of humans, the things we humans leave behind.

So let's dig deeper. Let's work out exactly what we mean when we say archaeology. For most people, when we think of archaeology, our imagination conjures up exotic scenes with Indiana Jones narrowly escaping treasure hunters and Nazis while hunting down the Ark of the Covenant. And I'll tell you what, man, it's not exactly an insult that some folks think of us anthropologists as adventurous globetrotters digging in the dirt, but actually there are better metaphors out there for what archaeologists actually do.

If you want to keep with the fictional characters, we'll avoid Indiana Jones and Lara Croft Tomb Raider, and instead we'll go with a couple less intrepid characters. Archaeologists are actually more of a mix between, say, Hercule Poirot of Agatha Christie fame, and Marty McFly from *Back to the Future*—that is to say, part detective and part time traveler. Why? Well, as time-traveling detectives, archaeologists try and figure out what humans were up to long after they're gone. The archaeologist tracks down, classifies, and analyzes the material remains of people who are no longer present to speak for themselves. And by analyzing these material remains, things like tools for example, we learn not only about the daily lives of our ancestors but also about how humans adapted over years on the path to our sole survivor status. So, in short, archaeology is the perfect subfield for seeking answers to why we're the sole survivors on the hominin family tree.

But, before we visit some archaeological sites to see how the creation of stone tools helped carry humankind into the 21st century, let's finish our abridged introduction to archaeology with a few specific examples of archaeologists at work. There are all kinds of specializations within anthropology and archaeology, but we can take a few moments to look at three exciting ways anthropologists do this thing called archaeology.

First, let's go to one of my absolute favorite branches of the archaeological family. It's called experimental archaeology. Now, experimental archaeologists are fun because they not only explore and interpret the archaeological record, but these anthropologists go a step further. They actually recreate the technology and tools of their research populations because they want to probe their lives and minds. It's one thing to find and study a stone tool, but I guarantee that you'll have a lot more respect for, and you'll better understand, the earliest stone toolmakers once you try to recreate their handiwork.

Later in this lecture, we'll explore the early history of human toolmaking. But first, let's get a glimpse into the breadth of archaeology by checking out three compelling specializations. First on our list, we'll visit Polynesia for a thrilling example of experimental archaeology. Have you heard of Thor Heyerdahl and maybe his *Kon-Tiki*? While living on a remote Polynesian island with his wife, Thor Heyerdahl researched regional cultural traditions as well as the local flora and fauna, and he noted that local plant and animal life actually migrated over the millennia in the same direction of regional ocean currents. So, Heyerdahl drew evolutionary paths connecting the spread and distribution of the region's animal life, and that got him thinking. He turned his gaze to the humans and to their Pacific migrations. Ultimately, he became engrossed in exploring the possibility of human contact between pre-Columbian Polynesians and American populations. And, like any good scientist, Heyerdahl set out to test his theory that the earliest Americans were in contact with Polynesian groups over 1,000 years ago.

So, in 1947, he built a balsa raft and he set sail from Peru—destination: Polynesia. Heyerdahl and his crew left most of their modern conveniences at home. They did bring a basic radio, some cooking supplies, a few cameras, and, believe it or not, even a parrot, but that was it. And, in short, Heyerdahl was practicing experimental archaeology. He was recreating a technology of the past in order to seek answers to a historical question.

The journey started in late April, with the stars and Pacific currents as their guides. And some 3.5 months and over 4,000 miles later, their raft, the *Kon-Tiki*, reached the Tuamotu Islands in French Polynesia. They actually made it. So, you can see the historic footage of this journey in the 1950 Academy Award-winning documentary *Kon-Tiki*.

Now, this remarkable journey doesn't prove definitively that ancient Peruvians visited the Polynesian Islands, but it does show us that pre-Columbian contact between South Americans and Polynesians is entirely plausible. Humans could have, technically speaking, made it from Peru to

Polynesia without equipment from the modern era. And here's the thrilling update. Despite decades of critiques, more recent genetic studies keep the door open to the idea of pre-Columbian transpacific relationships. In fact, recent analysis of the genomes of living Polynesian Islanders actually supports this idea. Their DNA includes a shared genetic legacy with Native Americans that dates back to just after Polynesians populated Easter Island and the Hawaiian Islands.

So, scientists have yet to agree on which group was first to take the transpacific journey, and only future research will bring us closer to an answer, but one fruitful place that we might turn to is our second archaeological specialization, and that's underwater archaeology. Think about it. Over 70 percent of the Earth's surface is covered in water, so if archaeologists are going to search for traces of humanity in every nook and corner of the world, some of them better look beneath the surface of the sea. So, tracing human migrations across the Bering Strait into the Americas, for example, is one project that needs help from underwater archaeology. As sea levels rise and fall over the centuries, the footsteps and remains of many of the earliest Americans are most certainly on the ocean floor. But we talked about Paleo-American studies in a previous lecture, so, instead, let's go to another exciting example of underwater archaeology research. That's the recovery and preservation of shipwrecks.

If we sail just off the coast of Beaufort, North Carolina, for example, we'll find the site where researchers discovered the pirate Blackbeard's flagship, the *Queen Anne's Revenge*. And besides navigation equipment, cannons, guns, and gold, archaeologists also unearthed things like kitchen utensils and personal possessions like game pieces and smoking pipes, all of which help us reconstruct what life was like for sailors in the early 1700s. But one thing we've definitively learned is that Blackbeard actually offered his crew a medical plan. Seriously. His ship doctors dealt with everything from bullet wounds and dysentery to scurvy, burns, and much worse. But don't be too jealous of this pirate perk, because shipboard healthcare in 1717 was anything but painless. The doctor's med kit included measuring devices, as well as a mortar and pestle for mixing up medicine. There was even a clyster pump for delivering enemas, and, worst of all, a mercury syringe for treating syphilis.

So, we'll leave underwater archaeology with Blackbeard in North Carolina for now, and move on to the third area that will round out our cursory glimpse at the specializations and applications of contemporary archaeology. Now, for all of you who love visiting our National Parks, our third example of archaeological work brings us to cultural resource management.

The idea of cultural resource management is an intentional parallel to natural resource management. Just as scientists preserve and study natural resources, some archaeologists apply their talents to preserving, studying, and even sharing a wide variety of cultural resources. For the Tennessee Valley Authority, CRM work included the management of over 9,000 archaeological sites, the relocation of cemeteries, and collaborative research and management of cultural resources on Native American lands. Another example of cultural resource management work takes us to the Nashville Sounds baseball stadium on the banks of the Cumberland River. And because the stadium was built in an area known for artifacts from early Native American settlements, cultural resource management archaeologists were on the scene to facilitate excavations and ensure that important archaeological artifacts were not destroyed or disturbed by the construction process.

And similar work is done all throughout the US, and even throughout the world. UNESCO, the cultural, scientific, and educational wing of the United Nations, actually manages the World Heritage Centre and its World Heritage List. It has over 1,000 extraordinary heritage sites. And one site, which archaeologists and others will be exploring and preserving for centuries to come, is Timbuktu in Mali, West Africa. Timbuktu is a center of Islamic scholarship, and it housed a treasure trove of ancient manuscripts, most of which have yet to be translated. Now, these remarkable volumes contain the intellectual work of generations of premier scholars from all over the world dating from the 15th century and even earlier.

So there we have it: cultural resource management, underwater archaeology, and experimental archaeology. These specializations give us a quick picture of the breadth of contemporary archaeology. Nonetheless, archaeologists do so much more. Some archaeologists focus on specific regions and time periods, like, say, Pharaonic Egypt or Classical Greece, while others look at themes like development of agriculture, or conquest, or even pottery production. And, in all of these endeavors, archaeologists continue to find evidence that helps us understand the question: how have humans survived when other hominin species have gone extinct? And nothing in the archaeological record speaks more directly to that question than human tools. So, for the final piece of our archaeological adventure, let's turn to a classic archaeological subject: stone tools. Let's visit some archaeological sites to dig up the early history of human toolmaking.

Our first destination is a dried-up riverbed in Kenya, and that's where, in both 2011 and 2012, anthropologists discovered some of the oldest hominin tools we've ever found to date. They found over 100 artifacts that show us that upright apes, or hominins, were making tools some 3 million years ago. And the next oldest tool we've unearthed dates to 2.6 million years ago, and was found in neighboring Ethiopia—note, still in east Africa. Now, this tool style, which we call the Oldowan tool kit, is a very early type of a manufactured stone tool that was widely adopted by our hominin ancestors. Archaeologists and paleoanthropologists found Oldowan tools throughout east Africa all the way down into southern Africa, and they also found, at these Oldowan sites, clear use of fire. Of course, if we're now butchering meat, it's probably time for a fire and some barbecue.

So, our *Homo habilis* ancestors were quick to develop and adopt this Oldowan tool production as part of their survival strategy, and it was likely one of the reasons upright-walking apes survived and eventually evolved into modern, tool-using *Homo sapiens* like you and me. And that's because Oldowan tools opened up new ways of feeding ourselves, among other things, and we're not talking about just banging a walnut open with a rock here.

In general, the Oldowan tool style is known as a chopper. It had a single, intentionally crafted blade, which was perfect for butchering. And with these new tools, *Homo habilis* broke through thick hides, got to the meat, and even cracked bones to get to the marrow. And, remarkably, *Homo habilis* lived in an era when Africa was in a drying trend, and that was, again, remember, 2–3 million years ago. And these tools are material evidence for the creative ingenuity of *Homo habilis* adapting to new ecologies and forging new food systems. We hominins sometimes know a good thing when we see one, and Oldowan toolmaking was like Wi-Fi; it was game changing. It was a technology that spread like wildfire, and it did usher in new waves of human innovation.

Our next destination on our stone tool adventure here is a site named Saint-Acheul in northern France. And that's where, in 1847, we discovered fossil evidence of the next great leap forward in toolmaking. Now, this new tool type is referred to as the Acheulean style, and that's been found in southern Africa, across east Africa, and even into the Mediterranean world, southern

Europe, and Asia as well. And, with its revolutionary design, this Acheulean tradition produced one of those all-in-one household gadgets—it slices, it dices, it chops, it grinds, it cracks, it digs, and a whole lot more. These hand axes were slightly larger than the Oldowan tools that they came to replace in the archaeological record, and, uniquely, they had a signature symmetrical teardrop shape, and they were symmetrically crafted on both sides.

Acheulean toolmaking essentially emerges with the arrival of *Homo erectus*. And, confirming the *Homo erectus* connection, archaeologists discovered Acheulean tools in Africa that date back to well over 1.5 million years ago, and then slightly after that in Asia and into Europe. And as specialized toolmakers honed their skills over the generations, Acheulean knappers— that's what anthropologists call toolmakers—eventually started what was a transition into our third and final destination on this stone tool safari.

As knappers became more specialized, they became experts at selecting and shaping the best materials into a remarkable tool kit. In the beginning of the Acheulean tradition, knappers started with what's called a core stone, and they used a hammerstone to chip away at the stone tool they had pictured in their mind. So, over time, knappers found more and more uses for the sharp flakes that splintered off from the core stone during the crafting of that larger hand ax. So the archaeological record shows that knappers were clearly in the business of making core stone tools and those flake tools I just mentioned just as *Homo sapiens* are emerging in the fossil record some 200,000 years ago.

It appears that the Acheulean tradition, that came to include the creation of smaller flake tools, kept humans alive almost into the modern age, but one last major technique in stone tools burst into the archaeological record some 12,000 years ago. And remember, that's right around the time that the most recent ice age drew to a close. As glaciers melted and sea levels rose, humans were releasing their newest stone technology. It's called the microlith. *Lith* is from the ancient Greek word *lithos*, meaning "stone," and that's how we get the term Mesolithic. *Meso*, or middle, plus *lith*, gives you Mesolithic, meaning "Middle Stone Age." Add the prefix micro instead, and we get microlith or "small stone." The hallmark technological innovation of the Middle Stone Age was this microlith, a toolmaking tradition that provides a thrilling example of *Homo sapiens* mastering the art of ecological adaptation in the face of existential crises. Let's take a closer look at the knapping techniques that spread with humans all across the planet just before the development of agriculture.

Now, when we talk about stone toolmaking, we're with the Flintstones in the Stone Age, and that's defined by an era in which humans relied on stone as their primary material for making tools. It's important to note that the Stone Age is split into three different eras. First, Oldowan and Acheulean toolmaking occurred in the Old Stone Age, or the Paleolithic. The tools our ancestors made back then helped them deal with their natural environment, which for most of the Paleolithic included the most recent ice age. The Northern Hemisphere was covered in ice for about 1.5 million years. But, nonetheless, turning our lemons into lemonade, our ancestors created tools for cooperatively hunting and processing the huge mammoth and other big game that roamed the earth during the Paleolithic.

But, as George Harrison warned us, all things must pass, and after 1.5 million years, so went the ice age. And melting ice sheets and rising sea levels weren't the only changes humans were going to have to get used to. As landscapes dramatically transformed from tundra to forests, savannahs, and mountains, the animal population changed, too. So the big game many of us relied on for so long for our survival followed the path into extinction, and humans needed a new tool kit to chase the more plentiful yet much smaller and flighty prey like deer, boar, and rabbits.

It was microliths that answered this need. The range of microliths that have been unearthed by archaeologists is absolutely extraordinary. These small tools were made of flint or chert, another sedimentary rock, and very highly skilled knappers manufactured points that were less than a centimeter wide. And by making an array of very small stone tools that could be affixed to arrows and spears, Mesolithic humans effectively transitioned out of the ice age and into a radically new world. In other words, we see once more how toolmaking allowed us to adapt to our environment, embolden our food system, and, again, sustain our ongoing survival.

So, we've seen humans making stone tools in the Paleolithic and the Mesolithic ages, but we haven't really talked about the Neolithic, or the New Stone Age. Actually, on the cosmic scale, we're not missing much. You see, by the time the Neolithic Age comes around some 10,000–12,000 years ago, humanity's Stone Age success story was already written in these small stones. And with the development of our stone tool traditions, our ancestors eventually created the microlith, and with that microlith in hand, humankind stopped merely surviving, and instead we started thriving. And I'll put it this way: prior to the microlith toolmaking, the human population was barely

100,000 people total. But with that microlith in hand, and with the ice age behind us, our human population exploded to several million in the span of just 5,000 years.

Think about that. We learned in the previous lecture that it took almost 7 million years to evolve from the first upright-walking apes into a small population of *Homo sapiens* using stone tools in Africa. But then, when *Homo sapiens* unleash this microlith some 12,000 years ago, our species literally grows from thousands to millions of individuals spread all across the planet. And that is no small feat.

Today, we took a closer look at archaeology as one of the four subfields of anthropology, and then we did some time travel back into the Paleolithic. Specifically, we visited the archaeological record to reconstruct the history of human toolmaking in the Stone Age. And we saw that humans refined our stone tool kits over the millennia, and that those adaptations certainly played a major role in *Homo sapiens*' exclusive status as the sole surviving hominin, and to our success in populating the planet.

For me, one of the most fascinating facts about the Stone Age tools is that they stand up to the test of time, even despite the great blades produced today with steel. One example would be the stone blades made by contemporary Konso women toolmakers in eastern Africa. They create specialized stone scrapers that, to this day, are studied from afar, and it's because they're more efficient than commercially manufactured alternatives. These Konso women are not making tools because they can't afford steel ones at the market; they make stone tools because they're efficient for common local tasks such as processing hides.

A more local example of the timeless sophistication of stone tools is obsidian blades. Humans have made razor sharp blades with obsidian for thousands of years, and they're not about to stop just because we figured out some digital printing here. In fact, some surgeons prefer to use obsidian scalpels for surgeries precisely because their blades can be 4–5 times sharper than surgical steel. And that means that these sharper blades also reduce scar width, healing time, and even inflammation. Not so bad for a Stone Age technology.

Of course, humans weren't content with a kit that included only stone tools, and eventually we would shift to tools made of bronze, iron, and other materials. But the big picture says it all: we went from Oldowan tools to iPads and nanotechnology in about 2.5 million years. And the ancient ingenuity of technological innovation is clearly evident in the craft of our ancestors' stone tools—the hand axes, the blades, and the flake tools that fed them and sustained the survival of our species.

LECTURE 8

Agricultural Roots of Civilization

P aleoanthropologists and archaeologists analyze the fossil record to reveal the myriad ways humans have cooperated to produce food across the millennia. That fossil record definitively tells us that our human ancestors were hunter-gatherers, of one variation or another, for 99.98% of our 7,000,000-year history. Even if we reduce the scope of our family tree to only modern humans—otherwise known as *Homo sapiens*—we've been hunter-gatherers for about 190,000 of the past 200,000 years. As odd as it may sound, farming, in the grand scheme of things, is about as modern as space travel and the Internet. This lecture looks at how agriculture came to be.

The Pre-Agricultural Past

At first thought, the idea of living a migratory, hunter-gatherer lifestyle might sound like a horrifying alternative to those of us who rely on modern ubiquities like carryout and grocery stores.

However, as renowned anthropologist Marshall Sahlins explains, the hunter-gatherer lifestyle had surprising advantages. Hunter-gatherer societies—which consisted of small, close-knit bands of fewer than 100 people—lived prosperously, had diverse and balanced diets, and even had longer and healthier lifespans than early farmers.

What's more, the small size of their bands made their communities adaptive and manageable. These small societies were decentralized and actually rather egalitarian with respect to leadership, work, and social class. Additionally, these small communities were migratory, which allowed them to go where the food was.

The gradual shift to agriculture was thousands of years in the making, and it emerged in different places at different times.

Parietal Art, Grinding Stones, and Genetics

We don't have histories recording the shift to agriculture written by our earliest farming ancestors, but we do have strong historical evidence. For example, in the mountains of southwest Libya, rock art—which anthropologists refer to as parietal art—documents that the regional domestication of cattle was well established some 7,000 years ago.

At the Tin Newen site, for example, early humans recorded all kinds of daily activities in their parietal art. We see thoughtful paintings of lots of cattle tended by a few human figures here and there. Further down the line is a scene of people seated in small groups like a prehistoric café.

Archaeologists have also analyzed pottery fragments and fossilized bones from the region to further substantiate this transition from foraging to agriculture and animal husbandry.

In Asia, the archaeologist Li Liu has provided another compelling regional example of how anthropologists date the origins of farming. When Liu analyzed grinding stones from a site near China's Yellow River, she found that they dated back some 23,000 years—not too far on the timeline from the coming domestication of plants.

Starch analysis and other techniques show us that for thousands of years, people in this region processed foraged foods like grasses, roots, and wild millet seed. Generations of these foraging activities gradually ushered in the domestication of wild plants like millet, which became a staple for ancient Chinese civilization.

Archaeologists have unearthed evidence that people in the Near East were growing cereals and figs as early as 12,000 years ago. In Mexico around the same time, people were growing squash and playing with teosinte, the wild version of maize. Meanwhile in China, we see the emergence of rice cultivation.

Genetics also helps us date the domestication of animals like cattle and goats. This dates to roughly the same timeframe, just a little before we begin to see archaeological evidence of farming.

Why Farm?

The agricultural transformation was a widespread response to the changing ecological, technological, biological, and cultural lives of humans. The shift began around 10,000 to 12,000 years ago.

One important factor was we evolved. Hunter-gatherers in the Middle to Late Stone Age were remarkably different from our earliest ancestors. The Middle Stone Age brain, for example, evolved to be 4 times larger than the brains of *Sahelanthropus*—the earliest hominin ancestor, which dates back 6–7 million years.

After millions of years, we became hunting and foraging machines. This led to surplus food, which in turn led to 2 choices: Build granaries for the surplus, or build permanent or semi-permanent settlements at the most productive hunting and foraging sites. Our pre-agricultural ancestors did both.

Progressively more hunter-gatherer populations began to like the idea of settling down occasionally to generate and store surplus food, tools, and all the other things that don't fit into a migrant's backpack.

Climate Change

Another major factor that accelerated our transition to agriculture was climate change. After the most recent ice age peaked around 20,000 years ago, humans watched their hunter-gatherer worlds change dramatically.

If we return to parietal art, our southeastern Algerian ancestors left us rock paintings depicting the Sahara as a lush, grassy expanse with giraffes, elephants, trees, streams, and lakes.

Animal bones and geological evidence confirm the Sahara was green as recently as 8,000 years ago.

Then, increased temperatures and humidity progressively dried up the once lush Sahara. Along major rivers like the Euphrates and Tigris, these changes created ideal conditions for the rise of agriculture.

Similar climatic pressures challenged pre-agricultural societies across the globe, and over the generations, humans gradually changed the way they made food. They became farmers.

Farming, Families, and Breakthroughs

Sedentary, agriculture-based living made it much easier to raise children. Raising an infant is exhausting enough—even in the 21st century and even with the very best baby monitors, diaper stations, and cribs. But imagine how hard parenting would be if we were all migratory foragers.

It's no surprise that when folks settled down and started farming, their populations grew. Now capable of generating some serious food surplus, human societies eventually grew into massive ancient civilizations in China, Egypt, Peru, and all over the world.

In short, the transition to agriculture inspired sedentary living and urban centers, and it sparked technological breakthroughs like writing, mathematics, medicine and much more.

But this shift wasn't an easy one. In his classic text, *Paleopathology at the Origins of Agriculture,* George Armelagos analyzes fossilized human remains to document early-agricultural diseases.

He found evidence that early farmers in the Illinois River Valley, when compared with the hunter-gatherers who preceded them, were rather stressed out and sickly: They had bone lesions, anemia, degenerative spinal conditions, and even lower life expectancy.

Despite these poor health outcomes, the relatively rapid spread of farming indicates that agriculture was a rather seductive alternative to hunter-gatherer life.

Poverty and Wealth

One more significant change was spurred by agricultural life: the emergence of poverty and wealth.

Today's tech geniuses reap benefits from the gadgets and patents they create, and that incentivized system was in effect for early farmers too. Super-farmers could become extremely rich.

Yet, for all its glitz and glory, farming was a very risky food production strategy. A single pest infestation, inadequate rainfall, or a loose group of cattle could all ruin a farming family or community in a flash.

We see this in parts of the world even today. For example, in Mali, a family endured several years of debt and scarcity only because one of their best farmers broke his leg.

Society tasked scientists with this next food revolution. Organized into national and international research institutes, funded by government and industry, and aligned with university research programs, agricultural scientists sought solutions to the feed the planet.

The Green Revolution

An existential food crisis emerged in the 20th century, and here in the 21st century, we're not past it just yet.

Prior to the 20th century, scholars like Thomas Malthus warned that exponential population growth would eventually outstrip food production. As a result, Malthus warned of a future of famines and disease.

Nearly 2 centuries after Malthus voiced his concerns, fears of runaway population growth destroying humanity found a new voice in Paul and Anne Ehrlich's classic, *The Population Bomb.* This influential 1968 book predicted that widespread starvation was imminent, barring another food revolution.

Thomas Malthus

In the US, for example, cereal farmers used scientists' improved seed, and the results were indeed revolutionary. By the start of the 21st century, US farmers more than tripled their cereal yields.

Early on, the 20th-century food revolution was optimistically named the Green Revolution. Farmers hoping to reap improved Green Revolution harvests had to change the way they farmed. For instance, they had to stop producing their own seed on the farm. Instead, they had to begin buying improved seed engineered by agricultural specialists.

After buying modified seed, farmers also needed to buy fertilizer, pesticide, herbicide, seed treatment, and so on. These petroleum-based products fueled the Green Revolution.

Unfortunately, as we settle further into the 21st century, many of the world's farmers, including the poorest of the poor, have yet to reap Green Revolution benefits like improved yields.

Green Revolution farming has benefited some more than others. Sub-Saharan African farmers, for example, grew only 1 ton of cereal per hectare in the 1960s, while US farmers harvested 2 tons per hectare.

To this day, in places like Mali, family farmers still get an optimistic average of around 1 ton of grain per hectare. Conversely, US cereal producers have harvested a stunning 6–7 tons per hectare since the early 1990s.

As extreme hunger and food insecurity persist into yet another century, humanity is again turning to scientists and other specialists in search of the future of food. Like the early humans who brought us agriculture, today's society is faced with its own set of unique challenges, including population growth, climate change, and global security.

Suggested Reading

Cleveland, *Balancing on a Planet.*

Cohen and Armelagos, *Paleopathology at the Origins of Agriculture.*

Shostak, *Nisa.*

Questions to Consider

1. When and why did humans start farming, and how did this agricultural transformation alter the human experience?

2. In what ways is the planting of seeds also a planting of civilization?

Agricultural Roots of Civilization

Think back to the last time you sat down for a large family meal. When we enjoy a traditional holiday feast on a day like Thanksgiving, we might easily imagine that humans have always eaten classics like potatoes, gravy, sweet corn, but you know what? That couldn't be further from the truth. Paleoanthropologists and archaeologists analyze the fossil record to reveal the myriad ways humans have cooperated to produce food across the millennia, and that fossil record definitively tells us that our human ancestors were hunter-gatherers of one variation or another for 99.98 percent of our 7 million-year history.

Put another way, even if we reduce the scope of our family tree to only modern humans, otherwise known as *Homo sapiens*, we've been hunter-gatherers for about 190,000 of the past 200,000 years. So, as odd as it may sound, farming—in the grand scheme of things—is about as modern as space travel and the internet. So, before we trace the origins of agriculture, let's first go deeper into the past in order to consider how humans produced food before farming came along.

Now, at first thought, the idea of living a migratory, hunter-gatherer lifestyle might sound like a horrifying alternative to those of us who rely on modern ubiquities like carryout and grocery stores. But, as renowned anthropologist Marshall Sahlins explains, the hunter-gatherer lifestyle had surprising advantages beyond free rent and no bills. In fact, Sahlins is famously cited for describing preagricultural humans as the original affluent societies.

These societies, which consisted of small, close-knit bands of fewer than 100 people, lived prosperously, and they had diverse and better balanced diets. They even had longer and healthier life spans than early farmers. Apart from avoiding delivery food and Netflix binges, foragers also avoided the high-starch, highly processed foods that most of us consume on a daily basis. Even our early farming ancestors, when they transitioned to agriculture, they suffered numerous negative nutritional outcomes, anemia and shortened life expectancy for one. But we'll get to that and the farming era soon.

So, beyond their variable and balanced diet, what other advantages allowed our hunter-gatherer ancestors to live longer lives than their early farming counterparts? Well, for one, the small size of their bands made their communities adaptive and manageable. These small societies were decentralized and actually rather egalitarian with respect to leadership, work, and social class. Think of it this way: if private property is restricted to only that which you can carry, none of us will have much more than anyone else.

Second of all, these small communities were migratory, and this allowed them to go where the food was. And yet, despite the upside of the hunter-gatherer life, the fact remains that humans did eventually switch to agriculture. But why is that? Well, I'll return to that important question in a few moments, but first let's place the origins of plant and animal domestication both chronologically and geographically.

The archaeological record shows us that humans first started farming approximately 12,000 years ago. In textbooks, we often see the phrase Neolithic Revolution applied to the origins of agriculture, and the name is apt, because this dramatic change in food production altered absolutely everything about our lives. And, that said, most anthropologists tend to think of this food revolution as more of a transformation, because the gradual shift to agriculture was thousands of years in the making, and it emerged in different places at different times.

It's archaeologists who tell us when and where humans started farming, but how? What's the actual evidence? Well, we don't have histories written by our earliest farming ancestors, but we do have strong historical evidence. For example, if we go to the mountains of southwest Libya, we have rock art—what anthropologists refer to as parietal art—that documents the original domestication of cattle in that region some 7,000 years ago. It's as if our ancestors sent postcards from the past to help us understand our origins.

The people who created the parietal art of southwest Libya, for example, show us that they herded and milked cattle, and they sometimes even turned their ancient milk into prehistoric yogurt and cheese. At the Tin Newen site, for example, early humans recorded all kinds of daily activities in their parietal art. We see thoughtful paintings of cattle, lots of them, tended by just a few human figures here and there. Then, further down the line, you see a scene of people seated in small groups, like a prehistoric café. Other archaeologists have analyzed pottery fragments and fossilized bones from the region to further substantiate this transition from foraging to agriculture and animal husbandry.

Now, as I mentioned, agriculture emerged in different places at different times. As such we can turn from northern Africa to Asia where Stanford archaeologist Li Liu provides another compelling regional example of how anthropologists date the origins of farming. Among other methods, she examines the tools of our ancestors. When Liu analyzed grinding stones from a site near China's Yellow River, she found that they dated back some 23,000 years, and that's not too far on the timeline from the coming domestication of plants. That said, starch analysis and other techniques show us that, for thousands of years, people in this region processed foraged foods like grasses, roots, and wild millet seed. So, generations of these foraging activities gradually ushered in the domestication of wild plants like millet, which became a staple for ancient Chinese civilization.

So, by analyzing parietal art and early tools for food processing—and we can even add fossilized remains of our ancestors—anthropologists reveal the geography and the emergence of agriculture. These archaeologists have unearthed evidence that people in the Near East were growing cereals and figs as early as 12,000 years ago. Around that same time, we also see people in Mexico and China starting their own agricultural adventures. In Mexico, they were growing squash and playing with teosinte, the wild version of maize. Meanwhile over in China, we start to see the emergence of rice cultivation. And genetics are also going to help us date the domestication of animals like cattle and goats, and what we find is that it dates to roughly the same time frame, actually just a little before we begin to see archaeological evidence of farming.

OK, so now we understand when and where our ancestors started farming, and we've seen how archaeologists use parietal art, prehistoric tools, and pottery analysis to trace these origins. But, before we move on, why, after 99.98 percent of our 7 million-year history, did humans switch food strategies? Why bother? Why didn't we just stick with the good old foraging days? I mean, think about this. As foragers, we worked fewer hours, we had a more balanced diet, our population was in check, and best of all none of us were poor or hungry—that is, unless all of us were poor and hungry.

So, OK, maybe we're romanticizing this hunter-gatherer lifestyle just a bit. It was actually exceptionally challenging and it's a demanding mode of food production, but the question remains: why did our human ancestors recently switch food strategies after millions of years? Why such a historic food revolution? Put simply, the agricultural transformation was a widespread human response to the changing ecological, technological, biological, and even the cultural lives of humans.

Now that's a pretty broad statement, so let's unpack it a bit by taking a closer look at a few specific reasons why some humans turned to agriculture beginning 10,000–12,000 years ago. Now, first and foremost, we evolved. Hunter-gatherers in the Middle to Late Stone Age were remarkably different from our earliest ancestors. The Middle Stone Age brain, for example, evolved to be four times larger than the brains of *Sahelanthropus*, the earliest hominin ancestor that dates back to 6 or 7 million years.

Essentially, our muscles, our skeletons, the nervous system, and locomotion were finely tuned over millions of years for hunter-gatherer success. Now this doesn't, in and of itself, explain why successful, big-brained humans switched to farming, but stick with me here. Before we get to agriculture, what about culture? Agriculture? Culture: our ability to communicate, to cooperate, to share, not to mention our inclination to give and receive a little help from time to time. These are foundations of humanity's survival.

To get a clearer picture, let's imagine living in a small band of hunter-gatherers 20,000 years ago. And, with our brains clicking and our bodies primed, humans like us have become more efficient as foragers and more skilled as hunters, and even our food processing techniques and technologies have vastly improved. In fact, because we now know where to camp along that annual migratory passage of antelope, we can cooperate and acquire far more meat than all of us could ever eat in one session. And for that matter, we also get far more hides and useful bones than we could possibly carry. Moreover, our foraging innovation now has us collecting much more wild millet than we could ever eat or take with us. Put simply, after millions of years, we became hunting and foraging machines.

Now, if you stuck with our earlier thought experiment, and you're still thinking like a 20,000-year-old ace forager, you'll feel an irresistible compulsion to build some granaries for our surplus, granaries that we could return to as part of migratory hunter-gatherer life. Or maybe we could start building permanent or semi-permanent settlements wherever we find our most productive hunting and foraging sites. And you know what? Our preagricultural ancestors did both, and that is how bigger brains and human culture contributed to the rise of farming.

Progressively, more hunter-gatherer populations were liking the idea of settling down occasionally to generate and store surplus food, tools, and all those other things that don't quite fit into a migrant's backpack. And it's my college students that relate really well to this transition toward permanent settlement, and I bet their parents do, too. Let me explain. Every August, college students pack their most important belongings so they can move in to campus life. Then, after only nine months, they repack and head somewhere else for the summer, only to return to campus three short months later. So, with a new dorm or apartment every year, along with three summers of living elsewhere, college students have annual migration patterns that discourage the accumulation of too much stuff.

And, like our hunter-gatherer ancestors, when a student pushes the limit on too much stuff, the people who will help him move about will certainly offer some sharp critiques. So, regardless of whether or not we chose semisedentary living, our direct ancestors had good and bad years, just like you and I. Our ancestors had to innovate and create new strategies to confront ecological, technical, and cultural challenges to food production. And when a regional antelope population decreased, for example, our successful ancestors moved on, and they found new hunting grounds, or they created new ways to make a living.

Now, we've just seen how our transition to agriculture and sedentary living was due in part to our evolving minds and bodies, our increasingly sophisticated tools, as well as our ability to adapt and innovate. But there was another major factor that accelerated our transition to agriculture, and that was climate change. After the most recent ice age peaked around 20,000 years ago, humans watched their hunter-gatherer worlds change dramatically. If we return to that parietal art, our southern Algerian ancestors left us rock paintings depicting the Sahara as a lush, grassy expanse with giraffes, elephants, trees, streams, and lakes. And animal bones and geological evidence confirm that the Sahara was indeed green as recently as 8,000 years ago.

But then, increased temperatures and humidity progressively dried up the once lush Sahara. But along major rivers like the Euphrates and Tigris, these changes created ideal conditions for the rise of agriculture. Similar climatic pressures challenged preagricultural societies all across the globe, and, over the generations, humans gradually changed the way they made food. Frankly, they became farmers.

Now, when we talk about hunter-gatherers, we saw how their food system influenced more than their dinner choices. Basically, the way they made food made their society, and it guided their technological, biological, and cultural development too. And the same can be said of agriculture. After nearly 7 million years of foraging, we suddenly started farming, and that food revolution changed our lives, all aspects of our lives, and forever.

First, sedentary living and farming allowed humans to generate serious surplus, encouraging larger constructions and settlements to produce and store that surplus. We also developed innovative tools and production strategies, which further bolstered our capacity to produce food. Additionally, sedentary living made it a heck of a lot easier to raise children. Raising an infant is exhausting enough, even in the 21st century with the very best of baby monitors, diaper stations, and cribs. But imagine how hard parenting would be if we were all migratory foragers. It's no surprise, really, that when folks settled down and started farming, their populations grew. No longer on the road, and now capable of generating some serious food surplus, human societies eventually grew into massive ancient civilizations. We see this in China, Egypt, Peru, and all over the world.

In short, the transition to agriculture inspired sedentary living and urban centers, and it sparked technological breakthroughs like writing, mathematics, medicine, and so much more. But this transition wasn't a smooth one. And to see why, let's turn to biological anthropology and linguistics for a revealing glimpse into our early agricultural life.

First, let's head to the Illinois River Valley near modern-day Peoria for some bioanthropology. Now, in his classic text *Paleopathology at the Origins of Agriculture,* George Armelagos analyzed fossilized human remains to document early agricultural diseases. He found evidence that early farmers in the Illinois River Valley, when compared with the hunter-gatherers who preceded them, were rather stressed-out and sickly. They had bone lesions, anemia, degenerative spinal conditions, and, as I've said before, even lower life expectancies. So, despite these poor health outcomes, the relatively rapid spread of farming indicates that agriculture was actually a rather seductive alternative to the hunter-gatherer life.

For example, take the Bantu, an African population bound by common linguistic and cultural traditions. In fewer than 2,000 years, the Bantu and their farming techniques took hold of most of Africa below the equator. Hunter-gatherers displaced by the Bantu either became farmers alongside the Bantu, or they fled to places like the Kalahari Desert to be left alone. And we can trace this migration using linguistics to map the evolution of Bantu dialects. And you know what? This map shows the Bantu and their language starting near modern-day Nigeria and Cameroon, and then they spread south to populate most of southern Africa.

Now, before we move on to the Green Revolution and modern food production, we should discuss one more significant change spurred by agricultural life, and that's the emergence of poverty and wealth, otherwise known as private property. Today's tech geniuses reap benefits from the gadgets and patents they create, and that incentivized system was in effect for early farmers, too. Just think about it: unlike the rest of us, super farmers will produce super surplus—they'll be rich, so to speak. At first, they may have surpluses due to experimental luck, or perhaps the rain and ecology in my neck of the woods just wasn't as fruitful as they were in yours, but one sure way to farm surplus was creative innovation, just like today. And for super farmers, the transition to agriculture is, in hindsight, a celebrated and transformative event.

Yet, for all its glitz and glory, farming was a very risky food production strategy. A single pest infestation, inadequate rainfall, or a loose group of your neighbor's cattle—they can all ruin a farming community or a family in a flash. We see this in parts of the world even today. For example, in Mali, I worked with a family who endured several years of debt and scarcity, only because one member of their family, one of their best farmers, broke his leg—actually, a donkey broke it for him. Anyway, this family lost critical labor power for one season, and the resulting debts undermined their food security for several years.

Like *yin* and *yang*, poverty and wealth always come as a pair. And, like most things in life, we can find some good and bad in the agricultural trajectory that has carried us into the 21st century. And, conveniently, agriculture has also carried us to the final part of today's lecture, the Green Revolution. We've seen that, after the last ice age, ecological changes threatened humanity's ability to live off the land, and that crisis triggered the agricultural revolution.

More recently, a second existential food crisis emerged in the 20th century. And here in the 21st, we're not out of the woods just yet. You see, prior to the 20th century, scholars like Thomas Malthus rang the alarm bells. Not so different from our hunter-gatherer ancestors, modern humanity's capacity to produce food surplus encouraged population growth. But that exponential growth, Malthus believed, would eventually outstrip food production, and as a result, he warned us of a future of famine and disease.

So, here we are, nearly two centuries after Malthus voiced his concerns. These fears of runaway population growth destroying humanity found a new voice in Paul and Anne Ehrlich's classic *The Population Bomb*. Now, this 1968 book predicted that widespread starvation was imminent. We had finally reached our agricultural limit, and, to avert a massive increase in the global death rate, humanity needed a second food revolution. But wait a minute. How do you do that? How do you make more food for more people, but with less arable land?

Well, society tasked scientists, not farmers, with this new food revolution. Organized into national and international research institutes funded by government and industry, and aligned with university research programs, agricultural scientists sought solutions to feed the planet. Each major crop had its own international research program and facilities, and, first through hybridization and then with bioengineering, scientists did the seemingly impossible: they boosted yields significantly. They found ways to make more food.

In the US, for example, cereal farmers used scientists' improved seed, and the results were indeed revolutionary. By the start of the 21st century, US farmers more than tripled their cereal yields in over three decades. Early on, the 20th-century food revolution was optimistically named the Green Revolution. Increasing food production is a monumental challenge, so if you happen to be a farmer, and if you hope to reap improved Green Revolution harvests, you're going to have to completely change the way you farm. But how?

Well, first, you'll have to stop producing your own seed on the farm; you'll be buying improved seed from agricultural specialists instead. And, by the way, if you're tempted to cheat the system by replanting any of that improved seed from the first generation of your improved harvest, you're going to want to reconsider because, since the 1990s, many of these engineered seeds carry what the world press literally calls a terminator gene, which produces sterile grain—good for eating, but bad for replanting.

Now, the next step in achieving your Green Revolution yields: shopping. After buying your modified seed, you'll also need to buy fertilizer, pesticide, herbicide, some seed treatment perhaps. And these petroleum-based products fueled the Green Revolution, and they'll allow your new seed to thrive. You see, all fields are not equal. To get the most from your improved seed, you'll need cash or credit to make conditions in your field just right. You'll definitely spend some money to get this seed in the ground, but the idea is to make up for that investment with the profits from your exponentially increased yields. And, really, this is not a bad strategy, as long as one can avoid unpredictable crop failure.

Currently, well over 80 percent of US soybean and cotton acres are planted with genetically modified varieties, and the first commercially produced genetically modified food actually hit the market quite a while ago, back in 1994. It was a tomato, by the way, which, among other qualities, has a fish gene that extends its shelf life. Unfortunately, as we settle further into the 21st century, many of the world's farmers, including the poorest of the poor, have yet to reap Green Revolution benefits like improved yields, and, if you're so inclined, the extended shelf life of the Flavr Savr tomato.

But alas, as we observed first in the agricultural transformation, green revolution farming benefited some more than others. Sub-Saharan African farmers, for example, grew only one ton of cereal per hectare in the 1960s, while US farmers harvested two. And to this day, in places like Mali, farmers still get an optimistic average of around one ton of grain per hectare. Nothing has changed for them. Conversely, US cereal producers have harvested a stunning 6–7 tons a hectare, and that's since the 1990s.

So, as extreme hunger and food insecurity persist into yet another century, humanity is again turning to scientists and other specialists in search of the future of food. And, like early humans who brought us agriculture, today's society is faced with its own set of unique challenges, including population growth, climate change, and global security.

Now, as we've seen in this lecture, anthropologists are uniquely positioned to explore and articulate the implications of our food production strategies both past and future. In fact, the persistence of extreme hunger and the quest for food security are the focus of my own career research and much of my teaching. And, as a cultural anthropologist, I've dedicated much of the last two decades to living and learning the quotidian lives of my remarkably generous Malian hosts.

Quite different from survey research and interviews, long-term participant observation reveals to the anthropologist the words, ideas, stories, art, music, and everything else that contains the hopes, knowledge, and histories of a community. This knowledge is an essential resource for bringing technology projects into the developing world, and, moreover, it can also help generate new knowledge and innovative solutions for reducing hunger. Let me give you just one example before we wrap up.

Ten years ago, when I launched a collaborative seed testing and production experiment with rural Malian farmers, I spent two years talking with them about their lives. Once, an elder named Bakari literally grabbed my arm and shook it vigorously as he told me that the sorghum plants his father once grew were as thick as my wrist. The sorghum his family grows today is barely thicker than a pinky. And he grows these new types of sorghum because he said those old ones just don't work anymore. There's not enough rain.

So, years ago, when his old varieties were failing, Bakari travelled north to seek out new seed. In the north there's less rainfall, and if his old seed no longer worked, he said the seed from the drier north certainly would. So, to make a long story short, he brought back and shared what is now one of the most popular types of sorghum in his community. The village, in honor of this effort, named the variety Bakari Kuruni Nyo, which means short Bakari's sorghum. Inspired by the considerable impact of his informal experiment, I became Bakari's farm apprentice, and I also acquired 23 new types of sorghum seed for us to test.

And from our experiment, I learned about local seed selection and exchange. I learned about the importance of seed names and ownership, farming techniques, labor organization, field hollers, and so much more. And, as an anthropologist, I'm uniquely placed to make sense of, to broker and spread, this information in a way that might ultimately benefit other farmers and scientists in other places all around the globe.

Renowned 20th-century anthropologist Ruth Benedict once said that the purpose of anthropology is to "make the world safe for human differences." So, in terms of the history of food production, what is it that anthropologists actually know or do that might possibly make the world safer? Well, simply put, we investigate and explain the biological, linguistic, and cultural histories of humankind, including the nature and consequences of our food systems. Decades ago, it was Bob Marley that reminded us: a hungry mob is an angry mob. So, whether you're a part of the angry mob, or if your belly's full, we can all make the world safer for human differences by finding ways, new and old, to feed the hungry.

Today, we called upon the four subfields of anthropology to help us understand how food production shapes our lives and society. We even considered how anthropology can contribute to innovative solutions to feed the world. So, when you share your next meal, why not pause for an anthropological minute to appreciate the 7 million-year history that put bread in your hand and tea in your cup.

Rise of Urban Centers

I n previous lectures, we saw how tool making and agriculture helped early humans adapt and endure as a species. This lecture continues our archaeological exploration by showing how our human ancestors invested in the future of humankind by building major cities and civilizations across the planet. To start, we'll take a moment to check out the earliest known cities on record, and then our archaeological adventure will give us an in-depth look at a few astonishing early, world-class cities that forever shaped the story of humankind.

Criteria for a City

What are the criteria for a city being considered as such? The influential archaeologist Gordon Childe gives us a checklist we can use, albeit loosely, to identify early cities across the archaeological record:

1. A city has a large, dense population with long-distance trade.
2. A city exists as a class-structured society.
3. A city has some system of governance and taxation.
4. A city boasts cultural resources including artwork, monuments, scientific knowledge, and the arts.

The actual emergence of anything resembling a city didn't occur until around 12,000 years ago. Anthropologists who study the emergence of cities generally note that humans started urbanizing several thousand years after the most recent ice age ended.

Ancient Mesopotamia

Archaeologists point to ancient Mesopotamia as the site of the first known urban revolution in the history of humankind. Mesopotamia is the ancient region between the Tigris and Euphrates rivers. There, in the 4th millennium B.C.E., a revolution began that would give rise to such powerful cities as Ur, Babylon, and Nineveh.

Jericho also fits the bill for one of the first human cities on record. With its 6-foot-wide protective walls, Jericho has been home to at least 20 unique settlements, the earliest dating back over 10,000 years ago. But it took another millennium before the droughts and cold conditions in the region gave way to more fertile grounds for the first cities.

With better conditions and an abundant ecology, people settled into permanent settlements and farming. One of the most remarkable archaeological discoveries of early Jericho life is the clear evidence for religious or cultural practices dealing with funerary practices.

The early urban Jericho people tended to bury their dead under the floors of their small, circular, clay houses. Remarkably, they occasionally preserved the heads

Ruins in Jericho

of the deceased, plastering the skull to preserve a rather lifelike, albeit spooky, appearance.

Jericho is more familiar to us than many other first cities because of its prominence in the Bible. The site is in the modern-day West Bank, about 15 miles northeast of Jerusalem. Archaeologists think people were attracted to the region because of its proximity to the Jordan River and its plentiful springs and palm trees.

Upward of 3,000 people lived in these earliest versions of Jericho, and they helped usher in the agricultural transformation with

domesticated barley and wheat production.

Byblos

Well before Egypt started building pyramids, ancient seafaring Phoenicians built Gebal, a small fishing community. By 3000 B.C.E., Gebal had grown into a full-fledged city called Byblos, another one of the first cities on record.

Byblos was ideally situated for the development of agriculture and international trade. It ushered in the powerful ancient Phoenician civilization and played a pivotal role in the development of the written word.

The archaeological record traces the early development of the Phoenician alphabet to Byblos. That ancient Phoenician alphabet is connected to languages the world over. English speakers, for example, have no trouble identifying a Phoenician L, M, or N. Even the K is just a rotated version of what we use today.

Aleppo

Aleppo is an ancient city that remains a major urban center in contemporary Syria. Aleppo, once a powerful geopolitical neighbor to pharaonic Egypt, dates back 8,000 years.

The region's archaeology shows us that nomadic camps roamed through the region as far back as 12,000 to 13,000 years ago. That's right at the cusp of the Neolithic transition from hunting and gathering to settled societies.

Aleppo eventually grew into a world city known for its legendary citadel, fortified palaces, impressively ornate mosques, and an extensive

network of *hammams*, which were essentially a massive bathing infrastructure.

Sadly, like untold numbers of ancient cities, Aleppo has endured colossal tragedy and destruction into the 21st century. The civil war in Syria, which began in 2011, has wreaked havoc upon the people and the architecture of Aleppo, imperiling one of humanity's first major cities.

Uruk

Some 5,000 years ago, ancient Uruk was a city on the Euphrates River, with a population of over 50,000 people. It was located about 150 miles south of modern Baghdad.

After the Euphrates River branched out some 10,000 years ago, the early inhabitants of the region built adobe-style houses near the river, attracted and sustained by the abundant ecosystem.

The region's natural resources inspired these early populations toward agricultural innovation, and they eventually built complex irrigation systems and canals, which helped transform this settlement into one of Mesopotamia's earliest and most famous world cities.

Uruk is also the scene of one of the greatest ancient literary works of all time: the *Epic of Gilgamesh*. Here's a synopsis:

- In the earliest of days, Uruk was ruled by a rather mean king: Gilgamesh. His reign of terror was so brutal that the people of Uruk prayed to the gods in hopes of peace.

- The gods heard their prayers and came up with a solution: They sent a primitive man to earth named Enkidu. He set out to defeat Gilgamesh, but befriended him instead.

- But then Enkidu died. His death crushed the king, who set out to seek Utnapishtim, the holder of the secret to eternal life.

- As it turns out, Utnapishtim was rewarded with immortality from a god named Enlil. It seems Enlil gave the gift of immortality for Utnapishtim's heroic effort of building an ark to preserve his family and all kinds of wild animals.

- As Gilgamesh left Utnapishtim to return to his kingdom back in Uruk, Utnapishtim took pity on Gilgamesh, and he shared one last secret. He reveals that there is a sacred plant at the bottom of the sea that could restore his youth.

- Gilgamesh goes underwater to retrieve the miracle plant. Despite having the source of immortality in his hands, he loses it moments later as he's bathing.

- With that final loss, Gilgamesh returns to his kingdom, where he is pleased by the remarkable wall his subjects have built to protect all of Uruk.

Stories like these may contain plenty of wonderfully magical and unrealistic elements, but they also contain fragments of the ancient Uruk worldview. Beyond interpretivist analyses of plot and characters, the content within the Gilgamesh story reveals hints about local ecology, values, and artistic or symbolic expression.

For example, in 2015, researchers at the Sulaymaniyah Museum discovered a new segment of the Gilgamesh adventure. This new morsel takes Gilgamesh and his pal Enkidu into a cedar forest where they meet up with monkeys and birds before they kill a demigod. It offers what we might otherwise consider to be an environmental morality tale.

In Uruk, the people developed both numbers and writing symbols. Archaeologists estimate that it took about 1000 years to go from scraping pictures in clay to a complex writing system that allowed for the composition of epic tales like the adventures of King Gilgamesh.

As the earliest settlers in the region gradually built sedentary lives and agricultural surplus, temples became some of the original sites for redistributing surplus food. As warfare among Mesopotamian city-states proliferated in the age of metals, religious power bases conveniently shared their surplus with warlords and others who had the capacity to protect Uruk from attackers.

Gilgamesh

Tiwanaku

The rise of cities was a global phenomenon. In modern-day Bolivia, not far from the shores of Lake Titicaca, are the remains of ancient Tiwanaku. Just over 2,000 years ago, the settlers at Tiwanaku were on track to building one of the most historic cities in South America.

From around 400 to 1100 C.E., Tiwanaku was a thriving capital city with adobe homes and a massive ceremonial complex with temples, pyramids, palaces, and brilliant archaeological monuments.

One of the most impressive monuments is the Gate of the Sun, which is a massive stone cut into an open gateway covered in intricate carvings. At the top is a great bas-relief image of a god wearing an imposing headdress. Remarkably, archaeologists have interpreted the complex pattern of these carved images as a farming calendar.

Another Tiwanaku feat was their underground water drainage system, and an absolutely brilliant network of irrigation canals that made Tiwanaku an agricultural powerhouse. They built a network of canals around some 50,000 raised bed fields to feed an entire empire.

This wasn't just about delivering water to plants. These raised beds also naturally protected crops from frost damage, a serious problem for Andean farmers.

Here's how it works: Water in the canals surrounds these raised beds, and it soaks the grounded roots, providing a steady source of water. But as the day goes on, the Sun heats the canal water as well.

When the cold mountain evening draws in the frost, the warm water releases its heat as a misty vapor

Tiwanaku, Bolivia

that literally covers the raised beds like a blanket protecting the delicate plants.

The symbolic, artistic, architectural, and agricultural legacy of this Tiwanaku was confirmed and celebrated in 2015. The occasion: After winning his 3rd term as president of Bolivia, Evo Morales invited the world back to the Sun Gate and the Tiwanaku ruins as the site of his inauguration.

As the country's first indigenous president, his success and his Tiwanaku inauguration help reclaim the glory of this ancient city as a striking metaphor for the ingenuity and power of Bolivians.

Suggested Reading

Jennings, *Killing Civilization.*

Leick, *Mesopotamia.*

Pauketat, *Cahokia.*

Questions to Consider

1. Where and when did some of the earliest known human cities emerge?

2. When and why did humans begin to create cities?

3. Why did ancient cities fall? Were there similarities in the conditions that led to their decline?

Rise of Urban Centers

When Mark Antony tried to think of what in the world he could offer as a gift for Cleopatra, he gave her a city. Yep, an entire city. The historic, walled city of Jericho. Not too shabby, right? What is it about cities? Why are so many of us so enamored with them? From Cleopatra to millions of people in the 21st century, most humans just love city life, but why?

It seems like a universal thing. In Bamako, for example—that's the capital city of Mali where I've worked for the past couple decades—it's population has grown over 10 times its size in only the last 50 years. And Bamako isn't alone. The World Health Organization estimates that around 60 percent of the global population will live in or near a city as early as 2030. What's going on here? Why is the city such an alluring place for us humans? Well, that's a question we'll address today.

In our previous lectures, we saw how tool making and agriculture helped early humans adapt and endure as a species. So today, we'll continue with our archaeological exploration and we'll watch our human ancestors invest in the future of humankind by building major cities and civilizations all across the planet.

To start, we'll take a moment to check out the earliest known cities on record, and then our archaeological adventure will give us an in-depth look at a few astonishing and early, world-class cities. Cities that forever shaped the epic story of humankind.

First, as we head back in time in search of the earliest cities on record, let's make sure we're clear about what we mean when we say we're searching for an early city. What're the criteria here?

We'll keep it in the anthro family, and go with a system devised in the 1950s by the influential archaeologist Gordon Childe. It's him that gives us this checklist we can use, albeit loosely, to identify early cities across the archaeological record. So what is it that we are looking for?

First, a city has a large, dense population with long-distance trade. Second, a city exists as a class-structured society. Third, a city has some system of governance and taxation. And last, a city boasts cultural resources including artwork, monuments, scientific knowledge, and even the arts.

Now, that's a fairly general checklist, but really, that's all we need. Because, despite the fact that hominins have been around for some 7 million years, the actual emergence of anything resembling a city, well that simply doesn't appear until around 12,000 years ago. Take a moment with that one. We've talked about our hominin ancestors living as hunter-gatherers for seven million years, and then, the ice age clears and we suddenly see sedentary communities, agriculture, and now urbanization, and all of this within just a few thousand years.

Leave our early ancestors out of it, even in terms of our relatively brief 200,000-year existence as *Homo sapiens*, settling down, living in cities and even agriculture, they're all really recent developments. So with that in mind, let's take a glance at the first known cities in the history of humankind.

Anthropologists, especially archaeologists, who study the emergence of cities generally note that humans started urbanizing several thousand years after the most recent ice age melted into the history books. Sedentary living became the latest prehistoric, post-ice age craze.

In our search of the earliest cities, archaeologists are going to point us to ancient Mesopotamia as the site of the first known urban revolution in the history of humankind. Mesopotamia literally means land between the rivers and it refers to the ancient region between the Tigris and the Euphrates. There, in the 4th millennium B.C.E., an urban revolution began—a revolution that would give rise to such powerful cities as Ur and Babylon and Nineveh. Let's take an archaeological tour of some of these remarkable first cities, and then we'll return to Mesopotamia to look more closely at ancient urban life.

Jericho easily fits the bill for one of the first human cities on record—at least if we're using Childe's criteria for distinguishing urban centers versus other types of settlements. With its six-foot wide protective walls, Jericho has been home to at least 20 unique settlements, the earliest one dating back over 10,000 years ago.

But it took another millennium before the droughts and cold conditions in the region gave way to more fertile grounds for the first cities. With better conditions and an abundant ecology, people gathered into permanent settlements and farming. One of the most remarkable archaeological discoveries of early Jericho life is the clear evidence for religious or cultural practices dealing with funerals.

For one, the early urban Jericho people tended to bury their dead under the floors of their small, circular, clay houses. And remarkably, they occasionally preserved the heads of the deceased, plastering the skull to preserve a rather lifelike appearance. And they even added physical details that recreated specific characteristics of particular individuals, as if they were pre-historic versions of the presidential busts we might see in our nation's capitol.

Plaster skulls aside, Jericho is more familiar to us than many other first cities because of its prominence in the Bible. The site is in the modern-day West Bank, about 15 miles northeast of Jerusalem. In the beginning, archaeologists think, people were attracted to the region because of its proximity to the River Jordan, and its plentiful springs and palm trees. But regardless, Jericho proved to be a fruitful place to usher in the Neolithic revolution, and that's just what generations of people did there. Upwards of 3,000 people lived in these earliest versions of Jericho, and they helped usher in the agricultural transformation with domesticated barley and wheat production.

Over time, the city's reputation and population ebbed and flowed like the tides. Regardless, as one of the very first cities on earth to thrive on an agrarian economy with a surplus, Jericho easily makes our list of one of the top three first cities of humankind. Now let's head north to the second city on our list— Byblos.

In the area we now know as Lebanon, the illustrious Byblos emerged only a couple thousand years after Jericho. Well before Pharaonic Egypt started building pyramids, ancient, seafaring Phoenicians built Gebal, a small fishing community, which by B.C.E. 3000, had grown into a full-fledged city called Byblos, another one of the first cities on record. So how does an ancient settlement of some Mediterranean fishermen and their families become one of the first world cities on record?

Byblos was ideally situated for the development of agriculture and international trade. It ushered in the powerful ancient Phoenician civilization. But what is it that makes it unique? Besides its huge population, and its role in developing world trade and agriculture, the main reason Byblos easily makes our top three first-cities list because it's the birthplace of the ABC's.

Let me explain. The archaeological record traces the origins and/or early development of the Phoenician alphabet to Byblos. Now, that ancient Phoenician alphabet is connected to languages the world over. And as English speakers, for example, we'll have no trouble identifying a Phoenician L, M, or N. Even the K is just a rotated version of what we use today.

So how about that? Our search for first cities just revealed details about the emergence of writing. And that makes sense. I mean, think about this relationship between the emergence of cities and writing. Once we move beyond the small, close-knit economy of a foraging band, everything gets more complicated.

Just Imagine how hard it would be to keep track of all your household income and expenses without writing words or numbers. That would be hard enough, but try keeping track of the integrated economies of the thousands of thousands of people living in Byblos. Impossible.

So over the years, we've uncovered a treasure trove of ancient human inscriptions, and one of the oldest of these archaeological finds is a historic carving on the stone coffin of King Ahiram of Byblos. The sarcophagus, dated to B.C.E. 1000 features a remarkable scene of a priestess giving a lotus flower to a king seated on a throne adorned with winged sphinxes. And then, on the lid rim is where we found a historic inscription that's 38 words long, and it warns future kings and conquerors to resist the temptation to uncover this coffin at the risk of suffering cataclysmic consequences.

With those engravings and bas relief images, this sarcophagus is now an invaluable artifact of an earlier era. An era when we were inventing the urban experience and beginning to teach alphabets.

Leaving ancient Byblos, let's head just a bit further north and east to the edge of Mesopotamia. Aleppo. Yes, we're talking about the ancient history of the major city that remains a major urban center in contemporary Syria. Aleppo, once a powerful geopolitical neighbor to Pharaonic Egypt, dates back 8,000 years and the region's archaeology shows us that nomadic camps roamed through the region as far back as 12,000 and 13,000 years ago. And, if you remember, that's right at the cusp of the Neolithic revolution.

Situated between the Mediterranean Sea and the upper Euphrates, Aleppo emerged right in the middle of several important trade routes. Location, location, location. Through trade and an urban transformation fueled by agriculture, Aleppo eventually grew into a world city known for its legendary citadel, fortified palaces, impressively ornate mosques, and an extensive network of hammams, which were essentially a massive bathing infrastructure. Think about that. How much would you really enjoy the city if it didn't have any showers?

Sadly, like untold numbers of ancient cities, Aleppo has endured colossal tragedy and destruction into the 21st century. The civil war in Syria, which began in 2011, has wreaked havoc upon the people and the architecture of Aleppo. And it's to such an extent, that the archaeological record of one of humanity's first major cities remains, to this day, in peril. As a symbol for all the unknown cities and civilizations that have vanished from the archaeological record, we'll celebrate ancient Aleppo and add it to our top three list of first cities.

So we've arrived 5,000 years ago in ancient Uruk—a city on the Euphrates River, with a population of over 50,000 people. It's actually just about 150 miles south of modern Baghdad.

After the Euphrates river branched out some 10,000 years ago, the early inhabitants of the region built adobe styled houses near the river, attracted and sustained by the abundant ecosystem. It was the region's natural resources that inspired these early populations toward agricultural innovation, and they eventually built complex irrigation systems and canals, which, in turn, helped transform this settlement into one of Mesopotamia's earliest and most famous world cities.

Fueled by the products of agricultural innovation, the people of Uruk developed non-farming trades, religious traditions, and even public architectural monuments like you'd see in any modern city. If you were to visit the region today, you'd still be able to see the ruins of a great tower dedicated to the sky god Anu.

Uruk is also the scene of one of the greatest ancient literary works of all time— the *Epic of Gilgamesh*. And though this isn't a course on literature, let's take a little time to discuss this epic, because as anthropologists, there's plenty we can learn from hearing these ancient tales. So, turn on those anthro-brains, and as I retell the story, why don't you think about what the details of this epic reveal about the lives and minds of the people of Uruk—the citizens who lived in one of the first global cities we know. Here we go.

Once upon a time, in the earliest of days, Uruk was ruled by a rather mean king—King Gilgamesh. His reign of terror was so brutal that the people of Uruk prayed to the gods in hopes of peace. The gods heard their prayers and came up with a solution, they sent a primitive man to earth named Enkidu.

Enkidu was a hairy beast of a man who lived with the animals, but over time and with a little help from a beautiful woman he was civilized. And once his civilized self learned of King Gilgamesh's atrocities, Enkidu rushed to end the people's suffering. In the midst of fighting, however, the two adversaries became great friends. Yeah, so much for that stop the evil king plan.

Well, not so fast. It seems that Enkidu's friendship was more powerful than his brute force. Through mutual death-defying adventures, the bromance between Enkidu and Gilgamesh blossomed. But then Enkidu died. His death crushed the king, and he set out to seek Utnapishtim, the holder of the secret to eternal life.

And after a series of failed adventures, Gilgamesh finally gets to Utnapishtim, but he messes up every chance he's given to earn immortality. Hilariously, at one point Utnapishtim challenges Gilgamesh to give immortality a try by not sleeping for a week, a challenge Gilgamesh quickly fails. Recognizing his failure, and as one last attempt, Gilgamesh finally asks Utnapishtim how he himself became immortal.

As it turns out, Utnapishtim was rewarded with immortality from a god named Enlil. It seems Enlil gave the gift of immortality for Utnapishtim's heroic effort of building an ark to preserve his family and all kinds of wild animals. The ark, like the biblical account of Noah, saved the animals and the ark builders family from perishing with the rest of humanity.

So that's why King Gilgamesh never fully shed his mortal roots. In fact, as he left Utnapishtim to return to his kingdom back in Uruk, Utnapishtim took pity on poor Gilgamesh, and he shared one last secret. He revealed that there is a sacred plant at the bottom of the sea that could restore Gilgamesh's youth. Of course, Gilgamesh goes straight to the source, and binding stones to his feet, he goes underwater to retrieve this miracle plant.

But, by now you've seen the way things work for this guy. Of course, despite having the source of immortality in his hands, he loses it moments later as he's bathing. And with that final loss, Gilgamesh gives up and jut returns to his kingdom where he is pleased by the remarkable wall his subjects have built to protect all of Uruk.

Now, stories like these may contain plenty of wonderfully magical and unrealistic elements, but they also contain fragments of the ancient Uruk worldview. The same way widely popular modern legends like the *Game of Thrones* reflect our contemporary world, the Gilgamesh adventure shows us what people thought and talked about in Uruk. Beyond interpretivist analyses of plot and characters, the content within the Gilgamesh story reveals hints about local ecology, values, and artistic or symbolic expression.

For example, in 2015, a newly discovered segment of the Gilgamesh adventure was discovered by researchers at the Sulaymaniyah Museum. This new morsel takes Gilgamesh and his pal Enkidu into a cedar forest where they meet up with monkeys and birds before they kill a demigod. I won't spoil all the details about this new exciting piece of the story, but remarkably it actually offers what we might otherwise consider to be an environmental morality tale.

So that's the outline of perhaps the earliest classic of world literature. But how do we know about the epic of Gilgamesh? Well, Uruk like Byblos was a monumental site in the history of human writing. What started first as a way to keep records of grain and livestock that were processed through the central warehouses, quickly developed from picture-like scribbles to abstract symbols and then numbers. In Uruk, they developed both numbers and writing symbols. Archaeologists estimate that it took about 1,000 years to go from scraping pictures in clay to a complex writing system allowing for the composition of epic tales like the adventures of King Gilgamesh.

Thanks to the plentiful and durable clay tablets found all throughout Mesopotamia, they help archaeologists reconstruct the city of Uruk as an enthralling world city with bulging populations, literature and the arts, social hierarchies, trans-continental trade routes, and even highly organized religious practices. Clay tablets from Uruk, for example, contained terrific lists of dozens of professions in the ancient city. There's the king, priests, labor supervisors, ambassadors, gardeners, masons, blacksmiths, farmers, weavers, potters, cooks, and so much more.

But how did such an early society manage so many people and so much food and trade? How did they do all that without Microsoft Excel? They didn't even have a slide rule? Well, as the earliest settlers in the region gradually built sedentary lives and agricultural surplus, temples became some of the original sites for redistributing surplus food.

By the time we're bringing and exchanging food at the temples of Uruk, food production and sharing have completely transformed from the idea of a band of hunters sharing a single antelope or mammoth. Eventually, societies built food systems managed by religious leaders who had to share the community's surplus with administrators, farmers, and artisans. However, as warfare among Mesopotamian city-states proliferated in the age of metals, religious power bases conveniently shared their surplus with warlords and others who had the ability to protect Uruk from those who sought to trespass its mighty walls.

Speaking of those walls, before we leave Uruk, let's listen to the words from the prolog of one of the versions of the *Epic of Gilgamesh*. Here the poet proudly describes the city he called home,

> Walk on the wall of Uruk, follow its course around the city, inspect its mighty foundations, examine its brickwork, how masterfully it is built, observe the land it encloses, the palm trees, the gardens, the orchards, the glorious palaces and temples, the shops and marketplaces, the houses, the public squares.

Uruk was definitely a gem of an early city. And many other such urban gems emerged in ancient Mesopotamia. Babylon, Nineveh, Akkad, Assur the list is truly remarkable for this relatively small stretch of land between two rivers. So, yes, Mesopotamia fully deserves its reputation as the site of the world's first urban revolution.

But rather than lingering in this one region, let's go further afield to explore the emergence of cities in other parts of the world. Because, as we've seen in other lectures, the end of the last ice age inspired or perhaps forced humans all over the world to modify their hunter-gatherer ways. And one increasingly popular adaptation strategy was a sedentary, agrarian lifestyle, and eventually the big city. Indeed, this was a global phenomenon—not restricted to Mesopotamia and the Mediterranean world. So let's head west, past Europe, across the Atlantic Ocean, and look at two ancient urban centers in the Americas.

The first of these sites is in modern-day Bolivia, not far from the shores of Lake Titicaca. That's where we find the remains of ancient Tiwanaku. Remember our discussion about the first people to come to the Americas? While Mesopotamians and the Old World were building some of the first cities on the planet, other branches of the human family were finally flowing into the Americas. And just as King Menes was ushering in the first years of Pharaonic Egypt, people in the Americas were beginning to play with sedentary living and farming.

In the Andes Mountains, for example, some early farming communities grew to populations of several thousand or more. The archaeological record shows that the people who lived in these towns gradually and collectively developed the foundations for serious urbanization. There was pottery, irrigated farming, religious sites, and expansive trade networks that eventually connected the coastal and mountain populations. They had every box checked on Childe's urban center inventory.

Early agrarian communities in what is now Peru, Ecuador, and Bolivia eventually thrived as sedentary farmers and communities. And as communities developed new, innovative solutions for food production, they quickly grew into larger towns. So, just over 2,000 years ago, as the Han dynasty of China was beginning its descent, the settlers at Tiwanaku were on track to building one of the most historic cities in South America.

From around C.E. 400–1100, Tiwanaku was a thriving capital with adobe homes and a massive ceremonial complex with temples, pyramids, palaces, and brilliant archaeological monuments. One of the most impressive monuments is called the Gate of the Sun, which is a massive stone cut into an open gateway covered with intricate carvings. At the top is this great bas-relief image of a god wearing an imposing headdress. Remarkably, archaeologists have interpreted the complex pattern of these carved images as a farming calendar.

Additionally, as great stone carvers, the ancient artisans of Tiwanaku developed some amazing interlocking stone blocks that could form impenetrable walls without even mortar or cement holding things together. Another Tiwanaku feat of infrastructure genius was their underground water drainage system and an absolutely brilliant network of irrigation canals that made Tiwanaku an agricultural powerhouse.

They built a network of canals around some 50,000 raised bed fields to feed an entire empire. This wasn't just about delivering water to plants. These raised beds also naturally protected crops from frost damage, which is a serious problem for Andean farmers. It's sheer genius. The water in the canals surround the raised beds, and it soaks the grounded roots, providing a steady source of water. But as the day goes on, the sun heats this canal water, and, then, when the cold mountain evening draws in the frost, the warm water releases its heat as a misty vapor that literally covers the raised beds like a blanket protecting these delicate plants.

The symbolic, artistic, architectural, and the agricultural legacy of this, one of the most important of all South American archaeological sites, was confirmed and celebrated for the world to see in 2015 when, after winning his third term as president of Bolivia, Evo Morales invited the world back to that Sun Gate and the Tiwanaku as the site of his inauguration. As the country's first indigenous president, his success and his Tiwanaku inauguration helped reclaim the glory of this ancient city as a striking metaphor for the ingenuity and power of Bolivians, ancient and modern, indigenous and otherwise.

So, while Tiwanaku flourished in the south, up in North America a similarly impressive first city was blossoming along the Mississippi, not too far from modern-day St. Louis. The economic and cultural center of the entire Mississippi valley, it was a place called Cahokia, and it was technically bigger than London in the year 1250. We'll learn more about Cahokia later in our lecture on apocalyptic anthropology. But for now, let me briefly describe this underappreciated American site.

Located in modern Illinois, as we said not far from St. Louis, Cahokia is an ancient civilization whose influence spread well beyond the Mississippi Valley. Like all first cities we discussed today, it was agricultural innovation and ecological adaptation that precipitated the urban experience in Cahokia.

And fed on maize and the benefits of an extensive Mississippi Valley trade network, one of Cahokia's most impressive architectural achievements was a remarkable pyramid that actually rivals the Pyramid of Giza. In terms of sheer size, the Cahokian pyramid was bigger.

So with that brief visit to Cahokia, it's time to wrap up our exploration of the great early cities that have forever shaped our humanity. The end of the most recent ice age transformed the earth enough that we humans had to find a way to adapt to survive through future generations.

Today, in the origins of the urban experience, we've seen how the emergence of first cities across the globe, from Cahokia to Uruk, we've seen how across cultures and geography, first cities bring with them some revolutionary tools to sustain our species. The urban experience, at its roots, was ironically all about farming. But we've also seen that we wouldn't have cities without technological innovation, tools, sedentary living, architecture, and even writing itself.

Anthropological Perspectives on Money

The past few lectures have introduced some of the game-changing things that humans did to become the sole remaining hominins. For example, we've seen how agriculture enabled humans to develop food surpluses and to transition to a sedentary lifestyle, which itself promoted a population explosion. And we've seen how the emergence of cities provided humans with opportunities to make remarkable strides culturally and technologically. This lecture looks at how we created numbers and money.

Numbers

Generally speaking, there are 3 phases to the creation of human numbers. First, biological anthropology teaches us about rudimentary number sense. Next comes the development of the early number concept. Finally, we get to counting with actual numbers.

The rudimentary number sense is phase 1, and it's not exclusive to humans. Other animals display rudimentary number sense too, and we humans still have it today. Rudimentary number sense doesn't use quantities. It's about sensing less or more—relative amounts.

An excellent example of rudimentary number sense is the solitary wasp. One type, the *Eumenes*, is sexually dimorphic; females are much larger than males. Remarkably, the mother wasp provides her developing larvae with food, usually small grubs.

The wasp provides exactly 10 grubs for each female larva. They provide half as many grubs per male larva. The wasp is not counting up from 1 to 5 and 10. It just knows 5 and 10. The wasp has rudimentary numbers sense.

The 2nd phase that leads to actual numbers as we understand them today is the early number concept. Think of this as a midpoint between modern numbers and rudimentary number sense. In this phase, we use clever devices and terms to get specific quantities.

An example: In the Andes Mountains, proto-accountants used a system of strings and knots called quipus to record important numbers like debts or annual harvests. Archaeological iconography and artifacts date this system to just under 5000 years ago.

Once agriculture seeded cities and empires, civilizations across the globe created number systems. From the Maya and Egyptians to the Chinese dynasties, numbers themselves appear to be as invaluable as tools and agriculture in terms of sustaining our species.

The Anthropology of Money

When we want to examine the earliest human societies, we largely depend on archaeology to reconstruct the remains of the day into human stories. It's a remarkable process that has taught us volumes about the origins of many early creations.

Two of the biggest names in the anthropology of money aren't archaeologists at all, but when they write about the origins of money, they head straight to the archaeological record. David Graeber's work *Debt: The First 5,000 Years* is perhaps the definitive book on debt, providing an exhaustive history of money and credit since the time of the ancient Sumerians.

And similarly there's Caroline Humphrey's extensive body of work on the history and cultural complexities of bartering. Both of them consult the archaeological record and much more to discover a massive truth that contradicts the traditional idea that bartering led to money, which in turn led to credit.

There never was an exclusively barter-based society, or at least not according to the empirical record. As a matter of fact, everywhere anthropologists have encountered bartering, it has always been in societies that concurrently use money.

Credit

As Graeber notes, the earliest archaeological evidence of financial transactions doesn't show us the money; it shows us debt (and therefore credit). If we go back to the first urban revolution in Mesopotamia, we find some of the absolute oldest financial records ever found. They are cuneiform tablets that recorded credit. Ancient Mesopotamians leaders created something called money of account.

Around 5000 years ago, city dwellers in Mesopotamia had established a fixed exchange rate to settle accrued debts. Citizens could take silver on credit, and then repay the debt with grain.

With a fixed rate set to exchange grains and silver, folks recorded debts on clay tablets, and then at harvest, accounts were settled. This is credit, or technically speaking, it's money of account, and it appears to be the earliest recorded financial transaction between humans. It happened independently in different sites with different materials.

It's suggested that people in early agrarian communities probably took seasonal loans in much the same way that many small-scale family farmers in Africa, Asia, and Latin America do to this day.

Coins

As societies grew, thrived, and interacted, precious metals became cross-cultural and timeless mechanisms to facilitate long-distance and local trade. Metals were more durable than paper or clay, and the fact that they can be scaled by weight and metal type makes it easier to trade, even across regional currencies.

Starting just under 3,000 years ago, evidence starts to show coins and precious metals as actual units of value. Archaeologists have unearthed coins all across Asia and into the Mediterranean.

In Turkey, which was once called Lydia, coins were found dating back to around 600 B.C.E. The Greek historian Herodotus advises us that Lydians were intrepid traders

who bridged Eastern and Western civilizations. As such, it's no surprise that they were early adopters of coin currency. Their currency was called the stater.

We can trace people trading specific weights of gold and silver bars in the names they used to differentiate between various denominations. The 96th denomination of a stater, for example, weighed of the weight of a full stater.

The earliest staters were composed of electrum, which is a naturally occurring alloy of gold and silver. But eventually Lydians modified their system by introducing pure versions of both gold and silver staters. In this new system, it took 10 silver staters to get 1 gold stater.

Paper Money

By the year 1007, the Song dynasty of China was minting over 1 million coins a year. Successful merchants were burdened by success because they had to haul and store all kinds of coins.

As part of the solution, authorities gave exclusive franchises to a few lucky enterprises that became deposit shops. Merchants could deposit money and goods at one of these shops, and they would get a receipt for what they left behind.

First, merchants traded these receipts in a proto-cash transaction. Then, once the central government claimed full control over the deposit system, China effectively produced the world's first paper currency that was issued and backed by a government treasury.

Centuries would pass before paper currencies emerged in Europe. That's exactly why Marco Polo, an anthropologist years before his time, was flabbergasted and inspired by the paper currency used throughout Kublai Khan's empire in the 13th century.

While coins made it to Europe very early, paper money didn't emerge until almost the 1700s, and it wasn't immediately embraced. Instead of paper, Europeans stuck with coins, and in the absence of coins, they used alternative devices like the tally stick.

King Henry I's England, circa 1100 C.E., required subjects to pay

King Henry I of
England

taxes using tally sticks, which
calculated debts and payments in
local currency denominations. The
sticks were segments of branches
or sometimes bones that people
carved and notched to document
a debt or payment.

Alternative Currencies and the Future

Alternative currencies are a fascinating phenomenon in the history of money. We're all familiar with coins and cash, but through the ages, people have also devised a range of alternative currencies. In the days before paper money was common, other money types like cowrie shells, clam beads, and leather sheets were used to facilitate exchange. Local currencies were part of what led to success in the American colonies, and they continue to help communities thrive into the current era.

One modern alternative currency example is the BerkShare of the Berkshire region in Massachusetts. Residents buy 100 BerkShares for $95. Then they spend this money in the community. Recipients can use it as full value at other businesses, or they can cash it in at the local bank for a 5% fee. Remarkably, some 20,000 residents purchased millions and millions of BerkShares from 2006 to 2016.

The history of money includes many different forms of currency. It's not just about coins and cash. And it seems that no single form of currency endures forever. It stands to reason that there are certainly some changes in store for how we use and understand money in the future.

From wearable, touchless payment options to phones, ATM cards, and web payments, our dollars are more virtual than ever. For years and years now, our paychecks have been electronically transferred into our personal accounts.

But ironically, one of the wealthiest empires in the Americas didn't have money. It was the Incas. The Incas had a community labor system in which males 15 and older worked for the state.

These workers collectively built thousands of miles of roads and palaces. They did so with amazing block work. In return for this fairly demanding group labor, the Incan government provided everything from housing and food to clothing.

Without money changing hands, and without markets, Incans collectively fulfilled their basic needs. All the while, they piled up hoards of gold and silver. But it's likely that Incan life was incalculably sweeter for the rulers versus the citizens.

Suggested Reading

Graeber, *Debt*.

Hart, *The Memory Bank*.

Humphrey and Hugh-Jones, *Barter, Exchange, and Value*.

Questions to Consider

1. How does anthropology's explanation for the history of money differ from the history presented in classical economics textbooks? Why?

2. Who were some of the first civilizations to make coins and cash? What did that early money look like?

3. Why do economic anthropologists and others predict the end of money as we know it? What possibly could replace a $100 bill?

Anthropological Perspectives on Money

For the past few lectures, we've been thinking about some of the game-changing things that humans did to become the sole remaining hominins on the human family tree. For example, we've seen how agriculture enabled humans to develop food surpluses and to transition to a sedentary lifestyle, which itself promoted a population explosion. And we've seen how the emergence of cities provided humans with opportunities to make remarkable strides, both culturally and technologically. Today, let's look at how we created numbers and the one thing we probably all like to count the most—money.

It may feel as if numbers and money have been rather permanent features of the human experience, but most of the Homo sapiens who have lived on this planet, they lived well before us, let alone American Express. And of all the humans who ever lived, most never touched money. Not a dime. They say winning a big sum of money can really change a person, but the anthropology of money is going to show us how creating money can really change a species.

So as we turn our anthropological lens to money, we'll get a peek at how economic anthropologists apply anthropology to understand how things like exchange, money, and credit transform the human experience, and how our economic behavior may have actually helped us retain our title as the last remaining hominin on the planet. So let's dig deeper into our humanity what is this thing that makes the modern world go round? What is money?

To start, try and imagine a time and place before money. What would it look like? How would you have to change your morning routine? To answer that question, Aristotle theorized that the earliest humans didn't even need money because they made everything they needed. But then, he did say, things changed with farming when new specializations emerged. Folks that were great at growing peanuts, for example, could barter with their surplus to get milk from cattle herders.

And it was the famous Scottish philosopher Adam Smith who agreed. In his classic The Wealth of Nations, Smith describes how there was a time before money in which barter was the way people exchanged things. Then money emerges in order to help make bartering easier.

Since Smith's time, we've continued adapting, making transactions easier by creating credit cards and mortgages. That's the classic story of money.

That's what you'll read in most economics textbooks. Barter into money, then credit.

Aristotle, Smith and just about every economist for the past couple centuries had relied on hypothetical history to explain the origins of money. So, what possibly can anthropology add to this story? It turns out, quite a lot. Where Smith and others go to thought experiments to explain the origins of money, anthropologists go to archaeology and the material record.

And the archaeological record indicates that we have the history of money all wrong. Our earliest artifacts show us that the idea of credit came well before actual money. That's the exact opposite of what most of us think. So let's reconstruct this very carefully. Let's see how anthropologists understand the origins and the history of money.

But wait, we skipped a beat here. We're jumping right into money, but we haven't even figured out when humans created numbers and counting. That's going to be the precondition for money; you got to have numbers. Using numbers and money is something that certainly makes us unique among the primate order and the animal kingdom as a whole. Other primates can be taught to count and use human numbers, but we humans are the only ones with calculators on their smartphones.

Let's take a brief moment to think about our human transition into mathematical apes. Generally speaking, there are three phases to the creation of our numbers.

First, biological anthropology teaches us about rudimentary number sense. And we'll come right back to that. Then, the second phase is the development of the early number concept, and today, archaeology will help us with that one. Then last, we get to counting with the actual numbers that you and I use them every day.

Rudimentary number sense is phase one, and it's not exclusive to humans. Other animals display rudimentary number sense too, and we humans still have it today. Rudimentary number sense doesn't use quantities. This is more about sensing less or more. If anything, it's relative quantities.

My favorite example of rudimentary number sense is the solitary wasp. One type, the Eumenes, is sexually dimorphic; females are much larger than males. Remarkably, the mother wasp provides her developing larvae with food, usually small grubs. Astonishingly, the wasp provides exactly 10 grubs for each female larvae. Not sometimes 11, sometimes nine. Always ten. And they do the same for male grubs but with half as many grubs per larvae. That wasp is not counting one, two, three, four. It just knows ten and five. Like a bird that returns to its nest and sees one of her eggs has been poached, the wasp has rudimentary numbers sense.

The second phase that leads to actual numbers as we understand them today, is the early number concept. Think of this as a midpoint between modern numbers and that rudimentary number sense we just talked about. In this phase, we use clever devices and terms to get specific quantities.

In South America, for example, we can visit with archaeologists and the living descendants of pioneers in numbers and counting. Have you heard of quipus? We'll in the Andes Mountains, proto-accountants named camayacs used a system of strings and knots to record important numbers like debts or annual harvests. Archaeological iconography and artifacts date this system to just under 5,000 years ago. Knotted numbers, but with strings.

Once agriculture seeded cities and empires, we see civilizations across the globe creating number systems. From the Maya and Egyptians to Chinese Dynasties, and just about everywhere you look, numbers themselves appear to be as invaluable as tools and agriculture in terms of sustaining our species and the emergence of cities and civilizations.

Anthropologists look at money in a number of ways. We think of money as a vehicle for exchange, as economists do, but we're also interested in deconstructing money, both linguistically and culturally. My friends in Mali often joke about the frenetic pace of life in the US, and many people who don't speak English know the phrase "time is money." And from my Connecticut perspective, "time is money" is not funny. It's not funny at all. It's a reminder that we've got to keep busy, no wasting time. But for my Malian friends, that's utterly absurd. It's laughable. The anthropology of money, with our four-field approach, can reveal some rather surprising facts about the history of money and human exchange.

Into the 21st century, there's a unique and enduring trend in the anthropology of money that explores the human relations involved in financial transactions. We're looking more at the collective human experience with money, rather than a purely fiscal phenomenon. Anthropologists are fascinated by the idea that you might sell your house to your child for $1, but you'd never dream of listing it on the market for that gift of a price. Actually, you might sell it for $1 if it was a terrible money pit, but if that was the case, you probably wouldn't sell it at any price to your kid. Money is money, a dollar is a dollar, but the way we use money depends on a heck of a lot more than some externally guided market.

So, let's take this question of the origins of money to anthropologists to see what they add to our understanding of something so central to human lives that we equate it with time. Time is money. So let's get to it. When we want to go to the earliest human societies, we largely depend on archaeology to reconstruct the remains of the day into human stories. And that's a remarkable process. It's a process that has taught us volumes about the origins of many early creations. We've seen archaeology reconstruct the origins of stone tools, for example, so now let's use it to track down the origins of money.

Two of the biggest names in the anthropology of money aren't archaeologists at all, but when they write about the origins of money, they head straight to the archaeological record. David Graeber's work Debt: The First 5,000 Years is perhaps the definitive book on debt to date, and it provides an exhaustive history of money and credit since the time of the ancient Sumerians.

And similarly, there's Caroline Humphrey's extensive body of work on the history and cultural complexities of bartering. Both of them consult the archaeological record and much more to discover a massive truth that contradicts the barter to money to credit story from economics. Humphrey puts it succinctly:

> Barter is at once a cornerstone of modern economic theory and an ancient subject of debate about political justice. In both discourses barter provides the imagined preconditions for the emergence of money; all available ethnography suggests that there never has been such a thing.

The ethnography she speaks of is the collective work of anthropologists and others who have documented people and cultures across the world, and a primordial barter society just isn't in the record. There never was an exclusively barter-based society, or at least not according to the empirical record. As a matter of fact, everywhere anthropologists have encountered bartering, it's always been in societies that concurrently use money. What is in the record, though, is plenty of coins, precious metals, and unsurprisingly, credit.

In fact, as Graeber notes in Debt, the earliest archaeological evidence of financial transactions doesn't show us the money, it shows us debt, credit. If we go back to the first urban revolution in Mesopotamia, we find some of the absolute oldest financial records ever found. They are cuneiform tablets that recorded credit. Ancient Mesopotamians leaders created what is referred to as money of account. It wasn't a coin or paper IOU kind of money. It was much actually much simpler than that.

Around 5,000 years ago, city-dwellers in Mesopotamia had established a fixed exchange rate to settle accrued debts. Almost like a universal credit card account for citizens, who take silver on credit, and then repay the debt with grain. With a fixed rate set to exchange grains for silver, folks recorded debts on clay tablets, and then at harvest, accounts were settled. This is credit. Technically, money of account, and it appears to be the earliest recorded financial transaction between humans. And it happened independently in different sites with different materials.

For example, in nearby Egypt, we see a similar history in hieroglyphics. These ancient writings, once translated, affirmed that in Egypt—just like Mesopotamia and elsewhere—credit, or money of account, preceded actual coin and paper currencies by a couple thousand years, depending on location. In short, money of account is our archaeologically confirmed origin of money.

It's suggested that people in these early agrarian communities probably took seasonal loans in much the same way that many small-scale family farmers in Africa, Asia, and Latin America do to this day. In Mali, the hard-working families that I work with use credit at the beginning of the growing season in order to buy inputs like fertilizers and pesticides. Those farm-based lines of credit then become due at harvest, when grains, cotton, or other crops come in. The only difference here is that Mesopotamians used barley to settle their debts, and my farming friends sell their harvest and use good old cash. So how do we go from this credit/money-of-account phase and move closer to the $100 bill we love to see in our pockets?

Well, the evolution of money unraveled quite differently depending on where you're looking. Regardless, across the human world, as societies grew, thrived, and interacted, precious metals became cross-cultural and timeless mechanisms to facilitate long-distance and local trade. For one, metals were more durable than paper or clay, and the fact that they can be scaled by weight and metal type makes it easier to trade, even across regional currencies. For example, a one gram silver coin from the Swahili coast could be received as 1 gram of silver just about anywhere there's people who value silver as much as they do back on the Swahili Coast.

Starting just under 3,000 years ago we start to see coins and precious metals not as cuneiform symbols or receipts for amounts of grain, but as actual units of value. Archaeologists have unearthed coins all across Asia and into the Mediterranean. We can go to China, India, or Turkey to see some of the earliest coins ever discovered. In Turkey, for example, which was once called Lydia, we have coins dating back to around B.C.E. 600. The Greek historian Herodotus advises us that Lydians were intrepid traders who bridged Eastern and Western civilizations. As such, it's really not a surprise that they were early adopters of coin currency. Coins gave the Lydians a single, stable currency that enabled them to transact business across cultures, languages, and geography.

We can trace people trading specific weights of gold and silver bars in the names they used to differentiate between various denominations. Across cultural boundaries and geography, early coins were named for their respective weights. The Lydian stater, for example, had six denominations. The third, the sixth, the 12th, 24th, 48th, and 96th. The 96th denomination, for example, weighed 1/96 of a full stater. The earliest staters were composed of electrum, which is a naturally occurring alloy of gold and silver. But eventually, Lydians modified their system by introducing pure versions of both gold and silver staters. In this new system, it took 10 silver staters to get 1 gold. Ultimately, whether as a coin or as a weight, the Lydian stater was worth its weight in gold, literally.

Over in India, and around the same time as the stater, there were silver, punch-marked coins. And in Zhou dynasty China, some of the first coins looked a bit more like monopoly pieces; they were these small decorative spades. And we can't leave out the knife-shaped coins found elsewhere in Ancient China. Then across China, around 2,300 years ago, rounded coins emerged as the standard.

So far we've been talking about precious metals and coins. But what about cold, hard cash? Have we found evidence of paper money in the ancient world? Well, by the year 1007 of the current era, the Song dynasty of China was minting over a million coins a year. Successful merchants were burdened by success because they had to haul and store all kinds of coins. Champagne problem, but a problem nonetheless.

So, as part of the solution, authorities gave exclusive franchises to a few lucky enterprises that became deposit shops. Merchants could deposit money and goods at one of these shops, and they would get a receipt for what they left behind. First, merchants traded these receipts in a proto-cash transaction, and then once the central government claimed full control over the deposit system, China effectively produced the world's first paper currency that was issued and backed by a government treasury.

Centuries would need to pass before paper currencies emerged in Europe, and that's exactly why Marco Polo, an anthropologist years before his time, was utterly flabbergasted and inspired by the novel, yet well-established paper currency used throughout Kublai Khan's empire in the 13th century. With an eye toward advising Europe into modernity, Marco Polo wrote:

> It is in the city of Khanbalik that the Great Khan possesses his mint. In fact, paper money is made there from the sapwood of the mulberry tree, whose leaves feed the silkworm. The sapwood, between the bark and the heart, is extracted, ground and then mixed with glue and compressed into sheets similar to cotton paper sheets, but completely black.

> Wheresoever a person may go throughout the Great Khan's dominions he shall find these pieces of paper current and shall be able to transact all sales and purchases of goods by means of them just as well as if they were coins of pure gold. And all the while they are so light that ten bezants' worth does not weigh one golden bezant.

Any foreign goods to enter his lands were sellable only to Kublai Khan. You received his paper money, and that was good for any item or services you needed, anywhere in the empire. Any payments to Kublai Khan had to be made in his specially created paper currency, which you had to get through him. It was brilliant. Polo was astounded by Khan's masterful economic system.

So, while coins made it to Europe very early, we don't see paper money until almost the 1700s, and it wasn't immediately embraced. Instead of paper, Europeans stuck with coins, and in the absence of coins, they used alternative devices like the tally stick. And in this, they were not alone

Marco Polo also wrote about the use of tally sticks in China, and tally sticks were accepted as evidence of economic transactions under the Napoleonic Code. But cutting to the chase, in the 1970s archaeologists found a tally stick in Africa that is over 30,000 years old. Called the Lebombo bone, it's a baboon leg bone with 29 notches on it. Now, it'll be impossible to definitively work out what that 30,000-year-old human was counting on that leg bone, but we know exactly how humans used tally sticks in the post-money era.

Let's say you live in medieval England. And suppose you loan £160 to a merchant. To make sure he pays you back, you take a hazelwood branch and you use it as a tally stick. First, you make several notches on the branch, each notch representing a certain amount. Maybe one notch is the breadth of a thumb, and that represents £100. And then you have three notches, each the breadth of a little finger. And they represent £20 each. Viola, you have a tally stick, showing £160.

Now, you split the stick into two lengthwise, through the notches, so that both the lender and the borrower can be given half the stick, showing the amount owed. Because every tally stick has a matching set of notches and a unique ring pattern at the split. These two pieces, and only these two pieces, will match up when then borrower comes to pay you back. That reduces the possibility that either side in that transaction can cheat the other.

And here's an interesting side note. Once the tally stick was split lengthwise, the matching halves were broken or cut apart in such a way that one piece was shorter than the other. And the borrower always got the shorter piece. That's where we get the phrase "short end of the stick" as a metaphor for the bad side of the deal. The longer piece, however, that went to the lender, and it was called the stock, which gives us our modern term stockholder.

Okay. Before we wrap up today, I'd like to take a quick glance at a fascinating phenomenon in the history of money, and that is alternative currencies. We're all familiar with coins and cash, but down through the ages, people have also devised a range of alternative currencies. In the days before paper money was common, other money types like cowrie shells, clam beads, and leather sheets were used to facilitate exchange. Local currencies were part of what led to the success in early American colonies, and they continue to help communities thrive into the modern era.

One of my favorite alternative currency is called Ithaca Hours. Residents of Ithaca, New York have been exchanging Ithaca Hours for decades. There are nearly 1000 ways residents can spend these Ithaca Hours, and some local employers actually pay part of their salaries in Ithaca Hours. The system has a proven successful model, and it's inspired over 50 community currencies to adopt it. You might use one Ithaca Hour to pay someone who washes your car, and then they'll use that same Hour for the meatball sub and iced tea he gets over there on Main Street for lunch.

Another terrific community currency that's called the BerkShare of the Berkshire region in Massachusetts. It's kind of like the Ithaca Hour, where residents buy 100 BerkShares for $95. Then they spend this money in the community. Recipients can use it as full value at other businesses, or they can cash it in at the local bank for a 5% fee. Remarkably, some 20,000 residents have purchased millions and millions of these BerkShares in the first decade, from 2006–2016.

So, as we wrap up, it's important to note that the history of money includes many different forms of currency. It's not just about coins and cash. And it seems that no single form of currency endures forever. So there are certainly some changes in store for how we use and understand money in the future.

From wearable touchless payment options to phones, ATM cards, and even web payments. Our virtual dollars are more virtual than ever. For years and years now, our paychecks have been electronically transferred into our accounts, and that's without a single piece of paper. We can even pay our Amex bill on-line with those funds. We're kind of back to the tally stick. Instead of a paper check or an envelope full of cash, we get new e-notches on our e-tally account. But doesn't the money rat-race, electronic or otherwise, overwhelm us all sometimes? Have you ever pondered the idea of a world without money? Well, ironically, one of the wealthiest empires in the Americas didn't have money. It was the Incas.

The Incas had a community labor system in which males 15 and older worked for the state. These workers collectively built thousands of miles of roads and palaces, and they did so with amazing block work with precision stone fitting that you can't even stick a knife into. In return for this fairly demanding group labor, the Incan government provided everything from housing and food to clothing. Without money changing hands, and without markets, Incans collectively fulfilled their basic needs, and, all the while they piled up hoards and hoards of gold and silver. That said, I bet Incan life was incalculably sweeter for the rulers versus the citizens.

LECTURE 11

Anthropological
Perspectives on Language

This lecture's mission is to explore the development and nature of human language. Like cities, farming, tools, and money, language has definitely played a starring role in our continued survival as a species. The study of language will bring us right back to the primates, but we'll also get our first real glance of the 3rd subfield of anthropology: linguistics.

Tamarin Monkeys

Deep in the rainforests of Central America, primatologists interested in the origins of language record how tamarin monkeys act as well as their vocalizations. Eventually, researchers can start to break their code to see how this monkey talk actually works.

In tamarins, primatologists have already recorded and catalogued almost 40 distinct types of vocalizations. For instance, the greeting chirp is called the B chirp.

Tamarins make other calls too, like whistles. When they combine vocalizations like whistles and chirps, they actually convey simple sentences to their families and friends. For example, they can tell each other that there's a predator around. This is revolutionary: There was a time that we thought communicating in sentences was solely a human trait.

Apes

A handful of apes have helped us explore the boundaries of human versus other ape-to-ape communication.

In the 1930s, Gua was a chimpanzee who lived with Luella and Winthrop Kellogg. The Kelloggs brought home Gua, a 7.5-month-old chimpanzee, to raise him alongside their own 10-month-old son, Donald. This duo spent 12 hours a day together and they were fast friends.

Remarkably, in the year they lived together, Gua actually was a quicker study on basic manual tasks like using a spoon. That said, Gua simply wasn't able to make progress with learning to speak. In fact, after about a year it was Donald who started imitating Gua's sounds. Shortly after that, the Kelloggs returned Gua to a primate research center.

In the late 1940s and into the 1950s, the psychologists Catherine and Keith Hayes tried a similar experiment, raising a young chimpanzee, Viki, as a human. Despite tireless efforts on everyone's part, Viki only learned to speak four words: mama, papa, cup, and up. Sadly, Viki died of viral meningitis at the age of 7.

Another pair of researchers, Allan and Beatrix Gardner, changed the way we think about primates and their linguistic potential. The

Gardners raised a chimpanzee named Washoe as a human, but not with a sibling. To gauge Washoe's linguistic abilities, the Gardners taught her American Sign Language (ASL). In all she learned some 250 words, which she later taught to her adopted chimpanzee son.

The 2nd chimpanzee who learned ASL was named Lucy. Lucy lived her human life with Maurice K. Temerlin. She learned ASL from Roger Fouts, and she was a natural, even displaying humor.

Lucy and Washoe made it clear that our primate cousins have the capacity to learn our human language. More contemporary research has revealed that we are only scratching the surface of the linguistic boundary between humans and the other apes. The famous gorilla Koko, for example, has learned over 1000 ASL signs.

However, the influential linguist Noam Chomsky warns us not to get our hopes up about future conversations with gorillas about the meaning of life. His primary critique is based on the idea that our primate cousins just don't have the unique linguistic legacy developed exclusively in the *Homo*

sapiens line. Their physiology, neurology, bodies, and brains preclude them from any advanced comprehension of human language.

Historical Linguistics

What exactly is human language? That terrific question brings us to linguistics, the 3rd subfield of anthropology. And within that subfield itself, there are 3 major specializations that help us understand the origins and nature of human language:

1. Historical linguistics: the evolution and extinction of language.
2. Descriptive linguistics: the mechanics of language.
3. Sociocultural linguistics: the relationship between language and culture

Historical linguists collectively build a beautiful linguistic tree that shows us how our languages have branched out and spread through various populations across the globe.

One tool that that helps to create our massive human language tree is glottochronology. Just as we can map the spread of humankind into the Americas, we can use glottochronology to map the diversification of human languages across the planet.

The gist of glottochronology is that 2 languages like English and German share linguistic origins, and therefore they still share some core vocabulary words. The longer they are apart, however, the more their core vocabularies will differ.

Linguists and others sometimes question the reliability of glottochronology. More recently, with more sophisticated computational methods, complex phylogenetic reconstructions help us see how languages, like farming and tools, have spread across the planet with intrepid *Homo sapiens*.

In addition to the evolution of language, historical linguistics also researches dead and disappearing languages. It's noting that of the

6,000 to 7,000 languages known today, half or more of these languages are predicted to go extinct by the end of the 21st century.

Many Native American languages—the languages of the Caddo, Menominee, Yuchi, Pawnee, and many other peoples—are endangered because there are only a few speakers left who can keep them alive. There are efforts, though, to improve this situation. For example, the Living Tongues Institute for Endangered Languages works to prevent Native American languages like Yuchi from dying out.

Descriptive Linguistics

The field of descriptive linguistics focuses on the conventions and mechanics that make language work. There are 5 key elements of human language:

1. Phonemes, which are the base sound units with no meaning.
2. Morphemes, which are the base units of meaning, like *un-* or *non-*.
3. Grammar, which covers rules.
4. Syntax, which covers sentence structure.

5. Semantics, which cover meaning

With these constituent pieces, there are limitless possibilities to what we can learn about the nature and origins of language. Linguists working in the Middle East, for example, studied a community named Al-Sayyid, which offers intriguing insights into our instinct for language.

This Bedouin community in Israel's Negev Desert has a large deaf population, the result of a local genetic mutation that leads to a higher prevalence of deafness. This village developed a distinctly local sign language, the Al-Sayyid Bedouin Sign Language, which is structurally distinct from both the spoken languages and the regionally dominant sign languages of Hebrew and Arabic.

Ultimately, the development and evolution of this unique, local sign language supports the idea that humans are born communicators. We're equipped to create languages when we need to.

Sociocultural Linguistics

Sociocultural linguists investigate the relationship between language and culture. Linguistic norms definitely differ across cultures, and those differences can shape the way we observe and perceive our daily lives.

In Mali, for instance, farmers who speak Bamanankan utilize a 24-hour cycle to track each day, but in day-to-day life, they use Bamanankan time categories. These categories are different from Western modes of speech, which tend to focus on specific hours and minutes.

The whole day may fit a 24-hour cycle, but the pacing of work teams, family meals, and daily life is still basically structured into 4, dynamic, unequal parts:

- Phase 1: Waking and starting work.
- Phase 2: A brief break, followed by more work.
- Phase 3: Return to the household for showers and tea.
- Phase 4: Eating and socializing before bed.

This system does not need daylight savings time because nighttime never changes. Night is night. Due to the seasons, sometimes it breaks closer to what the watch might call 6:00 pm, and at other times it's closer to 7:00 pm or 8:00 pm. That leads to a longer workday over the summer months and much longer nights 6 months later. Overall, this is definitely different from the cultural attitudes reflected by the phrase "9 to 5."

Another example of sociocultural linguistics is the the male-female communication gap. Linguist Deborah Tannen offers some insight into the roots of male-female misunderstanding. She looks at the ways boys and girls play in the US and argues that our socialization during childhood manifests itself in the way we communicate as adults.

Boys in the US, according to Tannen, are raised to value larger group play while girl play groups tend to be smaller and more intimate. And those smaller play groups can produce girls who've learned, better than boys, how to build ties with friends.

Ultimately, Tannen says that males learn to embolden friendship by talking about all the cool stuff

males do together. Women develop a different culture of friendship. Specifically, they are much better attuned to the subterranean messages, or meta-messages, of even the most basic conversations.

Tannen isn't saying that all women are small-group-oriented people, nor is she saying that all men can't express themselves in terms of their emotional and personal lives. She's simply using sociocultural linguistics to consider gendered differences in the way men and women tend to communicate in the US.

Takeaways

Darwin theorized that language, like biological organisms, evolves from simple to dramatically more complex forms. Essentially, he said that human language emerged from animal communication. That would explain, for instance, how bees can give each other directions, or how wolves manage to organize intricately sophisticated group hunting sessions.

However, human language is so complex that we can reasonably distinguish human language as something remarkably special in the animal kingdom. We can point to biological distinctions of *Homo sapiens* as language users. We share a uniquely human mutation of the FOXP2 gene, which has been directly linked to human language formation as well as spatial orientation.

We also have intricately evolved bodies: Consider our mouths, lungs, and brains. We can orchestrate all of these in microseconds to produce just the right sequence of noise to easily communicate any idea that crosses our minds.

Human language did nothing short of change the very nature of evolution itself. With language, humans collectively increased our capacity for information exchange. Information exchange dramatically speeds up the development of new technology—and technology extends our biologically evolved abilities.

Thanks to human language, we can now fly like birds and dive into the depths like whales. We can heat our homes in winter and keep them cool in summer. In short, human language has increased our ability to adapt.

But, of course, this adaptation through language and technology can have a dark side—as the very power we use to create can also be used to destroy. Perhaps by understanding the history, structure, and cultural aspects of communication, we'll collectively find a way to make language work for, rather than against, our continued survival.

Suggested Reading

Berwick and Chomsky, *Why Only Us*.

Burling, *The Talking Ape*.

Fouts and Mills, *Next of Kin*.

Tannen, *That's Not What I Meant!*

Questions to Consider

1. What are primate calls, and how do they provide a window into the evolution of human language?

2. How are anthropologists and others working to preserve endangered languages?

3. How can the language we speak impact the way we perceive the world around us?

Anthropological Perspectives on Language

How many of your credit card numbers have you memorized? Or how about all the bank accounts you've ever had? How many of those account numbers do you have committed to memory? Is that an unreasonable question? After all, few of us can even recite the number pi beyond a few digits, 3.14 something, something, something. Don't feel too bad.

Focus on something that you're a lot more familiar with. How about words. How many words do you think you know? 3,000? 4,000? Maybe 10,000? Well actually, according to a sophisticated, yet really fun website TestYourVocab.com, if you're like most adults in the US, you probably know some 20–30,000 words. That's a lot to remember. For each word you have, you have to keep track of how each one looks, how it sounds, and even what it means. How can our brains remember so many words when we'd all struggle to remember a half-dozen internet passwords? The quick answer is that there's something special about human language, and anthropology is going to help us figure out why.

Today our mission is to explore the development and nature of human language because, like cities, farming, tools, and money, language has definitely played a starring role in our continued survival as a species. Surprisingly today's mission brings us right back to the primates, but we'll also get our first real glance of the third subfield of anthropology, and that's linguistics.

When I talk about the origins of language with my students back at Fairfield University, I usually start with a rather goofy experiment. I ask for a volunteer to leave the classroom, and then I hide one of the volunteer's possessions without them knowing it. Last semester I hid an iPhone on top of the ceiling projector. Then I bring that volunteer right back in and explain the rules. I tell the whole class that everyone has to help this volunteer find his phone and that we'll cancel that day's reading quiz if he finds it in less than 45 seconds. But, they can't use facial expressions, they can't use body language, or anything other than a gorilla hoot. They act shy for a second, but with that free quiz on the line, they get loud fast. And sure enough, they always win. Without any training, they manage to make hoots in such a way as to lead their pal right to the goods every single time. Anthropologists think these hoots are actually keys to the origins of human language.

First, let's look at contemporary primates to see how they communicate, and then, we'll push some boundaries and try to actually talk with some apes. Now, remember, we're not looking at contemporary primates because they're earlier, less evolved versions of us humans. To the contrary, in Lecture 4 we learned that contemporary primates like South American monkeys aren't less evolved at all; they just took a different evolutionary path. Nonetheless, because we share a common ancestor, our monkey friends can help us unpack the roots of human language. So it's time to listen to some monkey talk, or primate calls to use the more appropriate term.

Deep in the rainforests of Central America, as we climb into an observation platform, we hear him. He's a sprightly tamarin monkey, and he chirps at us from over in the next tree. As primatologists interested in the origins of language, we record how these monkeys. We record them and their vocalizations because eventually, we can start to break down their code to see how this monkey talk actually works. And for this tamarin friend over there, primatologists have already recorded and cataloged almost 40 distinct types of these vocalizations. The chirp that we just heard, he made greeting chirp, and that's what we call a B chirp. And there are actually a half dozen or more additional unique chirps that each mean something different. There are chirps to signal nearby food and there's another high pitch chirp that sounds the alarm of a nearby predator.

These tamarins make other calls too, like whistles. And then, when they combine vocalizations like whistles and chirps, they actually convey simple sentences to their families and friends. For example, they can tell each other that there's a predator around, and they can include a second call that gives a better idea of how close, or where that danger actually lurks.

And this is revolutionary. There was a time that we thought communicating in sentences was solely a human trait. But, does this mean if we raise a chimpanzee or bonobo as a human, if we work hard to teach it human language, that he'll be able to talk with us about the meaning of life? Believe it or not, we've already tried this.

Back in 1951, an actor who would later become president of the United States, Ronald Reagan, starred in a comedy movie called *Bedtime for Bonzo*. In the film, he was a psychology professor who was raising a chimpanzee at home as a human. His mission, to test the nurture versus nature question. He wanted to see if a human upbringing can produce a human chimpanzee. As goofy as that sounds, the film really wasn't far from the truth. Actually, it was just the Hollywood version of some work that had begun just a couple year earlier with a series of high-profile apes living and learning with humans. I want to briefly introduce you to a handful of apes who helped us explore the boundaries of human versus other ape-to-ape communication. Because it's on these boundaries primate communication that we find the origins of our language.

First, let's start in the 1930s where we're going to meet Gua, a chimpanzee who lived with Luella and Winthrop Kellogg. In what sounds like a pitch for a family movie, the Kelloggs brought home Gua, a 7.5-month old chimpanzee, home to raise him alongside their own 10-month old son, Donald. This duo spent 12 hours a day together and they were fast friends. Remarkably, in the short year they lived together, Gua actually was a quicker study on basic manual tasks like using a spoon. That said, Gua simply wasn't able to make progress with learning to speak. In fact, after about a year it was Donald who started imitating Gua's sounds. Shortly after that, the Kellogg's returned Gua to a primate research center.

Then in the late 1940s and 1950s, psychologists Catherine and Keith Hayes tried a similar experiment, raising a young chimpanzee as a human. Her name was Viki. They dressed Viki up in white lacey dresses, she ate at the table, and they even threw birthday parties with cakes and candles. However, one of their biggest hopes as primatologist parents was to teach Viki to speak. And despite tireless efforts on everyone's part, Viki only learned to speak four words, mama, papa, cup, and up. Sadly, Viki died of viral meningitis at the age of seven.

What we were discovering with these early studies was that there were physiological and neurological reasons Viki couldn't learn to sing *Twinkle Twinkle Little Star*. What if non-human primates physically can't speak like us? Perhaps their physiology can't quite keep up. We've got an amazing flexibility of our human tongue and larynx? So if we accept the idea that Viki couldn't speak with humans, there's one more question. Does that mean she can't understand human language?

That brings us to our next primate. Researchers Allan & Beatrix Gardner asked that exact question, and they changed the way we think about primates and their linguistic potential. The Gardners raised a chimpanzee named Washoe as a human, but not with a sibling. Washoe lived a human life with toys, clothes, playtime, and, believe it or not, even chores. To gauge Washoe's linguistic abilities, the Gardners taught her American Sign Language. In all, amazingly, she learned some 250 words, which she later taught to her adopted chimpanzee son. Not only was Washoe the first non-human primate ever to learn American Sign Language, she also was the first non-human primate to teach ASL to a new generation.

The second chimpanzee who learned ASL was a hilarious primate named Lucy. Lucy lived her human life with Maurice K. Temerlin, and she was a natural. She made and served tea, she had a cat, she even drank a bit of gin, swore, and, believe it or not, she even enjoyed flipping through the pages of Playgirl magazine. Lucy learned ASL from Roger Fouts, who fondly reminisces about Lucy's wicked sense of humor.

One time when Fouts arrived for an ASL session with Lucy, he discovered that Lucy had relieved herself right on the carpet. He called her in and pointed at the pile on the floor. He called her in, and remember this is with signs, he pointed at the pile on the floor. "What's that?" he said. Lucy acted a bit coy, "What that?" So Fouts doubled down, "You know. What that?" Lucy said, "Dirty dirty," That was her word for number two.

So Fouts continued the interrogation, "Whose dirty dirty?" And then Lucy flat out lies, right into his face, and blames Susan Savage-Rumbaugh, a famous primatologist that was a graduate student at the time. Fout says, "It's not Sue. Whose that?" And that's where Lucy goes all in. "Roger," she said. What audacity. She actually blamed Fouts himself. But Fouts pushed back, "No. Not mine. Whose?" To which she finally fesses up, "Lucy dirty dirty. Sorry, Lucy."

Lucy and Washoe made it clear that our primate cousins have the capacity to learn our human language, and as their language abilities emerged, their unique personalities became apparent to us. And more contemporary research has revealed that we are only scratching the surface of the linguistic boundary between humans and other apes. The famous gorilla Koko, for example, has learned over 1,000 ASL signs, and as a result, she's had a voice in expanding ape research in the 21st century to prioritize a conservation mission.

The fascinating research into primate language learning has given us spectacular views into, as strange as this sounds, the mental life of apes, and what we've discerned is that the more our ape cousins speak our language, the more their personalities and behaviors feel human. That said, influential linguist, Noam Chomsky warns us not to get our hopes up about future conversations with gorillas about the meaning of life. His primary critique is based on the idea that our primate cousins, despite our shared genetics, just don't have that unique linguistic legacy developed exclusively in the *Homo sapiens* line. Their physiology and neurology, their bodies and their brains they preclude chimpanzees from any advanced comprehension of human language.

So what is it exactly about human language that makes it so unique? What exactly is human language? That terrific question is going to bring us square into linguistics, the third subfield of anthropology. And within that subfield itself, there are three major specializations that help us understand the origins and nature of human language.

First, there's historical linguistics. That's the evolution and extinction of language. We have descriptive linguistics where we study the mechanics of language, and last, third, socio-cultural linguistics, and that's where we're going to explore the relationship between language and culture. Today, let's lead with historical linguistics.

As I said a moment ago, historical linguistics investigates the evolution of language. In a way, when we were talking with the primates earlier, technically we were doing some historical linguistics, because, primatology gave us biological and even cultural insight into the development of the unique nature of human language. Like Darwinians constructing a biological family tree back to single cell organisms, historical linguists collectively build a beautiful linguistic tree that shows us how our languages have branched out and spread through various populations all across the globe.

One tool that helps us create our massive human language tree is glottochronology. Just as we can map the spread of humankind into the Americas, we can use glottochronology to map the diversification of human languages across the planet. On its surface, the idea is simple. Based on the idea that vocabularies tend to change at a similar rate across time and space, it becomes possible to contrast the core vocabulary of two languages to calculate their chronological separation.

We're not going to get into the technicalities of these calculations, but the gist is that two languages like English and German share linguistic origins, and therefore they still share some core vocabulary words. The longer they are apart, however, the more their core vocabularies will differ. Linguists and others sometimes question the reliability of glottochronology, but, more recently, with more sophisticated computational methods, complex phylogenetic reconstructions help us see how languages, like farming and tools, have spread across the planet with good old intrepid *Homo Sapiens*.

In addition to the evolution of language, historical linguistics also researches dead and disappearing languages. And it's worth noting that of the 6,000 or 7,000 languages known today, half or more of these languages are predicted to go extinct by the end of this century, maybe even earlier. Around 10 percent of those remaining languages, for example, are spoken by fewer than 10 people. Total. And in the US, believe it or not, is one of the nations with the highest number of endangered languages. Right up there with Brazil, Indonesia, and India.

So many Native American languages—the languages of the Caddo, the Menominee, the Pawnee, and many other Native American peoples are endangered because there are only a few speakers left who can keep them alive.

There are efforts, though, to improve this situation. For example, the Living Tongues Institute for Endangered Languages works to preserve North American languages like the Yuchi from dying out. The Yuchi people traditionally lived in the Tennessee River valley, but after suffering from the disease and conflict that Europeans brought to the Americas, they were eventually removed to Oklahoma. As of 2011, there were only five fluent speakers of the Yuchi language left.

Obviously, the future of any endangered language is dependent on young people learning and using it. That's why language preservation programs focus on educating children and young adults. A head-bobbing example of some the newer ways people have been integrating youth into language revitalization is endangered language hip hop. Like SlinCraze, a rapper from Norway. This guy raps in Sámi, which has fewer than 20,000 speakers. Closer to home, first Nations groups in the Americas are also preserving their ancestral languages through rhymes and a host of other creative avenues.

Now that we have a basic understanding of historical linguistics, let's go ahead and turn descriptive linguistics, which we'll focus on the conventions and mechanics that make language work. Just imagine as if everything that made a language work—its basics parts—were neatly packed into a box. It's descriptive linguistics that can help us open that box to see what's in there. And when we do that we find there are 5 key elements of human language.

The first element is phonemes, and these are essentially base sound units; there's no meaning there. In English, we have 44 of those, other cultures and other languages let's say Lithuanian, they have 59. The second unit here is morphemes. Morphemes are also base units, but these are base meaning units as opposed to base sound units, like the idea of un- or non-. When I say un- it's not a word, but you know what I mean. The last three are fairly simple. We've got grammar, which is the rules; we have syntax, which is sentence structure—where to put the nouns and the verbs and the adjectives; and last we have semantics, which delves deep into meaning.

With these constituent pieces, there are limitless possibilities to what we can learn about the nature and origins of language. Linguists working in the

Middle East, for example, studied a community named Al-Sayyid, which offers intriguing insights into our instinct for this thing called language. Al-Sayyid is sometimes called the village of the deaf because this Bedouin community in Israel's Negev Desert has a large deaf population, which is the result of a local genetic mutation that leads to a higher prevalence of deafness.

And the geographical and social isolation of this community explains why this village of the deaf developed a distinctly local sign language, the Al-Sayyid Bedouin Language, which is structurally distinct from both the spoken languages and the regionally dominant sign languages of Hebrew and Arabic. Ultimately, the development and evolution of this unique, local sign language supports the idea that we humans are born communicators. With our language instinct, we're equipped to create languages when we need to.

OK, so far we've looked at historical and descriptive linguistics. And, wrapping up our introduction of the three major specializations of linguistic anthropology, is socio-cultural linguistics. From a bird's eye view, sociocultural linguists investigate the relationship between language and culture. And trust me, there's a lot more to that than you might think.

Consider this example: In the US, you know exactly what would happen if we were at a party and all of you saw me break that antique lamp in the corner. It was an accident, I assure you, but I did knock it right off that table. Then the host barges in and asks, "What just happened?" With your fingers pointed right at me, you'd all give me up in a second, "It was Scott. Scott did it." I'd blush, feel even worse, and begin apologizing. But wait, it didn't have to go down like that.

You might be surprised to learn that some languages are a lot more forgiving than others. So let's move that party to Tokyo. In Japan, that same scenario might take a different route. When the host asks, "What just happened here?" Japanese speakers might not be so quick to give me up. Instead of saying that I knocked the lamp off the table, the more conventional response might be something that expresses a greater appreciation for the concept of an accident. Maybe something like, The lamp fell off the table. Nice.

Of course, this is a generalization, but the point is that linguistic norms definitely differ across cultures, and those differences can shape the way we observe and perceive our daily lives. In Mali, my farmer friends who speak Bamanankan utilize a 24-hour cycle to track each day just like you and me, but we often use strictly Bamanankan time categories. And these categories are different from Western modes of speech, which tend to focus on specific hours and minutes.

By contrast, when my friend Bakari says he'll come visit me at *wula fe*, I don't have a specific clock-time in mind. But I know he'll be around sometime late in the afternoon, probably on the way home from the fields, just before it's time for the adults' turn to bathe. The whole day may fit this 24-hour cycle, but the pacing of work teams, family meals, and daily life is still basically structured into four, dynamic and unequal parts.

There's waking up and starting work in the morning, and just as we switch to *tilen* or phase two, we pause for a meal and maybe a brief break. And you better get back to work through the end of phase two, because as phase three comes round, folks return to the household for showers and tea, and the final phase brings us into night time when we eat and socialize with family just before turning in for bed.

It's a simple, less than sensational observation, but it does go to show how our language can influence our sense of time and how we run our lives. In the village, we don't need daylight savings time because *su-fe* or night time never changes. Night is night. Due to the seasons, sometimes it breaks closer to what the watch might call 6 PM, and at other times closer to 7 or 8. That leads to a longer work day over the summer months, and a much longer night six months later. But it's definitely different from the cultural attitudes reflected by the phrase nine to five.

Another example of sociocultural linguistics is the classic men are from Mars, women from Venus thing; the male-female communication problem. It was linguist Deborah Tannen who offered some insight into the roots of male-female misunderstanding. As a linguist who herself went through a divorce, Tannen applied her anthropological understanding of language to deconstruct what it is that can lead two people who love each other so much to quickly jump into heated arguments. She looked at the ways boys and girls play in the United States, and she argued that our socialization during childhood manifests itself in the way we communicate as adults.

Boys in the US, according to Tannen, are raised to value larger group play while girl playgroups tend to be smaller and more intimate. And those smaller play groups can produce girls who've learned, much better than boys, how to build one-to-one ties with their friends. Ultimately she says that males learn to embolden friendship by talking about all the cool stuff we do together, us guys. That and a few less personal conversations about the weather. Women on the other hand, according to Tannen, develop a different culture of friendship. Specifically, they are much better attuned to the subterranean messages, or meta-messages, of even the most basic conversations. And it's not just things like body language or tone, it's the entire communicative experience.

An example a lot of us can relate to is when real estate agent Sam meets up with his wife Elana, who's a high school principal. They've both had a tough and really long day. Sam gets on some comfy clothes and says to Elana, "Hey honey, I'm going to go for a bike ride." To which, Elana, a very nice person, replies, "Well go ahead and go already." And already we've got the makings of an argument.

Tannen proposes that Elana, watching for levels of communication well beyond Sam's actual spoken words, she basically heard Sam say, "Leave me alone, I want to go on a bike ride." Even though Sam thought he was saying, "Honey, I know we had a hard day, both of us, and I need a to ride a bike. I'd love for you to come with me, that's why I'm mentioning it, but I don't want to make you feel like you have to come along."

We can all relate to gendered communication breakdowns in our own families as sisters, brothers, spouses, parents, and more. Tannen isn't saying that all women are small-group-oriented people, nor is she saying that all men are idiots who can't express themselves in terms of their emotional and personal lives. She's simply using sociocultural linguistics to consider gendered differences in the way men and women tend to communicate in the United States.

Today, we've checked out primatology and linguistics to understand human language from anthropological perspectives. When we consider our complex language as a unique feature of our humanity, it quickly reveals itself as one of the best things we developed, to assure our sole place, as the remaining hominins on this planet.

As we did with agriculture and tools, today we tried to pinpoint the origins of human language, but we found that the question is a bit more complex than we originally thought. Language is so much more than the words we write, say, or think. Darwin theorized that language, like biological organisms, evolves from simple to dramatically more complex forms. Essentially, he said that human language emerged from animal communication. That would explain, for instance, how bees can give each other directions, or how wolves manage to organize intricately sophisticated group hunting sessions. Like us, they communicate with each other. Especially when it comes to the basics of survival, animals signal each other through body language, sounds, and smells.

That older-than-human compulsion to communicate is something we share with other animals. But, our language is so complex, we can reasonably distinguish human language as something remarkably special in the animal kingdom.

We can point to biological distinctions of *Homo sapiens* as language users. We share a uniquely human mutation, for example, of the FOXP2 gene, and that dates to about 100,000 years ago in our evolution. This gene has been directly linked to not only human language formation but also spatial orientation. We also have intricately evolved bodies—our mouths, lungs, brains, and more. We can orchestrate these in microseconds to produce just the right sequence of noise to easily communicate any idea that might come across our mind. And this unique ability, this distinctively human language, may be our most critical, game-changing adaptation yet.

The basic hand signs and body language that we speculate were pre-verbal forms of human language—think quiet hunters pointing and coordinating a kill—that communication style may have a lot in common with how bees or wolves may talk. But since that hypothetical earliest form of human language emerged, we've completely transformed the phenomenon of human language. In so doing, we've irrevocably altered our minds, our bodies, and even our planet.

And our big takeaway is this: This thing we developed called human language not only helped us survive as the sole remaining hominin on earth but now we can see that it did nothing short of change the very nature of evolution itself. With language, unlike other animals, we humans collectively increased our capacity for information exchange. Instead of being governed by exchange and recombination of simply genetic information, with human language we bring a new cultural force to evolution.

By this I mean that information exchange dramatically speeds up the development of new technology, and this technology extends our biologically evolved abilities. Thanks to human language, we can now fly like birds and dive into the depths like whales. We can heat our homes in winter and keep them cool in summer. In short, human language has increased our ability to adapt.

But, of course, this adaptation through language and technology can have a dark side—as the very power we use to create can also be used to destroy. Perhaps by understanding the history, structure, and cultural aspects of communication, we'll collectively find a way to make language work for, rather than against, our continued survival.

Apocalyptic Anthropology

Anthropologists can survey the whole of human history and can examine the various ways that humans have envisioned the end of our species. But more than this, anthropologists can also look for critical lessons from the collapse of some of the greatest civilizations on Earth. Maybe these earlier civilizations can provide us some prescient counsel on how to best sustain our lives, our species, and our planet. That's this lecture's focus—a topic we might call apocalyptic anthropology.

Eschatology

Broadly, the term *eschatology* refers the study of humanity's destiny. Different religions have different takes on eschatology. Despite all the nuances, world religions tend to fall into 1 of 2 categories on this subject.

Some religious traditions like Buddhism and Hinduism have a more circular conceptualization of our destiny, while others like Islam and Christianity tend to be more linear, with a clear beginning and end.

This lecture will take a look at what Buddhists, Hindus, Zoroastrians, and the Hopi have to share when it comes to thinking about the apocalypse.

Eschatology: Buddhism

Many people grow up learning that the Buddha was a man who lived sometime between the 6th and 4th centuries B.C.E. and that his name was Siddhartha Gautama. But it may come as a surprise to learn that this famous Siddhartha was actually the 28th Buddha.

Buddhist traditions describe long, epic cycles in which the teachings of the Buddha eventually disappear from the earth without a trace. According to this teaching, we are in the period of the 1st sun, and eventually, a series of 7 new suns will flare up in the sky, turning our planet into a giant, fiery ball before it explodes into oblivion.

On the plus side, destruction is followed by rebirth—a new cycle of existence. And in each cycle, our lifespans can increase or decrease, depending on the collective conduct of humankind.

Eschatology: Hinduism

Hindu cosmologies include three major deities: Brahma, the creator; Shiva, the destroyer, and Vishnu, the universe's protector, who repeatedly returns to Earth to serve this duty. The incarnation that most non-Hindus may have heard about is Krishna, Vishnu's 8th incarnation.

Even with Vishnu's protection, our universe won't last forever. Brahma creates the universe with a shelf life of about 4.32 billion human years. The bad news is that it appears as if we're in Kali Yuga, the 4th and final phase of our current epic cycle.

Vishnu is expected to make a final visit before this particular universe comes to an end. Vishnu is set to return on a white horse as his 10th incarnation, Kalki. Kalki will raise an army and destroy absolutely everything. But Brahma will set the whole process in motion yet again. And with a new epoch in its infancy, humanity will be restored to a golden age in communion with the gods, and with lifespans of 100,000 years.

This cycle will repeat eternally. Even when Brahma dies after trillions of human years, a new Brahma is born to repeat yet another cycle.

Eschatology: Zoroastrianism

Zoroastrianism dates back to well over 3000 years ago in Iran, and it thrived as the region's religious tradition of choice until the 7th century and the spread of Islam.

Structurally, Zoroastrianism resembles aspects of the Judeo-Christian and Islamic tradition, sporting a single god who created the world. The Zoroastrian tradition describes a final cataclysmic battle in which good triumphs over evil. Then, the last messiah, Saoshyans, resurrects the dead, and restores their earthly bodies for final judgment.

Then, in a dramatic flash, the metals in the mountains and hills melt, flooding the earth and forming a molten river. All humans, living and dead, are then shepherded through this burning river.

The righteous tread through in comfort, while the iniquitous collectively perish as the glowing river carries them to hell, where all wickedness in the universe will be destroyed. Those who pass through the molten river receive delightful eternal lives.

Eschatology: The Hopi

In the late 1940s, Hopi elders in America's southwest learned of the development of the atomic bomb and its mushroom cloud. Upon hearing this news, they made a remarkable decision. After keeping it an internal secret for almost 1000 years, they decided to share the Hopi prophecy with the entire planet.

Brahma

The gist of the prophecy is this: Humankind has lived through 3 different worlds, and our current phase, the 4th world, may be drawing to a rapid close. At the end of the 3 previous worlds, the Great Spirit "purified or punished" humanity for corruption, greed, and turning from his ways.

The Great Spirit charged 3 helpers with the task of partnering up with the Hopi to bring peace on earth. If they fail, then it's all over. But if they succeed, humanity would be saved from the terrible day of purification.

Secular Eschatology

Despite the remarkable diversity we see among the world's religious traditions, an awful lot of them allude to a fiery end to humanity. This flame-ridden finale matches up well with how many scientists foresee the end of our planet.

Most scientists predict that in around 5 billion years, the Sun's core will run out of hydrogen, and it will expand to about 250 times its current size. And when that happens, we can expect the Earth to completely vaporize as the Sun swallows it.

But some scientists see the end of humankind as the rise of the machines. What if the physical sciences and computer sciences fused through an irrevocable new machine-human hybrid?

The technologist Ray Kurzweil foresees just such a convergence, and he describes this transition as the singularity. Kurzweil predicts that in 2029, machines will emerge that exhibit intelligent behavior that is indistinguishable from humans. Then, by 2045, computers will be 1 billion times more powerful than all the human brains on Earth combined.

When it comes to the singularity, there are 2 major camps. First, there are people like the technologist Elon Musk, who see the rise of artificial intelligence as the biggest imminent threat to the survival of our species. The other camp sees an evolutionary opportunity to merge with machines once and for all. Regardless of which camp one

favors, it's clear that our biological and technological evolution are becoming progressively intertwined.

Lessons from the Past

This lecture now turns to some of the world's great civilizations to see if their histories of collapse might offer a few lessons about how to avoid extinction.

First up is the great Mesopotamian city of Ur. Well positioned in the Persian Gulf, this small village emerged as one of the earliest cities born of the new agricultural age. But shifting climate patterns and an overuse of resources and land pushed people to migrate from the region. Ur was abandoned and left for ruins in the 4th century B.C.E.

Not too far from Ur, Egypt came to international prominence and regional dominance around 3100 B.C.E. But this didn't last forever.

The agricultural prowess of Egypt fed an immensely productive civilization with great military might. However, regional rivals developed improved metal technologies, namely the Hittites and their iron. As bronze makers, Egypt rose to

prominence, but without access to materials to embrace the Iron Age, their technological slip helped chip away at one of the most celebrated of ancient civilizations.

Similarly, with a reliance on non-local raw materials like timber, Egypt found it harder and harder to sustain its growth, and then its existence. As Egypt weakened, peripheral satellites of the empire rose, and eventually Egypt fell into the grip of a seemingly endless list of conquerors.

Another case comes from southern Africa, just inland from the Swahili Coast. There lie the ruins of Great Zimbabwe, an amazing kingdom that thrived until the 15th century. The ruins are a World Heritage Site with a stunning rock structure at the center.

Hundreds of years ago, Great Zimbabwe was a crucial hub in an international trade network that spread far beyond Africa's Swahili Coast into parts of India, China, and the Middle East.

But by the year 1500, Great Zimbabwe had dropped out of this trade network. Ceramics found at this remarkable site show fewer and fewer imports were arriving as the civilization faded. Some experts argue that drought and overgrazing of cattle took its eventual toll on the surrounding region, and that made it tougher and tougher to feed the thousands of people who lived in Great Zimbabwe.

Like Ur and Egypt, Great Zimbabwe may have collapsed as a victim of its own success. With a remarkable capacity for growth and food production, these civilizations ran like a runaway train—until they ran out of fuel.

The final stop on this lecture's survival tour brings us to another World Heritage Site, this time in North America. The site is a place we call Cahokia; it's not far from modern-day St. Louis, Missouri.

The climate record shows that the site experienced violent and frequent floods for millennia. Then, as these floods tapered off, people settled what would become the magnificent Cahokia. Humans were quick to dig in and cultivate the rich local soil. Populations grew and flourished, and by the mid-11th century, Cahokia had become the first known megacity in North America.

They created woodhenges positioned to mark solstices. The woodhenges were basically a sylvan version of the stone structures found at Stonehenge in the U.K. But with every house, temple, and woodhenge built with wood, Cahokians depleted the forests that supported the rapid rise of this civilization. And it was this deforestation that altered the Cahokia watershed, bringing back unforgiving floods.

The collapse of Cahokia is certainly a complex phenomenon, but there's no doubt that deforestation and the return of violent floods undermined everything that once made it the largest North American city on record, a title it held all the way until Philadelphia in the mid-1700s.

The Lesson

What does the collapse of past civilizations teach us about our status as the sole surviving hominins? For starters, the archaeological record clearly warns us that we need to be thinking seriously about climate change and our use of fossil fuels. We need to come up with new options and solutions.

Stonchenge

Past civilizations transformed the human experience, yet they all faded away, one way or another. They share the same story arc: Humans find a great place for innovative food production; they parlay their food surplus into a full-fledged, urban civilization; but ultimately, they fail to sustain themselves.

And while climate fluctuations and our natural resource use alone can't take the rap or credit for the rise and fall of civilizations, they certainly play a significant role. It's a role that we'd do well to better understand. After all, for 200,000 years now, humans have displayed an uncanny knack for adapting to a wondrous array of ecologies and environments. It's up to us to extend that record.

Suggested Reading

Bostrom and Cirkovic, *Global Catastrophic Risks*.

Kolbert, *The Sixth Extinction*.

Kurzweil, *The Singularity Is Near*.

Shanahan, *The Technological Singularity*.

Questions to Consider

1. How do world religions and secular technologists envision the end of humankind?

2. As 21st-century humans, what are some of the lessons we can learn from previous world civilizations that flourished for centuries but then crashed into oblivion nonetheless?

3. How might humanity's remarkable fascination with imagining the end of times help us survive, for at least a little bit longer?

Apocalyptic Anthropology

I t's clear that adaptive innovations like tools, agriculture, and language helped get our species into the 21st century and hurray for us. But hold that balloon drop. What's it going to take to sustain us for another two, let alone another 21 centuries into the future? That's a scary question. We are literally talking about the eventual end of humankind. But take heart, we have anthropology on our side.

As anthropologists, we can survey the whole of human history and we can examine the various ways that humans have envisioned the end of our species. But more than this, we can also look for critical lessons from the collapse of some of the greatest civilizations on earth. Maybe these earlier civilizations can provide us some prescient counsel on how best to sustain our lives, and our species, and, of course, our planet.

So that's our project for today—a task we might call apocalyptic anthropology. Our survey will include perspectives both religious and secular, but let's begin by seeing what some of the world's religions have to say about the end of humankind.

In the field of religious studies, the word scholars use to describe what we're up to today is eschatology. Broadly, that's the study of humanity's destiny. And different religions will have different takes on eschatology. But remember, as anthropologists, we're not sorting through these different perspectives to find a single, definitive answer on our destiny. Instead, we're going to look for a comparative insight across multiple ways of thinking.

First, we can observe that, despite all the nuances, world religions tend to fall into one of two categories on this subject. Some religious traditions like Buddhism and Hinduism have a more circular conceptualization of our destiny, while others like Islam and Christianity tend to be more linear, with a clear beginning and end. So let's take a closer look. Let's look at some circles and some lines. Let's see what Buddhists, Hindus, Zoroastrians, and the Hopi have to share when it comes to thinking about the apocalypse.

Let's look a Buddhism first. Many of you will have grown up learning that the Buddha was a man who lived sometime between the B.C.E. 6th and 4th centuries and that his name was Siddhartha Gautama. But it may surprise you to learn that this famous Siddhartha was actually the 28th Buddha. Like any world religion, there's a wide diversity of practices and ideas across Buddhism, but in general, the end of times story is pretty clear. You see, Buddhist traditions describe long, epic cycles in which the teachings of the Buddha will eventually disappear from the earth without a trace. Even the memory of the Buddha gone.

According to this teaching, we are in the period of the first sun, but eventually, a series of six new suns will flare up in the sky, turning our planet into a giant, fiery ball before it explodes into oblivion. Boom. The end. Then the cycle begins anew, and the Buddha will emerge yet again to deliver his message to a new cycle of existence. And while this may not feel like good news to us, t at least in Buddhist eschatology, humans are long gone by the time things get way worse.

This is how it goes down. First, a severe drought destroys all life, plants, and animals. And that's our exit. Then a second sun flashes and dries up all brooks and ponds. Then a third sun that dries the world's great rivers, a fourth dries up the great lakes, the fifth dries up the oceans and seas, and then that sixth one leads to volcanic explosions and fires. And finally, a seventh sun emerges, and earth bursts into a fiery ball, expands and then explodes. Gone.

On the plus side, destruction is followed by rebirth—a new cycle of existence. And in each cycle, remember Siddhartha was the 28th Buddha, our lifespans can increase. They can decrease too, but depending on the collective conduct of humankind, we can increase our lifespan exponentially. And that's a groovy silver lining. Well, if that's not to your liking, let's check out the Hindu apocalypse instead.

Hindu traditions, like Buddhist ones, envision a cyclical apocalypse. But let's start from the top. First, we have got to remember that Hindu cosmologies include three major deities. There's Brahma who is the creator, and Shiva the destroyer, and Vishnu, who is the universe's protector. Now, Vishnu, he is definitely on our side. It's said that he's already returned to earth nine times to help keep things rolling. On one occasion, he returned as a dwarf. He also has arrived as a turtle, a lion-human mix, and even a warrior. But the incarnation that most non-Hindus have heard about is Krishna, Vishnu's eighth incarnation.

Even with Vishnu's protection, our universe is not going to last forever. Brahma creates the universe with a shelf life of one day, which, thankfully for us, is about 4.32 billion human years. The bad news is that it appears as if we're in Kali Yuga, the fourth and final phase, of our current epic cycle. Worse yet, because life gets harder as we move closer to the end, it's clear what we might expect from this Kali phase, an era marked by spiritual emptiness, corruption, materialism, hypocrisy, and lots of fighting.

The full update is that Vishnu is expected to make one final visit before this particular universe comes to an end, and that really won't be such a bad thing for us. You see, Vishnu is set to return on a white horse as his 10th incarnation, Kalki. At the end of our present cycle, Kalki will—bad news first—raise an army and destroy absolutely everything.

But, wait for it like a rising sun, the supreme being, Brahma, sets the whole process in motion yet again. And with a new epoch in its infancy, humanity will be restored to a golden age in communion with the gods. We'll even have lifespans of 100,000 years. Now, this cycle is eternal. Even when Brahma dies after trillions of human years, a new Brahma is born to repeat yet another meta-cycle.

So the end of the world and universe is definitely a scary thing from our human point of view, but cosmically, there is something mysteriously comforting about not having to worry about a singular beginning or an end. But for those of us who find more comfort within the context of a beginning and end, maybe you'll prefer another ancient religion's eschatology. Let's visit Zoroastrianism.

Zoroastrianism dates back to well over 3,000 years ago in Iran, and it thrived as the region's religious tradition of choice until the 7th century and the spread of Islam. As one of the world's oldest religions, it shares many core connections with contemporary religions, including the idea of heaven and hell, and the golden rule. Structurally, Zoroastrianism resembles aspects of the Judeo-Christian and Islamic traditions. Zoroastrians believe in a single god who created the world. They received god's words through a prophet. They pray daily and gather for communal worship, and they even have a holy book that is divided in to an older section and a newer section.

The Zoroastrian tradition describes a final cataclysmic battle in which good triumphs over evil. Then, the last Messiah, Saoshyans, resurrects the dead and restores their earthly bodies for final judgment. Then, in a dramatic flash, the metals in the mountains and hills melt, flooding the earth and forming a molten river. All humans, living and dead, are then shepherded through this burning river, but people who live good lives, doing good deeds have nothing to fear. The righteous tread right through in comfort, while the iniquitous collectively perish as the glowing river carries them to hell, where all wickedness in the universe will be destroyed. Those who pass through the molten river receive delightful eternal lives.

Now, at the risk of depressing you with these rather dire prophecies, let's add one more, but this time, we'll talk about a prophecy that includes a glimmer of hope, not just in the afterlife, but right here on earth. Hope that we can keep things rolling.

Actually, the Native American prophecy I'm about to describe wasn't known to anyone but a small group of desert-dwelling farmers until around the 1940s, when Hopi elders in America's Southwest learned of the development of the atomic bomb and its mushroom cloud. Upon hearing this news, they made a remarkable decision. After keeping it an internal secret for almost 1,000 years, they decided to share the Hopi prophecy with the entire planet. With the world's balance at stake, elders appointed four special members of the Hopi Nation to bring their warnings and prophecy to all humankind.

In 1999 the last of these four special messengers, Thomas Banyacya, passed away. But before he did, he shared this prophecy and warning with millions of people, including a rather remarkable and peace-filled speech at the United Nations. He explained that the Hopi believe that humankind has lived through three different worlds and that our current phase, the fourth world, may be rapidly drawing to a close. He didn't project specific doom dates, but the rich, metaphorical Hopi prophecy strikes some eerie parallels with what's going on in our modern world.

Simply put, Banyacya said that, at the end of the three previous worlds, the Great Spirit "purified or punished humanity for corruption, greed, and turning from his ways. Banyacya explained that the Hopi were told how the Great Spirit charged three helpers with the task to partner up with the Hopi to bring peace on earth. And that if they fail it's all over. He described what would happen if we fail, "The one from the west will come like a big storm. He will be many in numbers and unmerciful. When he comes he will cover the land like the red ants and overtake this land in one day."

But as I mentioned a moment ago, there's a silver lining At the close of his UN speech, Banyacya enjoined us to this whole thing around. And I quote, this is directly from his speech: "Only by joining together in a Spiritual Peace with love in our hearts for one another, love in our hearts for the Great Spirit and Mother Earth, shall we be saved from the terrible Purification Day which is just ahead."

From Hopi to Hindu traditions, we've briefly glanced at four world religions. So now, let's turn to secular eschatology. And let's begin with an interesting observation.

Despite the remarkable diversity we see among the world's religious traditions, an awful lot of them allude to a fiery end to humanity. And surprisingly, or not, this flame-ridden finale also matches up quite well with how many scientists foresee the end of our planet. Kind of like the Buddhist idea of the earth burning up, most scientists predict that in around 5 billion years, the sun's core will run out of hydrogen, and it'll expand to about 250 times its current size. And when that happens, we can expect the earth to completely vaporize as it's swallowed by our sun. Of course, some scientists, like many religions, don't necessarily see this as an absolute end—in fact, some have argued about the possibility of an eternally reproducing multiverse. But let's turn from these intriguing parallels between religious and secular end-times to a vision of the future that seems unique to our scientific age.

Now, pop culture may embrace the *Walking Dead* as a humanized doomsday metaphor, but the empiricists among us aren't exactly fretting about the zombie apocalypse. Instead, some scientists are preparing for a less cosmic and fiery end because they see the end of humankind as the rise of the machines. The singularity. Think about that what if the end of humankind comes at the hands of our computers instead of a fiery apocalypse? What if machines eventually become intelligent? What if the physical sciences and computer sciences fused through an irrevocable new machine-human hybrid? From *Homo sapiens* to *Homo roboticus*?

Well, technologist Ray Kurzweil foresees just such a convergence, and he describes this transition as the singularity. It sounds far out, but Kurzweil predicts that in 2029 machines will emerge that exhibit intelligent behavior that is indistinguishable from us humans. Then by 2045, computers will be a billion times more powerful than all the human brains on earth, combined.

Kurzweil's complex predictions have come true before. His predictions proved accurate for the first computer to beat the world's reigning chess champion, and even for the emergence of the driverless car. Google, which hired Kurzweil full-time in 2012, successfully tested their first fleet of driverless cars by driving over a half million miles without being at fault for any accidents.

The complex processing required to drive a car in live traffic while avoiding accidents, is truly remarkable, but can a machine actually master language and complex thinking beyond human capabilities?

Well, these days intelligent machines do a majority of Wall Street trades. And besides beating us at chess, and telling us our bank balances, computers have even beaten Jeopardy champions at their own game. In fact, in 2016, an artificial intelligence program beat one of the world's top players at the game Go—which is generally considered to be one of the most complex, strategy-driven games ever invented.

So, when it comes to the singularity, there are two major camps. First, there are people like technologist Elon Musk who see the rise of artificial intelligence as the biggest imminent threat to the survival of our species. If computers surpass our processing and technological capacities, maybe they could eventually enslave us, or worse, maybe they'll drive us to extinction and simply replace us.

But not everyone is so fearful of the rise of the machines. Rather than the subjugation or extinction of our species, a second camp sees an evolutionary opportunity to merge with machines once and for all. And we're not talking about Google Glasses here, we're talking about the fusion of the biological and digital worlds. The emergence of *Homo roboticus*.

Regardless of which camp one favors, it's clear that our biological & technological evolution are becoming progressively intertwined. And it's also a safe bet that the more our world goes digital, we'll have to change the way we interact with machines, and probably in ways we can't imagine yet. Now, whether or not all this leads to our extinction or a new hybrid species, we'll have to see. But for an anthropologist, whether we're considering the eschatology of world religions or the predictions of scientists the important takeaway is this:

Long before we learned that the sun will expand and vaporize the earth in around 5 billion years, we humans have been imagining our collective demise. There's no shortage of interpretations on how and when our species may meet its end, and that's peculiar. I bet rabbits, for example, don't dream up bunny nightmares about the end of the world. That's us. Us bizarre *Homo sapiens*. Like a dream that helps wire our brains for future crises and events, thinking about the apocalypse may actually be one of the big things humans developed to sustain our species.

Oddly enough, maybe thinking about and planning for an apparently inevitable apocalypse, maybe that actually helps us sustain our species in the meantime. Maybe even long enough to figure out a way to deal with the earth's vaporized future.

In terms of religious and even secular thought on the apocalypse, they give us meaning and order to our lives. The classic tenet shared by many world religions, the golden rule, for example, that's a human behavior that, if sustained, can contribute to our survival as a species. I'm not ignoring the fact that religious traditions have inspired violence and countless wars, but I am saying that religious and secular explanations for the end give meaning and order to our lives, and in ways that keep most of us from running amok without any regard for our fellow humans and our future.

So, while physicists work out the 5 billion year event horizon that ends up with the sun burning up, technologists and artificial intelligence folks can hold the fort to prevent the robot revolution. Meanwhile, maybe we anthropologists can think about how *Homo sapiens* might just stick around long enough to have the luxury of getting us to another planet or universe. So why don't we revisit some of the world's great civilizations, to see if their histories of collapse might actually offer a few lessons about how to avoid extinction.

We'll start with the great Mesopotamian city of Ur. Well positioned in the Persian Gulf, this small village emerged as one of the earliest cities born of the new agricultural age. It's actually known as the home of Abraham of the Abrahamic faith traditions. But even the hometown of Abraham isn't immune to its own personal apocalypse. This early city's story is a story that we see over and over again. And it's a story we'd do well to think more about.

The short of it is that shifting climate patterns and an overuse of resources and the land pushed people to migrate from the region. Abraham, for example, moved northwest to Haran and then southwest from there into Canaan. But then, Ur was abandoned and left for ruins in B.C.E. 4th century. Ruins that wouldn't be uncovered and restored to the historical record for a couple thousand years.

One of human's first urban centers could have been wiped from history, yet through archaeology, we recreate their story. And if we're wise, we'll note that the shifting climate and Ur's unsustainable consumption of natural resources almost erased this important city from our history books.

Not too far from Ur, Egypt came to international prominence and regional dominance around B.C.E. 3100. In a way, Pharaoh's Egypt is our symbol for the ancient world that agriculture built. And through several thousand years of growth and decline, Egypt finally joined the ranks of most every civilization we've ever read about—it collapsed.

The agricultural prowess of Egypt fed an immensely productive civilization with great military might. But, eventually, regional rivals developed improved metal technologies, namely the Hittites and their iron. As bronze makers, Egypt rose to prominence, but without access to materials to embrace the iron age, their technological slip helped chip away at one of the most celebrated of ancient civilizations.

Similarly, with a reliance on non-local raw materials like timber, Egypt found it harder and harder to sustain its own growth, and, eventually, it's own existence. As Egypt weakened, peripheral satellites of the empire rose, and eventually, Egypt fell into the grip of a seemingly endless list of conquerors. Shortly after Cleopatra's reign, for example, Egypt became a tributary, mainly through grain production, of the Roman Empire.

Ancient Egypt is a civilization that many Westerners know about, but the next civilization I want to examine is probably less familiar. So let's leave Egypt and travel down to southern Africa, just inland from the Swahili Coast. That's where we're going to find the ruins of Great Zimbabwe, an absolutely amazing kingdom that thrived until around the 15th century. The ruins are actually a World Heritage Site, with a stunning rock structure at its center.

Archaeologists assert that this structure, a circular wall that's several meters thick, was built without any cement or mortar to hold the blocks together. These sophisticated builders knew how to make a structure that could stand for centuries, and without any bonding materials apart from gravity. The unsurpassed rock carving tradition still present in this region is celebrated across the world today. In fact, next time you're at the Atlanta airport, check out the amazing exhibit of Zimbabwe stone sculptures that dot the walkway between terminals.

Hundreds of years ago, Great Zimbabwe was a crucial hub in an international trade network that spread far beyond Africa's Swahili Coast into parts of India, China, and the Middle East. That said, the archaeological record reveals that, by the year 1500, Great Zimbabwe had dropped out of this trade network. Ceramics found at this remarkable site show fewer and fewer imports were arriving as this once great civilization faded.

Some experts argue that drought and overgrazing of cattle took its eventual toll on the surrounding region, and that made it tougher and tougher to feed the thousands of people who lived in Great Zimbabwe. Whether it was an ecological apocalypse or a taste for new sources of gold, there is still some debate about what eventually led to the abandonment of this world city.

Recent research tends to place the blame on us humans rather than the environment. It appears as if the physical environment wasn't really cruel enough to single-handedly blow out the Great Zimbabwe candle. Like Ur and Egypt, Great Zimbabwe may have collapsed as a victim of its own success. With a remarkable capacity for growth and food production, these civilizations ran like a runaway train that is until they ran out of gas. Is there a lesson there for us? Well, for me at least, renewable energy sources like solar and wind power never sounded so great.

The final stop on our survival tour brings us to yet another World Heritage Site, this time right back home in North America. The first epic urban civilization north of the Yucatan is a place we call Cahokia.

In the heart of the Mississippi valley, not far from modern-day St. Louis, a great civilization was born. For millennia the climate record shows that the Cahokia site experienced violent and frequent floods. Then, as these floods tapered off, people settled what would become the magnificent Cahokia. Humans were quick to dig in and cultivate the rich local soil, once the floods primed the region for human settlements.

Populations grew and flourished as maize producers, and by the mid-11th century, Cahokia had become the first known mega city in North America. It surprises people to learn that Cahokians built an earthen pyramid that rivaled the Great Pyramids of Egypt. They even also created wood henges positioned to mark solstices. The wood henges were basically a sylvan version of the stone structures we find at Stonehenge in the U.K. The abundant forest resources surrounding Cahokia provided building materials for the wood henge and the great earthen pyramid called Monks Mound, not to mention dwellings for the entire population of farmers who lived there. But with every house, temple, and wood henge built with wood, Cahokians eventually depleted the forests that supported the rapid rise of this civilization. And it was this deforestation that altered the Cahokia watershed, bringing back the unforgiving floods.

The collapse of Cahokia is certainly a complex phenomenon, but there's no doubt that deforestation and the return of violent floods undermined everything that once made it the largest North American city on record, a title it held all the way up until the 1700s with Philadelphia. Unlike Philadelphia, though, Cahokia vanished from history. And it wasn't until the 20th century that a highway project helped us rediscover what has now become one of only a handful of World Heritage Sites right here in North America.

So what's the big lesson here? What does the collapse of past civilizations teach us about our status as the sole surviving hominins? Well, the archaeological record clearly warns us that we need to be thinking seriously about climate change and our use of fossil fuels. And not just thinking about them; we need to come up with new options, new solutions. Past civilizations transformed the human experience, yet they all faded away, one way or another. They share that same story arc.

Humans find a great place for innovative food production, and then they parlay their food surplus into a full-fledged, urban civilization, but ultimately, they all failed to sustain themselves. Climate fluctuations and our natural resources alone can't take the rap or credit for the rise and fall of civilizations, but they certainly play a significant role.

In the sci-fi movie *Logan's Run*, humanity realizes that our available resources can't sustain an unchecked population, so they work out a solution.

Another spoiler alert here, ultimately they preserve their common resources by limiting human life spans to 30 years. They simply kill you on your 30th birthday. That's not anything that I'd propose for us humans in the 21st century, but we'd be wise to work out something else.

Anthropology, as the study of humankind over time and space, is well positioned to contribute lessons from the vast accrued experience of human history. It can also show how to anticipate our extinction, and to imagine our escape. If we can't keep *Homo sapiens* on the planet forever, knowing the clock is ticking, maybe anthropology can help us overcome our fears and even find happiness along the ride as we evolve into *Homo roboticus*. After all, for 200,000 years now, humans have displayed an uncanny knack for adapting to a wondrous array of ecologies and environments. And, my dear friends, it's up to us to extend that record.

Cultural Anthropology and Human Diversity

C ultural anthropology builds bridges of understanding. It connects people across all kinds of linguistic and cultural boundaries. And, the more we understand other ways of seeing and being in the world, the more we actually understand ourselves as individuals and as a species. This lecture visits a few anthropologists across the world to learn how anthropology reconciles the infinite cultural diversity of humankind with the biological fact that we are one race, *Homo sapiens*.

Edward Burnett Tylor

The English anthropologist was the first professional anthropologist to publish a comprehensive definition of culture. We find his definition of culture in his 1871 book *Primitive Culture*. There, Tylor asserts that culture is "that complex whole which includes knowledge, belief, art, morals, law, custom, and any other capabilities and habits acquired by man as a member of society."

His methodology is ethnology. Rather than travel the globe documenting exotic cultures, Tylor and ethnologists explored human cultural diversity by mining libraries and archives. Ultimately, Tylor deployed a comparative approach that blended ethnology, archaeology, and philosophy or reason. One of the central tenets of Tylor's work and of cultural evolutionism is that progress is our dominant narrative.

Edward Burnett Tylor

Lewis Henry Morgan

Another voice behind the primitive/civilized schematic was Lewis Henry Morgan. In the 1840s, the Seneca were engaged in a land dispute with the Ogden Land Company. Morgan helped the Seneca work out a solution to that dispute.

Morgan's fascination with Native American cultures compelled him to break out of the library and into the field. He traveled throughout the US to collect kinship data.

While in the field, Morgan gathered data from government officials known as Indian Agents, and he accumulated missionary data. Curating all these data streams, Morgan noted a distinct kinship pattern he referred to as classificatory kinship.

Essentially, kinship is the way a society organizes social relations in terms of family descent, but Morgan's classificatory kinship includes people who may or may not be genetically related. He noted classificatory kinship patterns across Native Americans and concluded that this form of kinship was a remnant of humanity's primitive state.

Ready to test his idea further, he sought out kinship data from all over the world using a 7-page questionnaire. He compiled his results and found that he could sort kinship patterns into distinct kinship systems that we still teach today. For example, it is common in the US to call one's mother's sister one's aunt, and her children one's cousins. But in some cultures, one would refer to the aunt as mother and her children as siblings, not one's cousins.

In 1877, Morgan published his findings in a major book called *Ancient Society*. The book was fiercely influential on the era's great thinkers and writers like Marx and Engels. In it, Morgan clearly defines cultural evolutionism as a process in 3 phases:

1. The primitive stage, associated with foraging communities.
2. The barbarian stage, brought on by the domestication of plants and animals.
3. The civilized stage, with state-level societies.

Morgan wanted to document earlier-stage cultures because, for him, they were a living window into our less-civilized past. And they were dying out.

Morgan's legacy includes that call for more anthropology, but his true contribution was his vast archive of ethnographic data from all over the world. Additionally, he integrated field research with cultural evolutionism.

However, in the present day, anthropologists are long over the idea that cultural evolution occurs on a progressive spectrum from savage to civilized.

Franz Boas

Why did cultural evolution fall out of favor? The answer takes us to the Arctic, on Baffin Island. In the 1880s, Franz Boas, the father of American anthropology, was there on a 15-month field study.

Unlike Morgan and Tylor, Boas didn't believe in culture grades. Instead, where cultural evolutionists saw savages and barbarians, Boas saw complex, sophisticated people artfully adapted to their unique environment.

Upon arriving on Baffin Island, Boas realized that as long as he was in the Arctic, he was the primitive, and his hosts were the sophisticates. After all, left to his own devices, Boas wouldn't last a night on this frozen island.

Boas saw sophistication in how the Inuit on Baffin Island sustained themselves in such a challenging place to live. What he saw was cultural relativity. Boas didn't do it alone, but he helped change the way we think about culture and each other. His career literally marks anthropology's transition out of cultural evolutionism and into cultural relativity.

How did cultural relativity win out in an era of entrenched cultural evolutionism? Boas did plenty of research and accumulated massive collections of anthropological artifacts. He wrote over 700 books and articles alone.

Perhaps his greatest legacy, however, was the scores of students he mentored. These students went on to build new anthropology programs across the United States, and their names are the giants of early American anthropology:

- Ruth Benedict, who wrote the 1934 public affairs pamphlet *The Races of Mankind*, which articulates cultural relativity from the Boasian perspective.
- Melville J. Herskovits, who set up the anthropology department at Northwestern University.
- Edward Sapir, who created the University of Chicago's famous anthropology program.
- Alfred Kroeber, who created UC Berkeley's famous museum and department of anthropology.

Bronisław Malinowski

Boas had a revolutionary counterpart in England: Bronisław Malinowski. Like Boas, Malinowski was a Physics Ph.D. from continental Europe. He was drawn to anthropology, and he brought his scientific mind and methods to the discipline and his teaching.

He was fervently against the idea of cultural evolution despite the fact that it was the established norm, thanks to the work of people like Tylor. He argued that cultural evolutionism was bad science.

Malinowski did some of his most important work in the Pacific. It was on the Trobriand Islands that Malinowski had his cultural relativity moment, and it's where he invented the concept of functionalism.

Just as Boas saw complex people doing remarkable things to sustain their Arctic lives, Malinowski saw the same rational or functional side to things that others viewed as primitive or exotic. Specifically, he was interested in the logic of local economics.

Here's a brief segment from the book in which Malinowski

defines functionalism, *Argonauts of the Western Pacific*. It shows not only Malinowski's beautifully descriptive language, but also his understanding that these so-called natives are anything but primitive:

> A most remarkable form of intertribal trade is that obtaining between the Motu of Port Moresby and the tribes of the Papuan Gulf. The Motu sail for hundreds of miles in heavy, unwieldy canoes, called *lakatoi,* which are provided with characteristic crab-claw sails. They bring pottery and shell ornaments, in olden days, stone blades, to Gulf Papuans, from whom they obtain in exchange sago and the heavy dug-outs, which are used afterwards by the Motu for the construction of their *lakatoi* canoes.

In the early 20th century, this was a new way to understand cultural diversity.

Malinowski instilled the functionalist method in his students. Tylor and the cultural evolutionists had culture, race, and biology all interlinked. But with functionalism, Malinowski separates these concepts. To see the world through functionalist glasses, Malinowski offers us 4 major tenets of the functionalist methodology:

1. Malinowski promoted science-based fieldwork. He said we had to be empiricists. We must follow the scientific model to produce testable and correctable knowledge.
2. Malinowski stressed the importance of cultural immersion as the only way to get a glimpse of daily life and the mindset of people who lead very different lives than anthropologists.
3. Malinowski warned us not to dissect culture traditions down to isolated practices. Instead, he urged students to "study the whole." For example, his Trobriand Islands work focused on understanding the local economic system, but his research included everything about daily life in these communities.
4. Malinowski advised on exactly what kinds of anthropological data to collect while in the field. He mentions 3 data types:

    ›   Type 1 is the institutional outline. Researchers should go in and chart out how the community operates. Who's in charge of what? Who does what? What's the social hierarchy?
    ›   Type 2 is called ethnographic utterances. Malinowski says to record how people actually talk.
    ›   Type 3 is the "imponderabilia of actual life." Put simply, Malinowski wants researchers to write down every single thing they see.

The rich and deep descriptive language in Malinowski's work was made possible by the depths of details in his field notes. In fact, he is a brilliant model for writing about culture. Blending his Western point of view with the perspectives he acquired through cultural immersion, Malinowski invented a new genre of writing: ethnographic realism.

Malinowski also tells us that cultural immersion includes some classic ethnographic tools including doing a census, mapping, language study, surveys, and interviews. Importantly, he recommends a field diary to ensure the separation of objective and subjective data and notes. The subjective (or *etic*) notes are for the diary. The objective (or *emic*) observations are for field notes.

Interviews and census taking are two classic ethnographic tools.

Malinowski and Boas envisioned new ways to decipher cultural diversity in the name of anthropology and science. And the legacies of these early methodologists run deep. They laid bare the 4 cornerstones of modern anthropology:

1. Cultural relativity.
2. Participant observation.
3. The reconciliation of etic and emic perspectives.
4. The critical importance of interdisciplinary research via the 4 subfields of anthropology.

Suggested Reading

Kroeber and Kluckholn, *Culture*.

Stocking, *Observers Observed*.

Tylor, *Primitive Culture*.

Questions to Consider

1. What is cultural evolutionism, and how did early anthropological scholars like E.B. Tylor and Lewis Henry Morgan classify human cultural diversity?

2. What did cultural relativists like Boas and Malinowski do to challenge and eventually discredit cultural evolutionism?

Cultural Anthropology and Human Diversity

H ave you heard of the Nacirema culture, and their outlandish rituals? I'm sure some of you have. Every year, in hundreds of anthropology classrooms all across the country, professors introduce their students to cultural anthropology by discussing the classic anthropological field study by Horace Miner.

It's easy to whip students into a fury with all this one; they giggle and quickly compile a list of bizarre Nacireman rituals, as they try to imagine living the Nacirema way. There are high priests who do strange mouth purification rituals that involve drilling holes into their teeth, they even have shrines in their homes with all kinds of potions to help their constant need for purification. Some women even go to the lengths of baking the impurities out of their heads by sitting in ovens built just that purpose.

Then I tend to shock them when I mention that I myself am of Nacireman descent. Some go quiet and red with embarrassment for their previous levity, while others just express confusion as they wait for more details. But I quickly allay the discomfort, telling them the laughter was harmless, and they didn't hurt my feelings. They just don't understand my culture.

So I ask straight up, "Besides me, do any of you know someone from the Nacerima culture?" I pretend to be utterly shocked that so few people raise their hand. Then the twist. In feigned frustration, I get to the root of it. Where do they live? Well, the reading says somewhere between Canada and Mexico. Some lights go on. Maybe a Native American culture? Not quite. First a few get it until I throw out the final question, "What's Nacirema spelled backward?" Their jaws drop. American. It was us all along.

The strange shrines are just bathroom medicine cabinet, the priests, they were dentists; and the head ovens were the air dryers you find in the beauty salon. Miner just told us about these familiar components or our life in very unfamiliar ways. So how can we avoid misrepresenting or misunderstanding other cultures when we read or write about them? How can we avoid this Nacerima mistake? The answer—cultural Anthropology.

From an outsider's perspective, our daily lives are bound to appear nonsensical. For example, my Malian farmer friends laugh in disbelief at the thought of so many of us Americans walking behind our dogs and collecting their business, and with our hands no less, albeit in little plastic baggies. But think about it. Can you see how a subsistence-level family farmer might think this American tradition is utterly crazy? Yet, from our point of view, from our cultural context, it makes a great deal of sense that we happily pick up our dog's feces. Our behavior is relative to our cultural context.

Through ethnographic field methods that cultural anthropologists seek to understand the world through the eyes of the people with whom they live and research, precisely to avoid ethnocentric miscues, like, for example, failing to see ourselves in the Nacirema. Cultural anthropology builds bridges of understanding, it connects people across all kinds of linguistic and cultural boundaries. And, the more we understand other ways of seeing and being in the world, the more we actually understand ourselves. Both as individuals and as a species.

Today we'll visit a few anthropologists to learn how anthropology reconciles the infinite cultural diversity of humankind with the biological fact that we are one race, *Homo sapiens*. Today, we're headed all over the world, from England and the Arctic to New York and even the Pacific Islands. But wait, one last anthro trick. As we visit each anthropologist, let's note some of their methodologies. Let's learn not just what they discovered, but how they did it. As scientists, anthropologists pay close attention to each other's research methodologies because we know that the way you collect your data is going to influence and shape your observations and conclusions.

So, let's start our journey in England to meet up with Edward Burnett Tylor. Tylor is a great first stop because, in anthropology, he's earned many famous firsts including the fact that he's the first professional anthropologist to publish a comprehensive definition of culture. He's also the first Professor of Anthropology at Oxford, a position he assumed in 1896. But, 15 years before that, he wrote the first anthropology textbook, titled *Anthropology: An Introduction to the Study of Man and Civilization*.

Tylor is basically the undisputed father of British Anthropology. But back to task, why are we visiting Tylor? Well, he has that original definition of culture. And, he's also important because he was a leading proponent of the idea of cultural evolutionism, the idea that culture change, like biology, is an evolutionary process that transforms humans from primitives to sophisticates.

So let's start with Tylor's definition of culture, which we find in his 1871 book *Primitive Culture*. There, Tylor asserts that culture is "that complex whole which includes knowledge, belief, art, morals, law, custom, and any other capabilities and habits acquired by man as a member of society."

Now, where does he get this definition? Well, his methodology is what we consider ethnology, which is often characterized as classic Victorian armchair anthropology. Rather than travel the globe documenting exotic cultures, Tylor and ethnologists explored human cultural diversity by mining libraries and archives from the comfort of their desks.

Ultimately, he deployed a comparative approach that blended ethnology, archaeology, and philosophy or reason. And one of the central tenets of Tylor's work and of cultural evolutionism is that progress is our dominant narrative. Collectively we are on an arc from primitive to civilized, culturally and biologically.

One of his passions was the anthropology of religion, which he understood from an evolutionary point of view. As an element of culture, primitive religion starts with primitive people trying to solve life's big puzzles, like what's going on with this death thing? Then it evolves from magic and mystery through polytheism into monotheism, and then finally science.

So understanding culture, from Tylor's early anthropological point of view basically means that we see human diversity as evidence for the bio-cultural evolution of humankind. According to this perspective, we can sort world cultures into one of three grades, there's the savages, the barbarians, and the civilized.

Tylor wasn't alone as a cultural evolutionist. The armchair explorer and scholar James Frazer, who wrote the anthropologically famous book, *The Golden Bough*, was also a well-published advocate for the continuum from magic to religion and science. Cultural Evolutionism was a dominant thread in British Anthropology at the of the century, and it was also a popular perspective right back here in the states. So, let's go to the Seneca Iroquois Reservation in New York, to hear from Lewis Henry Morgan, one of the major voices behind this primitive-civilized schematic.

The reservation is located just outside Rochester, where in the 1840s the Seneca Nation was engaged in a land dispute with the Ogden Land Company. It's in this context that we meet a young lawyer named Lewis Henry Morgan, who helped the Seneca work out a solution to that dispute.

Morgan's fascination with Native American cultures compelled him to break out of the library and into the field. Unlike Tylor, Morgan was a tireless field researcher, and he traveled all throughout the US to collect kinship data. His interest in kinship blossomed into one of his most noted specializations. To this day, anthropologists still refer to Morgan as the father of kinship studies.

While in the field, Morgan gathered data from government officials known as Indian Agents, and he accumulated missionary data. Curating all these data streams, Morgan noted a distinct kinship pattern he referred to as classificatory kinship.

Essentially, kinship is the way a society organizes social relations in terms of family descent, but Morgan's classificatory kinship includes people who may or may not be genetically related. He noted classificatory patterns across Native Americans and concluded that this form of kinship was a primitive survival. A remnant of humanity's primitive state.

Ready to test his idea further, he sought out kinship data from all over the world. And with assistance from the Smithsonian Institute, he created a 7-page questionnaire with dozens and dozens of questions. With that, he compiled his results and found that he could sort kinship patterns into distinct kinship systems that we still teach to this day. For example, it's common in the US to call your mother's sister as your aunt, and her children are then your cousins, but, would you believe that in some cultures, we'd refer to our aunt as mother and therefore her kids would actually be our siblings, not our cousins.

In 1877, he published his findings in a major book called *Ancient Society*. This book was fiercely influential on the era's great thinkers and writers like Marx and Engels. In it, he clearly defines cultural evolutionism as a process in three phases. First, there's the most primitive, what he called the savage stage, which he associated with foraging communities. Then there's the barbarian stage brought on by the domestication of plants and animals. Then last, there's the ultimate civilized stage with state-level societies. Here he is in his own words,

> The latest investigations respecting the early condition of the human race, are tending to the conclusion that mankind commenced their career at the bottom of the scale and worked their way up from savagery to civilization, through the slow accumulation of experimental knowledge.

Morgan wanted to document these cultures because, for him, they were a living window into our less-civilized past, and, they were dying out. People actually called this line of research salvage anthropology. Here's Morgan again,

> The ethnic life of the Indian tribes is declining under the influence of American civilization, their arts and languages are disappearing and their institutions are dissolving. After a few more years, facts that may now be gathered with ease will become impossible of discovery. These

circumstances appeal strongly to Americans to enter this great field and gather abundant harvest.

So Morgan's legacy definitely includes that terrific call for more anthropology, but his true contribution was his vast archive of ethnographic data from all over the world. Additionally, he integrated field research with cultural evolutionism.

From Morgan's point of view, he advocated a comparative method that was much more at home in the humanities than in biology, and it embraced the enlightenment narrative of progress, with Victorian civilization defining the highest level of cultural evolution. And while this point of view was mainstream among scientists and scholars at the of the century, the idea of cultural evolution eventually falls out of favor. So what was cultural evolution's fate?

Well, Tylor's definition of culture stands to this day. And his use of subsistence strategies to distinguish cultures is still used today. That said, we anthropologists are long over the idea that cultural evolution is teleological; that it's on a progressive spectrum from savage to civilized. But why? Why did cultural evolution fall out of favor? Why did anthropology finally move on from that savage to civilized trajectory? The answer takes us to our next destination, and that's the Arctic.

It's here in the Arctic on Baffin Island where we meet up with Franz Boas, the father of American anthropology, and one of the most influential figures in the debates that pushed cultural evolutionism out of favor.

So, in the 1880s young Boas was out on Baffin Island on a 15-month field study. Before you hear from him, let me just prep you on who this guy really is. First of all, he was a physics Ph.D. A bona fide scientist from Europe. His research focused on color perceptions of Native Americans in Northern Canada. And he brought his physics mind to anthropology, which explains why he stressed empirical methods that, like in a lab, could be tested over and over again. Like Morgan, Boas did field research to collect data on human diversity.

His dedication to cultural immersion inspired his antipathy for cultural evolutionism. Because, unlike Morgan and Tylor, Boas didn't believe in culture grades. To him, there is no superior or more evolved cultures. Instead, where cultural evolutionists saw savages and barbarians, Boas saw complex, sophisticated people artfully adapted to their unique environment. And he got this point of view the moment he stepped foot on Baffin Island. Put yourself in his shoes. You, a physics Ph.D., one of the leading intellectuals of your era, you step off that steamboat with a couple suitcases and plenty of paper for writing notes. Tylor would certainly put Boas in the civilized category, but Boas wasn't buying it. When Boas arrived on that island, he was humbled, the way I was humbled, the first day I arrived in my Malian host village.

With one look around, Boas realized that as long as he was in the Artic, it was he that was the primitive, and his hosts were the sophisticates. After all, left to his own devices, Boas wouldn't last a night on this frozen island. He didn't know how to feed himself, clothe himself, how to get around, even his local language was elementary at best. Boas, by arriving on this island had become an infant who needs to learn absolutely everything, from how to say good morning to how to do some ice fishing with spears.

Boas saw sophistication in how the Inuit on Baffin Island sustained themselves in such a challenging place to live. What he saw was cultural relativity. He wasn't primitive because he didn't know how to protect himself against polar bears, he just wasn't born in a place where you learn that lesson. And the same goes for the Inuit. They're not primitive because they don't know quantum mechanics, they just were born in a place where it was more important to learn other things.

Early on, one of the lessons I gleaned from the letter diaries he kept with his fiancée in Europe was just how lonely and challenging anthropology field work can actually be. He wrote about happy times, but he also experienced significant troubles such as the time his hosts blamed him for someone's terrible illness. But overall, his 15-month research project was a smashing success, and in just a few short years Boas parlayed his anthropological prowess into a career running the anthropology department at Columbia University.

And his remarkable career helped change the world. He didn't do it alone, but he helped change the way we think about culture and each other. His career literally marks anthropology's transition out of cultural evolutionism and into cultural relativity, a perspective we still hold tightly to this day. Thanks to this work, there's no more ranking cultures from civilized to primitive. Now, we're all modern. Different in appearance and culture sure, but we're all modern. We're one race.

But how did he do that? I mean, government, industry, and eugenics all brought cultural evolutionism into the mainstream. How did cultural relativity win out in an era entrenched in cultural evolutionism?

Well, he did plenty of research and accumulated massive collections of anthropological artifacts. He wrote over 700 books and articles alone. Perhaps his greatest legacy, however, was the scores of students he mentored. These students went on to build new anthropology programs across the United States, and their names are the giants of early American anthropology.

We'll hear more from some of his students in future lectures, but for now here's a short list to give you a picture of the depths of his remarkable team. First, there's Ruth Benedict, who wrote the 1934 public affairs pamphlet *The Races of Mankind*, which articulates cultural relativity from the Boasian perspective. Then there's Melville J. Herskovits who set up the anthropology department at Northwestern University. And over at the University of Chicago, Edward Sapir created the famous anthropology program there.

There's also Alfred Kroeber who created UC Berkeley's famous museum and department of anthropology. Boas's fingertips are all over the growth of American Anthropology. He even reached beyond our discipline to groups like the NAACP and Howard Law School, to work with them, as allies in redefining race in the US.

Working to the end, Boas died in 1942 while eating lunch at Columbia, the place where he launched American anthropology. And you know what? He'd be pleased with his legacy, because, in his words, "Equal rights for all, equal possibilities, to have done even the smallest bit for this, is more than all science taken together." It's a physicist who said that.

In the next lecture, we'll hear more about Boas and how his students pushed anthropology into the future while leaving cultural evolutionism in the dust. But for now, let's leave the story here. Through research, teaching, and partnerships beyond the university, Boas clung to the scientific method and empiricism, and he launched four-field anthropology in the US And while he was busy training new generations of American anthropologists, Boas had a revolutionary counterpart over in England. And it sure wasn't Tylor. Let me introduce you to Bronislaw Malinowski.

Remarkably, Malinowski's story has some crazy parallels to Boas and his path. Like Boas, Malinowski was a Physics Ph.D. from continental Europe. He was drawn to anthropology, and he brought his scientific mind and methods to the discipline as well as to his teaching. He was fervently against the idea of cultural evolution despite the fact that it was the established norm, thanks to the work of people like Tylor. He argued that cultural evolutionism was bad science, and he taught students who went on to revolutionize anthropology in England and beyond.

Malinowski finished his career at Yale where he's buried, but he established his career at the London School of Economics where they still have the Friday Malinowski Lecture series. To hear more about his revolutionary approach to anthropology, let's go to Pacific Islands where he did some of his most important work. Because, it's here on the Trobriand Islands that Malinowski had his cultural relativity moment, and it's also where he invented the concept of functionalism.

Just as Boas saw complex people doing remarkable things to sustain their Arctic lives, Malinowski saw the same rational or functional side to things that others viewed as primitive or exotic. Specifically, he was interested in the logic of local economics. In terms of his approach, cultural practices that appear exotic only become familiar and rational when you look at things from a local point of view.

Here's a brief segment from the book in which Malinowski defines functionalism, It's called *Argonauts of the Western Pacific*. It shows not only Malinowski's beautifully descriptive language but also his understanding that these so-called natives are anything but primitive.

> A most remarkable form of intertribal trade is that obtaining between the Motu of Port Moresby and the tribes of the Papuan Gulf. The Motu sail for hundreds of miles in heavy, unwieldy canoes called lakatoi, which are provided with characteristic crab-claw sails. They bring pottery and shell ornaments, in olden days, stone blades, to Gulf Papuans, from whom they obtain in exchange, sago and heavy dug-outs, which are used afterward by the Motu for the construction of their lakotoi canoes.

The cultural relativity we see in this quote can feel rather obvious to our modern minds, but in the early 20th century, this was a new way to understand cultural diversity. Just like Boas, Malinowski sprung into fame as a practicing anthropologist and a teacher. But what exactly is this functionalist method he instilled in his students?

We'll it all comes down to how we define culture. Tylor and the cultural evolutionists had culture, race, and biology all interlinked. Basically, culture equals biology equals race. But then with functionalism, Malinowski separates these three concepts, allowing for what we now know as cultural relativity and the rejection of culture grades.

To see the world through functionalist glasses, Malinowski offers us four major tenets of the functionalist methodology. First and foremost, like Boas, Malinowski promoted science-based fieldwork. That was his physics background coming in. For anthropology to produce reliable knowledge, Malinowski said we had to be empiricists. We must follow the scientific model to produce testable and correctable knowledge.

Second, like Boas, Malinowski stressed the importance of cultural immersion as the only way we can get a glimpse of daily life and the mindset of people who lead very different lives than anthropologists.

Third, Malinowski warned us not to dissect cultural traditions down to isolated practices. Instead, he urged students to "study the whole" as he would say. For example, his Trobriand Islands work focused on understanding the local economic system, but his research included everything about daily life in these communities. He argued that economics or any individual piece of culture is inextricably connected in meaningful ways to everything else. Religion, art, clothing, canoe building, funerals, meals, celebrations, everything. If we look at local economies without seeing everything else, Malinowski says, we're not getting an accurate picture of local economics.

Fourth and last, he gave us some counsel on exactly what kinds of anthropological data we should collect while in the field. He mentions three types. Type 1 is what he called the institutional outline. What he's telling us here is to go in and chart out how this community operates. Who's in charge of what? Who does what? What's the social hierarchy?

Type 2 is called ethnographic utterances. This one is easy. He says to record how people actually talk. So, for example, when someone in Mali greets me in the morning with the phrase *I ni sogoma*, Malinowski warns me not to record that as good morning because that's not exactly what my friend just said. *I* means you. *Ni* means and, and *sogoma* means morning.

If I skip ahead and just write that I think this means good morning, I miss out on what was actually said. I lose a layer. Now, I can write that I translate this expression as good morning, but first, I had better write it as I heard it. The ethnographic utterance of my host community. Their actual words. Just like the rest of science, when we offer our anthropological conclusions and observations, we need to show our work too. How did we come to that conclusion?

The third and last type of data he suggests we collect is something he terms the imponderabilia of actual life. Put simply, he wants us to literally write down everything you see. Every single thing.

When I teach this idea to my students I ask them about their class notes, and I warn them that Malinowski would not be impressed with the level of detail in their notes. Why? What's missing? Well, there's my shirt. How many of you wrote down that I'm wearing a button-down shirt. And did you count how many times I walked across the front of the room? Did you write down what time I answered that first question from the class? No? Why not?

Well, my students remind me that all those things are unimportant. They know they won't be tested on my body language and my clothing, so they don't write it down. But that's because they're functioning within a culture they're very familiar with. They know students aren't tested on where the teacher stands during lectures, so they ignore that to focus on the important things. But, as Malinowski points out, when you get to a new and unfamiliar culture, you have no idea what's important and what's not. So to be thorough and comprehensive, write down every detail you can.

When you read Malinowski's work, the rich and deep descriptive language was made possible by the depths of details in his field notes. In fact, he's a brilliant model for writing about culture. Blending his western point of view with the perspectives he acquired through cultural immersion, Malinowski invents a new genre of writing—ethnographic realism.

In addition to the functionalist methods we just discussed, Malinowski also tells us that our cultural immersion, or as he refers to it, participant observation, includes some classic ethnographic tools including doing a census, mapping, language study, surveys, interviews, and importantly a field diary to ensure that you separate your objective and subjective data and notes. The subjective or etic notes are for your diary. The objective or emic observations are for your field notes.

So there you have it. A quick glance at how pioneers like Malinowski and Boas envisioned new ways to decipher cultural diversity in the name of anthropology and science. And the legacies of these early methodologists run deep, as their footprints endure to this day because they laid bare, the four cornerstones of modern anthropology: cultural relativity, participant observation, the reconciliation of etic and emic perspectives, and the critical importance of interdisciplinary research via the four sub-fields of anthropology.

If we embrace these lessons, and the idea of cultural relativity, we'll avoid that Nacirema trap most of my students fall into when they start learning about cultural anthropology. After all, if we're going to take the time to really understand new cultures from within, we might as well do this work in a way that our research subjects agree that it's an accurate representation of their daily lives and thoughts.

Field Research in Cultural Anthropology

I n the previous lecture, we saw how Franz Boas and Bronislaw
Malinowski used their scientific approach to usher in the era of cultural
relativity. It was nothing short of revolutionary to move from the
dominance of racially charged cultural evolutionism to cultural relativity—
but what happened next? This lecture's task is to examine how new
generations built on the foundations laid by Boas and Malinowski. We'll also
see what new thinkers can teach us about doing ethnographic field research.

Alfred Kroeber

Alfred Kroeber was in his senior year when Franz Boas arrived at Columbia as its 1st professor of anthropology. Kroeber's early intellectual curiosities focused on literature and languages, which is what opened the door to his career as an anthropologist.

One of Boas's popular courses, on Native American languages, attracted Kroeber's attention. This eventually led to an anthropological adventure and 2 years of field research amongst the Arapaho.

He earned his Ph.D. in 1901 on Arapaho symbolism, and Boas's influence is clear. First of all, Kroeber was an empiricist. He embraced the idea that anthropological research should be rooted in the scientific method. Additionally, he carried the torch of cultural relativity and taught by example when it came to cultural immersion.

One of Kroeber's most enduring accomplishments is the creation of the flagship anthropology program at the University of California, Berkeley.

Zora Neale Hurston

One of Boas's notable students was the novelist Zora Neale Hurston. As an African American in the early 20th century, Hurston defied prevailing racial and gender conventions as a student of anthropology at Columbia University.

Boas had a keen interest in researching folklore as a window into cultural diffusion and assimilation. He relished folklore students like Hurston.

Hurston investigated the cultural diffusion aspect of folklore in the Boas tradition, but she did something else: She added an interpretivist approach to the anthropology of folklore.

Hurston not only documented and mapped out the folklore of the American South and the Caribbean. Her new approach produced 2 new and invaluable insights into the anthropological potential of an interpretivist approach to folklore.

Specifically, she noted that once we enter the realm of interpreting and deconstructing folklore, we open a fascinating window into a culture. Moreover, she noted that the otherwise muted voices of female perspectives and voices are rendered visible through folklore.

She also characterized folklore as an adaptive strategy. For instance, the stories of Br'er Rabbit could have been used by slaves raising children in the 18th century. One story involves Br'er Rabbit hiding from a powerful bear in a briar patch.

Such stories always had a trope that served as a thinly veiled lesson on how the powerlessness of slavery could be overcome with cunning and smarts rather than physical force.

Hurston's work provides several methodological lessons:

1. Hurston was masterful at recording not only the thoughts and worldviews of people, but she captured the sounds and rhythms of their speech. This is shown in books like *Their Eyes Were Watching God*, which transport the reader deep into Hurston's worlds.

2. Hurston was relentless in working out ways to gain the trust and partnership of her research communities.

3. Creative approaches to data collection are fun for both anthropologists and informants. For example, while collecting folktales and oral traditions in Florida, Hurston organized a "Big Lies" contest. This got the locals to open up with fantastic tales, one example being about people who were so small they could walk between raindrops during a storm.

Audrey Richards

Audrey Richards was one of Malinowski's star graduates. Richards spent her early years in India during her father's time in the British Colonial Service in Calcutta. She grew up to be an admired field researcher and a tireless and inspiring anthropology professor.

Where Malinowski taught, before going to the field, students were required to write a thesis based on existing literature. Richards embraced the 4-field approach and picked a topic that was as biological as it was cultural: nutrition.

She studied in Zambia with a group that practiced matrilineal descent and kinship. In her words, Richards set out to "make an intensive study of the social institutions, customs and beliefs of the [Bemba] tribe ... with special reference to the part played by women in tribal and economic life... the nature and importance of the family system and the marriage contract... and problems connected with the rearing and education of children."

Richards was essentially planting a seed that eventually grew into a specialization called medical anthropology.

Richards in Zambia

Richards took to field research and loved the anthropology life in Zambia. She rode around on her bicycle, living in Bemba communities and helping women with daily tasks. She also sat and talked with community elders.

Before arriving in the communities, she'd spent a month studying the local language at a remote mission. She was a skilled hunter, and her colleagues and friends described her with words like "fearless" and "inspiring."

Malinowski's empiricism was omnipresent in Richards's original and revolutionary approach. She recorded what individuals ate and weighed food to calculate its nutritional value. Additionally, she calculated the nutritional value of wild and gathered foods.

Her first field study was followed by another in 1933, and in 1939 she published her work in the book *Land, Labour and Diet in Northern Rhodesia.* In the book, she describes the agricultural system of her hosts. She also analyzes the gendered division of labor as well as land use.

Two takeaway ideas stand out:

1. She demonstrated that anthropologists need to seek out indigenous knowledge.
2. In the behavior of their research communities, anthropologists observe complex, local, ecological management systems.

Richards's Legacy

Richards's legacy is multidimensional. She helped preserve and embolden the empiricism and cultural relativity of Malinowski's functionalism. In fact, she wrote a comprehensive history of functionalism, teaching students the practical value of anthropology for colonial policymakers and anyone who studies development and social change.

Additionally, she demonstrated the primacy of fieldwork and the scientific method as principles of anthropological research. Her work opened a new field of medical (or nutritional) anthropology. She also championed bringing women into anthropology and then into the field.

With her focus on the nexus of food, nutrition, and health, Richards was ahead of her time. She promoted practical applications for anthropology, which translated into baby-weighing nutritional studies across Africa and the world.

Richards would be pleased at the vibrancy and growth of the Society for Applied Anthropology. Years before that society was born, Richards advocated a national, applied anthropology program in Britain that could teach basic anthropological theories and methods to government officials and other nontraditional anthropology students.

While rooted in nutritional science, Richards made anthropology stronger by embracing and emboldening the interdisciplinarity of the field.

Edwin Evans-Pritchard

This lecture will close with a visit to one more star in the Malinowski line. His name was Sir Edwin Evans-Pritchard. Evans-Pritchard was born in the UK a year after Kroeber started his position at the University of California, Berkeley. He studied under Malinowski, and he too found new ways to deploy his teacher's methods.

Evans-Pritchard promoted the humanities side of anthropology. Evans-Pritchard collected field data and used his observations to interpret local histories and social structures. His more interpretive approach, and his fascination with social structure, emerged as a new brand of functionalism, now known as structural functionalism.

He refined this approach in the field while studying groups living in East Africa. In his early study in the Sudan, he considered how beliefs and practices in witchcraft among

the Azande people reinforce social cohesion.

In his 1937 book on the subject, he argued that Azande witchcraft functionally operates as a safety valve that redirects conflict and tension. It stabilizes society for the good of the order.

Another one of his major studies emerged in his 1940 book, *The Nuer*, about a group of pastoralists who live in East Africa. Amongst the Nuer, he continued his inquiry into social structure. He wanted to see if his ideas would hold up in a community of pastoralists the way they did among the Azande farmers. He asked: How do less stratified societies work? What does their political organization look like?

Evans-Pritchard saw that cattle, not cash, made Nuer societies work, so he followed the cattle. He learned to relate the chief values and social structure of the Nuer by understanding their relationship and practices with cattle. He characterized the social idiom of the Nuer as a bovine idiom.

Evans-Pritchard creatively diverged from the more empirical work of others like Malinowski and Richards. His great ethnographic details render his books instructive for anyone who wants to write or read about studying culture and social structures.

Suggested Reading

Hurston, *Mules and Men*.

Malinowski, *Argonauts of the Western Pacific*.

Richards, *Land, Labour, and Diet in Northern Rhodesia*.

Questions to Consider

1. As highly influential figures in the history of anthropology, what did Zora Neale Hurston and Edwin Evans-Pritchard build upon the ideas and methods of their teachers?

2. How did Audrey Richard's field research methods presage the emergence of medical anthropology?

Field Research in Cultural Anthropology

Today I'd like to start with a remarkable story. It involves the Yahi people of Northern California who hunted, fished and were avid foragers well into the 1800s. But then gold was found in California, and by 1848, the Gold Rush was in full force, ushering in scores of settlers, who gradually pushed the Yahi out of existence. Prospecting settlers killed off most of the Yahi, and those who escaped, hid in the mountains, living as their ancestors did, for over four decades. Kind of like the Dogon cliff dwellers of Mali, they withdrew to preserve their way of life in the face of external forces

Within decades, the Yahi had been reduced to fewer than half a dozen people. And it all came to an end when a group of engineers discovered their hidden camp. These men took as souvenirs all the Yahis's tools, weapons, and cooking equipment. So, alone and without the tools he needed to survive, one of the last remaining members of the Yahi came out of the woods on August 29, 1911. After several years of fending for himself in the mountains, this man, somewhere in his 40s or 50s, had reached his limit.

Destitute, hungry, and exhausted, he showed up at a slaughterhouse in Oroville, California, where he was quickly apprehended and jailed by the local sheriff. The sheriff felt he had no other place to put this disheveled man. After all, the prevalence of cultural evolutionism influenced how the sheriff understood this remarkable survivor of a hunter-gatherer people. To the sheriff, the man was a primitive, or perhaps a barbarian at best. A human who was less evolved, and ill fit for a so-called civilized world. So, he did not pass Go, he proceeded directly to jail.

Meanwhile, some 140 miles south, news of this wild man from the woods reached an anthropologist named Alfred Kroeber. Kroeber was a cultural relativist trained by Franz Boas, whom we met in our last lecture. Like his teacher, Kroeber railed against grand evolutionary schemes that ranked people on the scale from savage to civilized. A Native American specialist whose career was largely focused on collecting and documenting disappearing native cultures, Kroeber sprang into action. He brought the last Yahi back to UC Berkeley, where he had set up a formidable department of anthropology.

Long story short, Kroeber named the man Ishi, which translates as "man" in Yahi. Ishi reported that his cultural tradition prohibits people from speaking their own name. So, without family or friends, there was nobody left on earth who could say his actual name. Like the Yahi people themselves, Ishi's true name is forever gone. And that's what motivated Kroeber. He, like Lewis Henry Morgan in our previous lecture, was a salvage anthropologist. They both were determined to document and preserve Native American knowledge and cultural traditions before they disappeared forever.

Ishi lived in an apartment at the UC Berkeley Museum of Anthropology where he taught thousands of visitors about his life and culture. The experiences he related served as the basis for the 1992 movie *The Last of His Tribe*, which stars Jon Voight as Kroeber. One of Boas' first Ph.D. students, Kroeber, provides a terrific example of how, as an apprenticeship discipline, we anthropologists build upon the work of our teachers.

So today, we're continuing the story we started before. We saw how Boas and Bronislaw Malinowski used their scientific approach to usher in an era of cultural relativity. And it was nothing short of revolutionary to move from the dominance of racially charged, cultural evolutionism to cultural relativity, but what happens after that? When new generations of students in the Boas and Malinowski tradition make their mark on anthropology, how does that change our understanding of culture?

Our task today is to see how new generations build on the foundations laid by Boas & Malinowski. And secondly, as we trace their intellectual lineage, we'll see what these new thinkers can teach us about doing ethnographic field research.

Here's the plan. First, we'll visit with Alfred Kroeber and 20th-century novelist Zora Neale Hurston, who were both heavily influenced by Boas. And then we'll head to Africa to see two of Malinowski's stars. Finally, we'll wrap things up, and see the emergence of an entirely new approach to studying culture something that will challenge the very premise of anthropology and everything we know about culture.

First, Alfred Kroeber. Kroeber was in his senior year when Franz Boas arrived at Columbia as its first professor of anthropology. Kroeber's first intellectual curiosities focused on literature and languages, which is what opened the door to his career as an anthropologist. One of Boas' popular courses, *Native American Languages* attracted Kroeber's attention, and eventually led to an anthropological adventure and two years of field research amongst the Arapaho.

He earned his Ph.D. in 1901 on Arapaho symbolism, and the Boasian influence is clear. First of all, Kroeber was an empiricist with a capital E. He embraced the idea that anthropological research should be rooted in the scientific method. Additionally, he carried the torch of cultural relativity and taught by example when it came to cultural immersion, or what we also refer to as ethnographic research and participant observation. This method allows us to see and corporally experience the daily lives of our study communities.

One of Kroeber's most enduring accomplishments is the creation of the flagship anthropology program at UC Berkeley the same exact year that he graduates in 1901. His pioneering stewardship of West Coast anthropology quickly garnered the attention of major donors like the Hearst family, and in 1909 he was named Director of Berkeley's Museum of Anthropology, a position he held until 1947.

But for me, one of Kroeber's best accomplishments was his proclamation that anthropology had accomplished his teacher's goal, the Boas vision that one day people would understand culture in a way that unified us as one human race, rather than savages, barbarians, and sophisticates. Here he is in his own words,

> The most significant accomplishment of anthropology in the first half of the 20th century has been the extension and clarification of the concept of culture. The outstanding consequence of this conceptual extension has been the toppling of the doctrine of racism.

He continues, "We have learned that social achievements and superiorities rest overwhelmingly on cultural condition."

Now, Kroeber is one of many Boas superstars, and he's one of my favorites, but my absolute favorite was the brilliant novelist, Zora Neale Hurston. As an African American in the early 20th century, Hurston defied prevailing racial and gender conventions as a student of anthropology, at Columbia University. She talked about anthropology as cultural exploration and vindication, so let's look closer at her research to see what that really means.

Hurston collected folklore, and that's what connected her to Boas. Boas had a keen interest in researching folklore as a window into cultural diffusion & assimilation. Like geneticists might use DNA to trace the spread of humankind, Boas used folklore to trace the spread of culture. And he relished folklore students like Hurston.

Growing up in Eatonville, Florida, just a couple decades after the Civil War, she grew up listening to the great stories and oral traditions that graced local porches and gatherings. As someone with direct connections to southern communities like Eatonville, Hurston was much more likely to gain the trust of her study communities than Boas. And when Hurston collected folklore she was brilliant. She investigated the cultural diffusion aspect of folklore in the Boas tradition, but she did something else. She did something that Boas didn't really get into.

Hurston, a keen student of culture, added an interpretivist approach to the anthropology of folklore. You see, Boas wasn't so keen on ideas you can't test and correct, he was an empiricist at the core. But Hurston not only documented and mapped out the folklore of the American South and the Caribbean. Hurston's new approach produced two new and invaluable insights into the anthropological potential of interpretivist approaches to folklore. Specifically, she noted that once we enter the realm of interpreting and deconstructing folklore, we open a fascinating window into a culture, one we've never seen before. Moreover, she noted that the otherwise muted voices of female perspectives and voices in the historical and anthropological record, are suddenly rendered visible through this thing called folklore.

Bigger yet, she characterized folklore as an adaptive strategy. Remember that term? In anthropology, we tend to use it synonymously with modes of production. Adaptive strategies like foraging and agriculture are easy enough to grasp, but how can folklore itself be an adaptation strategy. How does a folktale, for example, help someone survive from day to day?

Well, let's revisit a childhood classic. A folktale you probably heard growing up. B'rer Rabbit. B'rer Rabbit was a clever little guy who was featured in any number of tales. B'rer was short for brother. There was B'rer rabbit, B'rer bear, B'rer all-kinds-of animals. So, one day B'rer Rabbit was coming home from a long day of gathering clover for his family. And he was just around the corner, and ready to greet his family, when out of the blue, bang, Brother Bear jumps from behind a tree and grabs poor Brother Rabbit by the neck. He was a downright, mean old bully of a bear, and he swung Brother Rabbit around and around and around getting ready to serve this guy up as dinner.

And spinning around, Brother Rabbit knew his fate was sealed, but then he had an idea. He spoke up, "Please, please, please. I don't mind this spinning, I don't even mind being your dinner, I accept my fate, Brother Bear. Please, please please, whatever you do, just don't throw me into that nasty briar patch over there."

Well, Brother Bear was indeed a cruel bully bear, and when he heard his victim screaming about that thorny briar patch, he did what all bullies would do, and he threw Brother Rabbit right into the thick of it. Crash. But wait what's a briar patch to a rabbit? That briar patch is home, not torture for a rabbit. And, it's a darn good one since Brother Bear's big clumsy paws can't get through those nasty thorns.

So how would Hurston break this down and see evidence that folklore can be an adaptation strategy? Well, first of all, B'rer Rabbit tales, the briar patch and others, they always had this same trope that served as a thinly veiled lesson on how the powerlessness of slavery can be overcome, not with physical force and brawn, but with cunning and smarts. And, if I was raising children as a slave in the 18th century, I can imagine telling kids some new version of this story every single night. Despite all, They'll be up against, I'd want them to see hope, and a hope they can cultivate by being as witty and quick as Brother Rabbit. This tale shows us how to carry on, even though we face the monstrosity of the slave system.

So, with this interpretivist analysis of folklore as an adaptation strategy, what makes Hurston an anthropologist? Well, for one, she trained with Boas and did anthropology field research in Florida and Haiti. That said, she wasn't so conventional when it came time to writing books about her observations. She wrote a few classics packed with great folklore scholarship, but she gained more fame than her mentor as one of the most acclaimed American novelists of the 20th century. She wrote using local dialects, and her fiction was steeped in years of anthropological observation.

I first discovered Hurston through my terrific undergrad English professors, but it wasn't until I became an anthropologist that I came to recognize her importance as an innovative field researcher in the Boas tradition. These days, when I read Hurston's more anthropological musings, I value four unique methodological gems.

First, in both her fiction and non-fiction, Hurston was masterful at recording not only the thoughts and worldviews of people in places like Eatonville, but she captured the sounds and the rhythms of their speech. Hearing authentic voices as you page through books like *Their Eyes Were Watching God*, transports you deep, deep into Hurston's worlds. So, if anthropologists are responsible, among other things, for helping the reader see the world through someone else's soul, then Hurston is an anthropologist par excellence.

In addition to her linguistic prowess and imagery, Hurston's example taught me to be relentless in working out ways to gain the trust and partnership of your research community. Heck, when she first arrived on site, she regretted her initial choices. First, she wore a fancy dress from Woolworth, and she arrived, driving herself in a really nice car. Locals couldn't figure out if she was a cop or some other snoop. Eventually, she finally gained their initial trust by telling people she was a bootlegger.

So, when I read her methodological reflections here, I took note. Coming in to my host village in Mali with a satellite phone, computers, and drones of video cameras, that will only distance me from those I came to live with. And as ethnographic field researchers, our job is the exact opposite. We're out to reduce the distance between us.

One last lesson I gleaned from Hurston was that creative approaches to data collection are not only fun for the anthropologist, but they're fun for our informants too. Trust me, that does impact the quality of data you collect.

So, here's a great example. She's sent down to Florida to collect folktales and oral tradition in Florida, and, when she starts asking folks to share folk stories and tales, people didn't want to talk with her. To get these people talking, Hurston organized a Big Lies contest, because locals enjoyed throwing around these so-called big lies. Suddenly—with this contest—her informants opened up. They told stories about people they knew who were so small, that, when it rained, they never got wet because they could walk between the raindrops. And if you didn't know somebody that small, you might brag about knowing a guy so ugly, that after he took a swim, they had to skim that ugly from the river for over a week. You get the picture. The point was the creative approach of inventing a locally salient contest, rather than a structured questionnaire paid off big.

But, most important to Boas himself, the generation of anthropologists that he trained picked up the mantle of his push for racial justice. And Kroeber himself celebrated the fight against scientific racism as the chief achievement of American Anthropology. But, tracing the full impact of the generation of anthropologists trained by Boas is a task far beyond the scope of a single lecture. So today, we'll use Kroeber and Hurston, as symbols of a new generation of American Anthropology in the Boas tradition.

And that's how it all goes down on the west side of the Atlantic, but what happened to British Anthropology of Bronisław Malinowski? Let's see how his side of the cultural relativist revolution evolves into the mid-20th century.

First, it is my pleasure to introduce you to Audrey Richards, one of Malinowski's star graduates. Audrey Richards spent her early years in India during her father's British Colonial Service in Calcutta. Richards grew up to be an admired field researcher, a brilliantly and hilarious colleague, and a tireless and inspiring anthropology professor. At the London School of Economics, where Malinowski taught, before going to the field, students were required to write a thesis based on existing literature.

Richards embraced the four-field approach and picked a topic that was as biological as it was cultural—nutrition. She studied in Zambia with a group that practiced matrilineal descent and kinship. In her proposal, Richards set out to, quote make an intensive study of the social institutions, customs, and beliefs of the Bemba tribe with specific reference to the part played by women in tribal and economic life. The nature and importance of the family system and the marriage contract and problems connected with the rearing and education of children.

Richards, without precedent, is essentially planting a seed that eventually grows into a specialization called medical anthropology.

So let's visit her in one of her favorite places, the field. And as her train departs on the long journey towards Africa, Audrey Richards unwraps a farewell gift of colored pencils from her mentor Bronislaw Malinowski. He reportedly met her at the station to give her this gift, along with a reminder that brown was the best color for economics, and red was best for political organization. A final lesson on coding field notes never hurts. Thank you, Prof. Malinowski.

So, Richards took to the field research idea and loved the anthropological life in Zambia. She rode around on her bicycle, living in Bemba communities and she helped women with daily tasks. She also sat and talked with community elders. But, before arriving in the communities, she spent a whole month studying the local language at a remote mission. Then she didn't turn back. She was an amazing hunter, and her colleagues and friends described her with words like fearless and inspiring. Those who appreciated her quick sense of humor relished her famous trick, which was lighting a match with her toes.

Malinowski's empiricism was omnipresent in Richard's original and revolutionary approach. For one, she recorded what individuals ate, and she weighed food to calculate its nutritional value. Additionally, she even calculated the nutritional value of wild and gathered foods. Her first field study was followed by another one in 1933, and in 1939 she then publishes her work in the classic *Land, Labour, and Diet in Northern Rhodesia*. Briefly, in this book she describes the agricultural system of her hosts, she analyses the gendered division of labor, as well as land use. And her big take home was two ideas that are central to my own work.

One, she demonstrated that we needed to seek out indigenous knowledge as anthropologists. And two, she helps us work out the idea that what we see going on in the behavior and knowledge of our research communities is actually a complex, local, ecological management system. Richards is known for her work on nutrition and even taught that as a biological process, it was more fundamental than sex. We don't have the time to get into that one now, but let's remember that Richards also studied anthropological topics like kinship, economics, and rituals, but she always brought things back to food.

Kinship, economics, and rituals, she argued, are all organized around the making, eating, or distribution of food.

Today, it's common for contemporary anthropologists to focus on food as their window into a community. I'm a prime example. I study it all—the music, the proverbs, kinship, everything. But I tend to seek this data as a way to better understand food. For most of my career, I've partnered with farmers to do a lot of projects, but we always bring it home to sorghum and filling bellies.

Now, Richard's legacy is multi-dimensional. First and foremost she helped preserve and embolden the empiricism and cultural relativity of Malinowski's functionalism. In fact, she wrote a comprehensive history of functionalism, teaching students the practical value of anthropology for colonial policy makers and anyone else who studies development and social change.

Additionally, she masterfully demonstrated the primacy of field work and scientific method as a founding principle of anthropological research. And in so doing, by example and as a teacher, Professor Richards trained dozens and dozens of adoring students, who she referred to as her butterflies and lambs. In the words of Malinowski himself, Richard's work was pioneering and scientific, and it opened a new field of medical or nutritional anthropology. As a pioneer, she also championed bringing women into anthropology and then into the field.

With her focus on the nexus of food, nutrition, and health, Audrey Richards was ahead of her time. She promoted practical applications for anthropology, which translated into baby weighing nutritional studies, all across Africa and the world. Richards would be pleased at the vibrancy and growth of the Society for Applied Anthropology in the 21st century. Years before that society was born, Richards advocated a national, applied anthropology program for Britain that could teach basic anthropological theories and methods to government officials and other non-traditional anthropology students. While rooted in nutritional science, Richards like Malinowski and other cultural relativists made anthropology stronger by embracing and emboldening the interdisciplinarity of our field.

And before we run out of time, let's do one fast and brief visit to another star in the Malinowski line. His name was Sir Edwin Evans-Pritchard. Evans-Pritchard was born in the UK a year after Kroeber started his position at UC Berkeley. He studied under Malinowski, and he too found new ways to deploy his teacher's methods.

Parallel with Zora Neale Hurston's breach from sheer empiricism to shades of interpretivist analysis, Evans-Pritchard flat-out promoted the humanities side of anthropology. Unlike his physicist mentor, Evans-Pritchard collected field data, but he used his observations to interpret local histories and social structures. His more interpretivist approach, and his fascination with social structure emerged as a new brand of functionalism, now known as structural functionalism.

He refined this approach in the field while studying groups living in East Africa. In his early study in the Sudan, he considered how beliefs and practices in witchcraft among the Azande people actually reinforce social cohesion. In his 1937 book on the subject, he argued that Azande witchcraft functionally operates as a safety valve that redirects conflict and tension. It stabilizes society for the good of the order.

Seen from this point of view, Evans-Pritchard uncovers the rational functions behind rituals and beliefs that those cultural evolutionists once interpreted as primitive survivals.

Another one of Evans-Pritchard's major studies emerged in his 1940 book, *The Nuer*, about a group of pastoralists who live in East Africa. Amongst the Nuer, he continued his inquiry into social structure, and he wanted to see if his ideas would hold up in a community of pastoralists they way they did among the Azande farmers. He asked, how do less stratified societies work? What does their political organization look like?

Like a Washington Post reporter following the Watergate money, Evans-Pritchard saw that cattle, not cash made Nuer societies work, so he followed the cattle. Literally, in his book, he celebrates the choice as a methodological tip for anyone who would follow up on his Nuer research. If you want to understand these people, he said, "Cherchez la vache!" Seek the cow! And, his primary conclusion was that he learned to relate the chief values and social structure of the Nuer by understanding their relationship and practices with cattle. He characterized the social idiom of the Nuer as a bovine idiom. But, you get the point. Let's not beat a dead cow.

As someone who dove head first into the more interpretive side, the humanities thread within anthropology, Evans-Pritchard creatively diverged from the more empirical work of others like Malinowski and Richards. And I'd be remiss to mention that his great ethnographic details render his books instructive for anyone who wants to write or read about studying culture and social structures.

So to close, each of the four anthropology legends we visited today preserved core principles from their teachers, but they also probed further and broadened the way we think about culture, cultural diversity, and even folklore and social structure.

The initial explorations into interpretivist work in the early 20th century, however, mark a new and significant shift in the way we think and work as anthropologists. As new generations of anthropologists made their own unique contributions to our collective enterprise, interpretivist approaches gain popularity and opened new questions to carry us forward, such as: Is it even possible to scientifically see the world from someone else's point of view? Could we be just tricking ourselves with these anthropology studies?

Deeper into the second half of the 20th century, the interpretivist mode gained immense popularity with great work in psychological anthropology, linguistics, mythology, and so much more. But for now, let's reconsider how far we've come since the origins of the 20th century.

It was in the twilight of Kroeber's career, that he tracked down and classified definitions of culture throughout the anthropological record. The growth of the discipline, even when we see it through the dated, mid-20th-century eyes of Kroeber, can be seen in the growth of our definitions for culture. Around the turn of the century, he found less than a dozen unique definitions, but by the 1950s, he found there were over 160 definitions.

Anthropology loves answers, not an answer. Our cultural relativity and interdisciplinarity allow us to forge unlikely perspectives that bridge unlikely approaches and answers—take Richards's biocultural work for example. We consider it a success that our definitions of culture continue to evolve and flourish. That's one of the great lessons of anthropology—all understanding of our humanity is temporal incomplete, yet, more complete than before.

Kinship, Family, and Marriage

There is no universal rule about the way humans make families: Humans approach families in all sorts of ways. Anthropology brings a bird's-eye view to make sense of this diversity. This lecture examines a Malian village, some Tibetan farmers, and some people from Amazonia, because we need to investigate why different cultures create families in so many different ways. Throughout, we'll see compelling cross-cultural examples of how different cultures have different ideas about how to structure a family.

Different Family Types

The nuclear family is the kind of family Americans see all over television, from *The Dick Van Dyke Show* to *Family Guy*. In its simplest form, a nuclear family consists of 2 parents and their children. But if we set the nuclear family as the standard definition for family, such ethnocentrism will surely lead to our disappointment.

Generally speaking, in the US, when people start a new family, they tend to start a new household in a new location, separate from their parents. This is referred to as neolocal residence.

For subsistence-farming societies, on the other hand, unilocal residence tends to be the prevailing norm. In one farming community in Mali, for example, males marry females in surrounding villages.

The wedding festivities begin when the bride-to-be, her family, her community, and a host of drummers literally walk to her new home. Grown males may have wives and children of their own, but they live in the same compound they've always lived in. This is unilocal residence.

Family Research

Lewis Henry Morgan did anthropological research on family and kinship before anthropology was taught in American universities. With inspiration from work like Morgan's, early anthropologists were trained to collect kinship data, and to theorize on kinship to understand the social structure of traditional societies.

By the 1940s, kinship was all the rage among anthropologists. For example, Edward Evans-Pritchard wrote *The Nuer* based on his fieldwork with an East African pastoralist group.

He studied the social structure and quotidian life of these cattle-oriented people, and he made the general argument that kinship, or the way we do family, is basically a dynamic rulebook for social relations, albeit an unwritten rulebook. It connects us with each other, defining our relationships with the folks we call family as well as the rest of the world.

From the interpretivist point of view, anthropologist Claude Lévi-Strauss was interested in family and kinship in terms of

marital-based alliances rather than bloodlines or direct descent.

We can put this alliance lens to work in the Malian community of Dissan. While men remain in their birth communities, upon marriage, women in Dissan leave their birth village to live with their husbands. This pattern of marriages weaves a thick matrix of long-established family relationships that transform individual communities into a self-reinforcing and symbiotic cooperative.

Virtual Visit: Tibet

Next, this lecture will take a virtual anthropologic visit to Tibet, where we'll visit the Limi. The Limi are family subsistence farmers who have cultivated this challenging, rocky land for generations. This society practices fraternal polyandry, which is multiple brothers marrying one wife.

As we get into the participant observation spirit, we work

alongside our hosts preparing the fields for the coming rains. Our hosts tell us how difficult it is to grow food here. They have only limited lands that can produce food—and only then through grueling manual labor.

As we sit in the glow of a solar lantern writing our field observations, we need to write down everything we observed and learned today. But how does this information address the primary research question: Why is it that brothers share a wife?

When asked, our hosts kept saying that fraternal polyandry is just how people do things around here. Furthermore, if every brother had a wife, there would be way too many kids and not enough food.

They love kids but they fear having too many. Plus, they have to deal with food insecurity, manual farming, limited land, limited inputs and resources, and no real other work to be had in this remote region. The more we look at this fraternal polyandry from a Limi perspective, the more it begins to make sense.

Ultimately, with the help of our host community, we not only got a glimpse into their lives, values, and strong work ethic, but we also figure out why fraternal polyandry is the most rational and appropriate local strategy for sustaining life and building a family with new generations to come.

Fraternal polyandry would keep birthrates lower because not everyone has kids concurrently. Secondly, it prevents the breakup of family farms. In Limi, when the household head passes away, the household holdings stay together as a cooperative unit. In the US, by contrast, siblings might split up inheritance and property and then continue on distinct paths.

Virtual Visit: Venezuela

Our next visit takes us to Venezuela to visit the Bari people. They live in Amazonia. Here, our mission is to answer the question: How can it be that some children in the Amazon reportedly have multiple biological fathers?

The anthropologist Stephen Beckerman has spent decades visiting and studying the Bari. He explains that we're visiting with an indigenous group who fish, hunt, and grow manioc and bananas. Thanks to Beckerman's presence, a few elders are willing to talk about their fatherhood beliefs.

The Bari elders explain that for a Bari child to be born, a single insemination was not ample enough, and that healthy newborns required what they call multiple sperm washes. Biological parents will likely continue to fertilize their pre-born baby, and it's common that a second male will provide secondary so-called washes.

The idea of multiple biological fathers clearly violates everything we've been taught about how babies are born. But with our anthropological lens and mission, we can see some terrific reasons for this cultural idea and practice.

Beckerman explains that the Bari have long suffered from attacks and raids that sometimes decimate entire communities. The Bari themselves are a peaceful group, but, since the arrival of the conquistadors in the 16th century, generations of Bari people have been stricken with massacres and other attacks from outsiders.

In the 1940s and 1950s for example, regional cattle ranchers regularly massacred Bari villages in a campaign to claim their land. And in the 21st century, it is oil and coal companies who are maneuvering to seize Bari lands.

The point is that they've adapted to a life of recurring attacks and stress. Beckerman interviewed 897 Bari women who reported on over 900 pregnancies. Women have an average of 8 pregnancies, but stillbirth and infant mortality rates were quite high.

Unlike all other arrangements, women who bring in a secondary father actually had lower incidents of stillbirths and miscarriages. Having a secondary father is, at least here among the Bari, a proven way to increase the likelihood that a child will survive into puberty.

The Bari are not alone. There are indigenous groups with similar practices and beliefs across South America, India, and Papua New Guinea.

Ecological resources and geography can shape how we create families.

Though their understanding of human biology may be off, they still get the important question right. They know what to do to increase their children's chances of reaching adulthood.

Diversity of Ideas

People across the world think about and practice kinship in all kinds of different ways. The diversity of kinship approaches tells us there isn't a gold standard for how to do marriage or family.

In fact, we've discovered the exact opposite. Our ecological resources and geography really can shape the way we successfully create and sustain our families. And there isn't one single form of family that works for all socioeconomic and cultural contexts.

Suggested Reading

Evans-Pritchard, *The Nuer*.

Hrdy, *Mothers and Others*.

Lévi-Strauss, *The Elementary Structures of Kinship*.

Questions to Consider

1. What is kinship, and why do anthropologists dedicate so much attention to studying kinship, marriage, and family across cultures?

2. How do environmental and economic conditions help explain the presence of fraternal polyandry among the Limi of Tibet, as well as the phenomenon of partible paternity in the Venezuelan Amazon?

Kinship, Family, and Marriage

We cultural anthropologists just love to categorize humans and everything we do. We take a morsel of the human experience, and we start thinking about how that morsel varies across cultures and history. It's great fun because with anthropology you can take something so familiar, like family for example, and then, in an instant, you see it and yourself in a whole new light. On the surface, family is so familiar to us that it's literally the base of the word familiar itself. Family feels like it's innate or natural but is it? Today, let's have some fun and see how the anthropological lens can deepen our appreciation of this thing we call family.

Consider this to get us started. All across the world, across all known cultures, what's one universal about the way we humans make families? There's only one. When I ask this question to my students, they jump right in maybe you're thinking what they're thinking? Something about two or more people living together? Sure, but wait. My friends in Mali have families where there are more than two parents. And, to be honest, they'd be rather surprised to hear they're not a family.

Then, some students always mention something like married people co-raising children but, I know plenty of terrific single parents, and I also know great families that don't have any offspring.

By this point, it's usually the romantics amongst us that bring up love. Family is a group of people who love each other. That's nice and all, but again, there's plenty of exceptions to that rule. While I was blessed with a terrific, loving family, I have pals who tell me they weren't as fortunate.

So what is it? What's the one family universal that applies to all cultures across the world? The answer well, it's that there is no universal about the way humans make families. We're all over the place. We do family in all kinds of ways. But, with anthropology, we can use our bird's eye view to make sense of this diversity. Today we'll visit a Malian village, some Tibetan farmers, and even some people in Amazonia because we need to investigate why different cultures create families in so many different ways. So let's begin with what we know: the nuclear family.

The nuclear family is the kind of family we see all over the television, from *The Dick Van Dyke Show* to *Family Guy*. And, despite the fact that we now see more non-nuclear families in popular culture and in our daily lives, the nuclear family endures. In its simplest form, a nuclear family consists of two parents and their children. And on television, they usually throw in a pet.

But, if we set the nuclear family as the standard definition for all families, such ethnocentrism will surely lead to our disappointment. Because, no matter where you come from, when we see a new or unfamiliar kind of family or marriage, we have something to learn, not fear. And if we follow the methods of Franz Boas, Bronislaw Malinowski, as well as their students, if we embrace cultural relativity, we'll begin to see family and marriage in entirely new ways.

Now, the non-nuclear family. Let's think about this. What's a basic difference between early farming families versus 21st-century American families? Do early farmers form nuclear families? Well, not exactly. There's plenty of differences, but here's a big one, unilocal versus neolocal residence. Neo, new location. You see, in industrialized economies like here in the US, we're basically a neolocal residence society. You grow up, get married, you grab your spouse by the hand, and you both return to your childhood room in your parent's house where you'll live for the rest of your life. Oh, wait a minute, no. That's not what we do.

When we start a new family, we tend to start a new household, and in a new location, separate from our parents. Neolocal residence. Now, we do practice unilocal residence in times of crises, but, each generation, for the most part, settles a new home separate from, yet often with some help from parents and other loved ones.

But for subsistence farming societies, on the other hand, unilocal residence tends to be the prevailing norm. In the farming community I work with in Mali, for example, males in our village marry females in surrounding villages. And the wedding festivities begin when the bride-to-be, her family, and the entire community, as well as a host of drummers literally walk her and her belongings to her new home. My younger adoptive brothers, for example, now have wives and children of their own, but they live in the same compound we've always lived in. Unilocal residence. One family, one residence.

So, by studying the different ways people make family, we can see that the way you make a living, the way you feed and sustain yourself well, that significantly influences the way you create and think about your family.

My families are great examples. My adoptive Malian family is actually huge. Not including me, there are 47 people in the Sangare household, with four generations living together. And, our collective compound is so large that we eat together in a half dozen different crews. It's terrific. I sit down at a common bowl of porridge with my brothers, and I look over my left shoulder and see groups of other family members seated around their common bowls. But, we all worked together in the same fields and kitchens to put that porridge in our bowls.

When I tell people about my Malian family, sometimes they ask why these people, living in poverty, have such large families. And from a US perspective, that makes sense. Over here, life is expensive, and the thought of raising ten kids through college is enough to frighten most anyone. But over there in Mali, in communities where people eat only what they grow, this logic doesn't hold. Having less children isn't the way to reduce and conserve household resources. Ironically, it's the exact opposite.

Curious about this phenomenon, I did a study in Dissan, my host village, and I used my census data of all village households along with a farming questionnaire I had completed with each household prior to that growing season.

First, in Dissan, which is a rural, consumption-based farming community, the mean size of family households was 11. So, I split the community in two, those with 11 or more family members, and those with fewer than the mean 10 or less. I ran the numbers and found a significant correlation with household food security, and having at least 11 people in your family. Smaller households reported significantly more episodes of food insecurity. It seems that having fewer family members in Dissan means you'll have trouble putting food on the table. And this does not come as a surprise to Malians. But again, back in the states where we make our money at work, not by growing it, families I know often struggle to keep up with the expenses of maybe one or two kids. Ultimately, the way we make a living, the way we put food on the table that actually shapes how we make families.

Let's dig deeper and so some simulated field research in Asia and South America, but first, we need to see how we anthropologists actually study families. When it comes to researching family and marriage, anthropologists have a special term we like to use. That's kinship. Influential anthropological minds have been studying kinship since the dawn of our discipline. And even earlier than that.

It was proto-anthropologist Lewis Henry Morgan, for example, who did anthropological research on family and kinship well before anthropology was even taught in American universities. Morgan was fascinated with Native American cultures, and he even lived with the Iroquois for a short time. Native American kinship systems attracted his anthropological eye, and his extensive comparative study across Native American cultures grew into a much larger project involving cultures from all over the globe. Morgan argued that kinship is a window into social evolution. When he sees my host family in Mali with 47 members who collectively work the land, Morgan sees more of a primitive form of family, from which our so-called modern nuclear family has evolved.

Finally, with inspiration from work like Morgan's, early anthropologists were trained to collect kinship data and to theorize on kinship to understand the social structure of traditional societies. And by the 1940s, kinship was all the rage among anthropologists. Let's take a quick look at a couple of them.

First, there's Edward Evans-Pritchard, who wrote *The Nuer* based on his fieldwork with this East African pastoralist group. Now, Evans-Pritchard studied the social structure and quotidian life of these cattle-oriented people, and he made the general argument that kinship, or the way we do family, is basically a dynamic rulebook for social relations, albeit an unwritten one. And it connects us with each other, defining our relationships with the folks we call family, as well as the rest of the world.

And from the interpretivist point of view, anthropologist Claude Levi-Strauss was interested in family and kinship in terms of marital-based alliances rather than bloodlines or direct descent. We can put this alliance lens to work again in my Malian host community to get an idea of how our village alliance operates. You see, while men remain in their birth communities, upon marriage, women in Dissan leave their birth village to live with their husbands. And similarly, young women in Dissan leave our community to start families in neighboring villages.

This pattern of marriages weaves a thick matrix of long-established family relationships that transform our individual communities into a self-reinforcing and symbiotic cooperative. I remember a heart-warming episode during my dissertation field research, and it's a perfect example of how these alliances mutually sustain their members

As an apprentice to one of my Malian farming teachers, a great man named Bakri Jakite. I refer to him as ciwara, a mythical local farming hero who first taught humans how to farm grain. Well, Jakite and I were alone in his three-hectare field, and we were planting the whole thing by ourselves, by hand. In the intense sun, I was struggling, but I kept on because Bakri had lost all of his field labor that year. He was alone and he still needed to farm to feed a family of 10.

You see, his oldest son, Drisa was preparing their fields a month earlier when the family donkey kicked and broke his leg. So there we were, Bakri pressing on like a superhero, and me just trying to help the best I could, but struggling in the heat and sun.

About two hours into the morning, and I'll never forget this, I went to the tree where I had hung my canteen was hung because I needed some water. And that's where I saw it. It was like a movie. There on the far horizon, on the edge of the forest, was a line of 20 men, all of them with farming tools in hand. And without fuss or fanfare, or even a conversation they had heard about the trouble. And, I kid you not, they walked some seven kilometers to Dissan, and they helped us finish our planting in a single day. Afterward, we sat under a tree drinking tea and crunching on some peanuts. When I asked why they had come, they explained that, despite the fact that they were from a different village, they were one family with Bakri's household through generations of marriage. Now, if you ask me, those are some terrific in-laws.

Anyway, collectively, and in many different ways, early kinship-oriented anthropologists, established the value of kinship as a window into the way different societies organized themselves, including how we make and sustain our families. And with that quick kinship briefing, let's do some virtual field research to discover for ourselves what cross-cultural studies on families can reveal.

So first, it's off to Papua New Guinea where we find cultural anthropologist Bruce Knauft. Knauft's Gebusi research is a wonderful example of what we're going to do when we try out some virtual ethnographic research in Tibet and Venezuela.

Knauft lived the daily life of the Gebusi, learning terrific lessons along the way. For example, he once learned a super efficient way to fell trees. The Gebusi showed Knauft how they only partially chop down a constellation of trees. Then, they completely chop down one final tree, and remarkably, like falling dominoes, it does the rest of the work, lickity-split.

More in line with our discussion on family Knauft dedicated a lot of his research to document and explore Gebusi kinship. And as he became more and more familiar with how the Gebusi organized families, he suddenly discovered a connection between local kinship practices and the unusually high local homicide rate. What he found wasn't funny, but it does point to a similar, yet less extreme ethos we see in US culture back home. It's the dreaded in-laws. As it turns out, when Knauft maps out marital relations and local kinship, he can see that if you're looking for the culprit responsible for someone's death, you'd be wise to check out the in-laws as your prime suspects. And not just the mother and father-in-law, all of them.

Now, rather than unpacking that compelling conclusion, I'll leave you to do that on your own so we can do some virtual research for ourselves. Let's use our anthropological lens to work out some unique ways that different people form families in the world around us. And, for our first virtual ethnographic research mission, we're off to Tibet where we need to answer our primary research question Why do brothers share a single wife? Think about that. Would you marry someone if you also had to marry all of their siblings?

When I ask my students to consider that, some giggle, but most wrench their faces in curious disgust. And they have wide eyes, begging for an explanation. Some of us may joke about excessively interfering family members, but if you were born into the Limi community in rural Tibet, that's not a joke, it's the norm.

Now, rather than writing this cultural practice off, let's use anthropology to work out why these people practice fraternal polyandry. Polyandry is plural marriage with multiple husbands, so fraternal polyandry is exactly as it sounds multiple brothers marrying one wife. But why?

Now that we're among the Limi of Tibet, I've already facilitated our entry and introduced everyone to our village hosts. These farming families are terrific and generous. They greet us warmly and we quickly embrace the beauty of these people and their community. So let's just get right to work. But wait, what do we do first, now that we have the elder's permission to do some anthropology field research? Well, one of the most noted methodologists in the history of the discipline, Malinowski, encouraged us to see through these differences, and instead seek perspectives that reveal the rational actions and strategies behind what to us outsiders appear as mysterious and exotic practices.

So let's do just that with the Limi, family subsistence farmers, who have cultivated this challenging, rocky land for generations. And as we get into the participant observation spirit, we work alongside our hosts preparing the fields for the coming rains.

And, as we viscerally experience this day-in-the-life of a Limi farmer, our hosts tell us how difficult it is to grow food there. They have only limited lands that can produce food, and access to other modes of production like a gig at the local coffee shop or a shift at the factory, those aren't even options. The sweat on your brow and the sore back and legs are only a small dose of the magnificent effort that these farmers put forth to make sure their families are fed. It's remarkable.

As we sit in the glow of our solar lantern writing our field observations, we need to write down everything we observed and learned today. But how does this information address our primary research question: why is it, that brothers share a wife? When we asked, our hosts kept saying that fraternal polyandry is just how people do things around here and that if every brother had a wife, there'd be way too many kids and not enough food.

So what do you think? Is that a hint? Well, break it down. Talk this out in your field notes. First, we know their adaptation strategy. How they make their living and sustain their families? They're subsistence farmers. Yes, barley farmers who also dabble in animal husbandry. In this village, we saw a lot of sheep and goats. And two, we know they were very talkative about the extremely difficult growing conditions, including the lack of arable land. And last, their rationale for fraternal polyandry was that it kept birthrates at sustainable levels. In their words, it prevents having too many kids.

We asked, don't they like raising kids? And they laughed at such a silly question. They love kids but they fear having too many. Plus, there's food insecurity, manual barley farming, limited land, limited inputs and resources, and no real other work to be had in this remote region.

You see it, right? Here in this community if you want a family and kids, if you want a Limi life right here in rural Tibet, the first thing you have to work out is how to mitigate ridiculously challenging environmental and economic resources. It turns out, the more we look at this fraternal polyandry from a Limi perspective, the more it begins to make sense and when we consider our theory further, it cements right into place.

Ultimately, with the help of our host community, we not only got a glimpse into their lives, values, and strong work ethic, but we also figured out why fraternal polyandry is the most rational and appropriate local strategy for sustaining life and building a family with new generations to come.

For one, this practice limits reproductive rates to sustainable levels for a place with few resources. Rather than three couples producing up to three kids every two years or so, fraternal polyandry would keep birthrates lower because not everyone is having kids concurrently.

Secondly, something you may not have thought of is that this approach prevents the break-up of family farms. Here in Limi, when the household head passes away, the household holdings stay together as a cooperative unit. In the US we might split up our inheritance and property and then continue along on our own distinct paths as siblings. But here among the Limi, the inheritance stays intact as a whole. And that's a good thing because these lands are so difficult to farm, that, if three brothers split up their household fields to raise separate families none of them could do it. It's all or one here among the Limi.

And put it that way, frankly, fraternal polyandry is a beautiful expression of humanity and the strength of this thing we call family.

So now we have just a little more time to do one more virtual research mission. And for this one, we're off to Venezuela to visit the Bari people who live in Amazonia. Here our mission is to answer the question, how can it be that some children in the Amazon reportedly have multiple biological fathers?

The idea of multiple dads isn't such big deal. We've had stepfathers and adoptive fathers for ages, that's normal. But when it comes to multiple biological fathers, few of us believe that we actually have more than one biological father. Well, I know a place where that isn't the case. Let's go visit a Barí community along the Amazon to figure out this two biological fathers thing.

Now, when we arrive we meet up with Stephen Beckerman who spent decades visiting and studying the Bari. He explains that we're visiting with an indigenous group who fish, hunt, and they even grow manioc and bananas. And we sit on a woven grass mat as a few elders sit in their hammocks ready to talk with us about multiple or partible paternity. Because Beckerman is here, and because he's earned the trust of our hosts, we can get right to the point. And when we ask the elders if some Bari people have more than one father, they laugh with wide smiles and say, of course, they do.

When we asked how that was possible, they clearly explained that for a Bari child to be born, a single insemination was not ample enough, and that healthy newborns required what they call multiple sperm washings. So, biological parents will likely continue to fertilize their pre-born baby and it's common that a second male will provide secondary so-called washes.

They clarified that it is important for a husband and wife to handle the very first fertilization, but then, bringing in a second father is generally seen as a great strategy to raise a child and build a family. But, when we ask why, when we push on, they keep with the multiple washes of sperm idea, and they say that sometimes they simply don't want to wear out their husbands. But what else is going on here?

I mean, biologically, we can't rationally adopt their concept of partible paternity. The idea of multiple biological dads clearly violates everything we've been taught about how babies are born. But with our anthropological lens and mission, we can see some terrific reasons for this cultural idea and practice.

Beckerman explains that the Bari have long suffered from attacks and raids that sometimes decimate entire communities. The Bari themselves are a peaceful group, but, since the arrival of the conquistadors in the 16th century, generations of Bari people have been stricken with massacres and other attacks from outsiders. In the 1940s and 1950s for example, regional cattle ranchers regularly massacred Bari villages in a campaign to claim their land. And in the 21st century, it's oil and coal companies who are maneuvering to seize these same lands.

The point is, they've adapted to a life of recurring attacks and stress. Beckerman interviewed 897 Bari women who reported on over 900 pregnancies. The results are illuminating, but, the short of it is that women have an average of eight pregnancies, but stillbirth and infant mortality rates were actually quite high. And, unlike all other arrangements, women who bring in a secondary father actually had lower incidents of stillbirths and miscarriages. It seems the additional support and food of a secondary father is a sure way to carry a pregnancy through to a successful birth.

So I get it. Without microscopes or statistical mathematics, the Bari people figured out the same thing that Beckerman did with science and anthropology. Having a secondary father is, at least here among the Bari, a proven way to increase the likelihood that your child will survive into puberty. Through oral tradition, the Bari cultivated and carry forward this cultural knowledge through the idea of partible paternity. And, they're not alone. There are actually indigenous groups with similar practices and beliefs across South America, India, and Papua New Guinea. At worse, give them an F for the biology of human reproduction, but they still get the important question right. They know exactly what to do to increase their children's chances to make it to adulthood, and it works.

So let's wrap up this cross-cultural exploration of family and kinship. Today, in remote parts of Tibet and Venezuela, we observed how kinship patterns and local ecology are functionally intertwined.

The Limi of Tibet taught us that fraternal polyandry makes a heck of a lot of sense if you want to create a family in a society living with scarce arable land and other resources. And the Bari of Venezuela showed us how partible paternity can actually exist despite that well-known fact that the fertilization of a human egg is a singular event. Thank you cultural relativity and kinship.

People across the world think about and practice kinship in all kinds of different ways. What the diversity of kinship systems across cultures tells us, is that there really isn't a gold standard for how to do marriage or family. In fact, we've discovered the exact opposite. Our ecological resources and geography really can shape the way we successfully create and sustain our families. And there isn't one form of family that works for all socioeconomic and cultural contexts.

Today, we saw compelling cross-cultural examples of how different cultures have different ideas about how to structure a family. And, our anthropological perspective enabled us to appreciate the functional logic that underlies these differences.

Think about that. Anthropology gave us a fascinating glance into the lives of people who otherwise we wouldn't be thinking about or relating to. And by helping us make sense of these practices that, at first glance, may seem rather odd, anthropology also enables us to find remarkable pathways that connect the cultures of the world.

Sex, Gender, and Sexuality

The past several lectures have examined human diversity and cultural traditions. Today's anthropologists continue to test and correct their understanding of human diversity. Anthropologists started with cultural evolutionism, but now understand biology and culture as distinct facets of humanity. This lecture takes a similar approach with sex and gender. It uses anthropology to see how biological sex, gender, and sexuality are unique threads of humanity.

Intersex Individuals

A hermaphrodite is an organism that has both male and female sex organs. Hermaphrodites comprise a relatively small category within a larger group of people who, biologically, have both male and female sexual traits.

The term *intersex* is used to describe this group, whose biology is not exclusively female and male. This is strictly in reference to physiology—not sexual orientation.

Some people are born intersex. According to Anne Fausto-Sterling, the rate is 1.7%, which is a higher incidence than albinism.

In the US, cultural and medical practices have classified intersex as a deformity rather than diversity. As such, it's often the case that when intersex children are born, doctors and parents typically go into emergency mode. They make a decision: male or female. Shockingly, the decision to go male or female is commonly made immediately, based on the presence or absence of a functional penis. Many parents will opt for their child to have a so-called corrective surgery.

Studies have shown that people who received this sexual assignment surgery are at risk for gender identity issues later in life. Operating on the body can clear up biological ambiguity when it comes to sex organs, but those organs do not control one's sexuality or gender identity. Rather than reinforcing a non-biological duality of male and female, perhaps we should expand our understanding of other sexes.

Intersex Categories

Under the general term intersex, there are various categories based on different biological characteristics. The biologist Joan Roughgarden wrote the book *Evolution's Rainbow*, in which we can see that intersex and non-binary sexes are more than common throughout the entire animal kingdom. Roughgarden discusses several types of intersexuality.

First, there are hermaphrodites. Despite the fact that this may be the most recognized or familiar form of intersex, it's actually the rarest of them all. The ratio is 1 in 85,000. A hermpaphrodite usually has some combination of testes, ovaries, or ovotestis, which contain both ovarian and testicular tissue.

Hypospadias are more common. Hypospadia is considered to be minor if the vent is off-center on the tip of the penis. To one degree or another, this occurs in over 40% of males. But then, around 1 in 2000 males are born hypospadias major, which means that the vent is anywhere but the tip of the penis.

Even more common than hypospadias is chromosomal variation. This outcome happens about once in 1000 births. These varieties of intersex occur as the result of an additional chromosome. Instead of an XX female or an XY chromosome male, the combinations XXY, XYY, XXX, or even XXYY might occur.

The takeaway from this discussion: Biological sex isn't really a binary male-female concept.

Sexuality and Gender

Sexuality and gender are not strictly biological phenomena. That's why cultural anthropology can help us see the distinct qualities of biological sex as opposed to one's sexuality and gender.

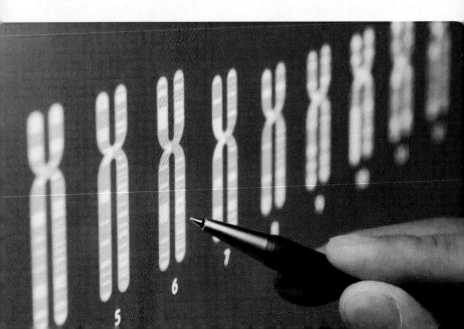

Darwin and his contemporaries may have argued that we're ultimately all one race, *Homo sapiens*. But they were clear that half the human race wasn't built for thinking or anything else outside of the domestic sphere. They believed that women's role in human reproduction is so physically and mentally demanding that it limits their potential as scholars, doctors, and presidents.

A brilliant French scientist named Clémence Royer challenged this idea. She was hired but then fired by Darwin as the translator of his masterpiece, *On the Origins of Species*. When she had a disagreement with Darwin, she let it be known. After her firing, she continued her scientific career.

She gave a remarkable and courageous speech to the Société d'Anthropologie de Paris in 1874. She was ahead of her times and most of anthropology with her succinct take on gender. Here's what she said:

> Up until now, science like law, made exclusively by men, has too often considered woman as an absolutely passive being, without instincts or passions or her own interests; as a purely plastic material capable of taking any form given her without resistance; a being without the inner resources to react against the education she receives or against the discipline to which she submits as part of law, custom or opinion. Woman is not made like this.

Royer's critique of European science in the late 19th century opened the door to a more comprehensive study of humanity—an anthropology that saw women in their fullness as human beings.

The Zuni

Undeniably, Western gender traditions strongly separate men's roles from women's roles, thus developing a gender binary that is closely tied to the biology binary. But cultural anthropologists have found and documented world cultures that don't have a strict male-female gender binary.

One example comes from the Zuni of the American Southwest, where scholars like Will Roscoe have studied Native American cultures and the presence of two-spirit people. Practices and norms certainly differed between distinct Native American groups, but the presence of two-spirit people was rather common.

Two-spirit people are sometimes referred to as a 3rd or 4th gender. They clearly contradict the male-female gender binary. Typically, the two-spirited person is celebrated as a gifted member of society who may perform trades or skills that are associated with the so-called opposite sex.

According to the literature, two-spirit people were not necessarily homosexual, but some were. They were not dressing or acting with the intent to deceive fellow community members; rather, they were known and admired publicly for possessing both male and female qualities.

For example, among the Zuni, when a potential two-spirit child emerges, a ritual served as a test to discern who this person truly was, regardless of biology. In one version, a bow and arrow is placed next to a basket, and the young child is asked to pick the one they like the most. If a young girl picks up the bow and arrow, she is celebrated and accepted as a two-spirited person, and she will likely go

on with life learning skills traditionally taught to biological males.

Native Americans aren't the only ones who see more than 2 genders. India has the hijras, and in the Balkans are people called the sworn virgins. In the Pacific are the fakaleitis of Tonga and the mahu of Hawaii. And there are others. What all these cultures and people have in common is a separation of biological sex, gender, and sexuality.

Sexuality

If gender isn't sexuality, and if biological sex doesn't determine sexuality, then what exactly is sexuality? Neuroanthropologists tell us that sexuality does in fact have biological foundations. Multiple studies confirm statistically significant correlations between human sexuality and several biological indicators, including the nerve cells in the hypothalamus.

The hypothalamus is located at the center of the brain and produces hormones. Inside the hypothalamus are suprachiasmatic nuclei (or SCN). The SCN, which controls our circadian body rhythm, is closer to some of the biological roots of our sexuality.

Research by noted neurobiologist Dick Swaab and others reveals a relationship between SCN size and homosexuality. Simply put, the SCNs of homosexual men are larger than the SCNs in heterosexual men. They're about twice as large in terms of physical size and the total amount of neurons.

If the SCN of homosexual men are around twice the size of their heterosexual counterparts, how will the SCNs in homosexual men compare with heterosexual females? As it turns out, female SCNs are much smaller than even heterosexual males, meaning women have smaller SCNs and therefore fewer SCN neurons.

Swaab and others like the neurobiologist Simon LeVay also identified another compelling connection between our brains and our sexuality. They identified a nucleus that is understood as the sexually dimorphic nucleus. It's called INAH-3, which is short for the 3rd interstitial nucleus of the anterior hypothalamus.

Like the SCN, the INAH-3 is significantly larger in males than females, and even larger in heterosexual males than in homosexual males and heterosexual females. Overall, the INAH-3 is smaller in people who are sexually attracted to males.

One last connection comes from the hypothalamus, which helps us with our olfactory system. Scientists looked into how the hypothalamus responds to the smells of estrogen (as found in female urine) and testosterone (from male sweat). The results were definitive.

The hypothalamus of people who were sexually attracted to women (heterosexual men and homosexual women) responded to the estrogen. The hypothalamus of people attracted to men, including both homosexual men and heterosexual women, responded to testosterone.

Twin Studies

Roughgarden's exploration of sexuality and biological sex, *Evolution's Rainbow*, provides a great bird's-eye view of twin studies that have looked at sexuality. Collectively, these studies ask the question: Is there evidence for the biological foundations of sexuality?

The studies compare identical versus fraternal twins. The thought is that if homosexuality is in fact biologically based, identical twins who are both gay would be more common than fraternal twins who are both gay.

When researchers tested this theory, they found it was accurate. A 2015 University of California study presented at the American Society of Human Genetics examined the genomes of 47 sets of identical twins. Thirty-seven of the sets had 1 homosexual twin and 1 heterosexual twin. The remaining 10 sets, just over 20% of the study population, all identified as homosexuals.

Those results are quite revealing. For instance, the fact that over 20% of the twin sets had 2 gay twins indicates that there is something biological going on here.

However, not all identical twins mirror each other's sexuality. In twin sets with at least 1 gay twin, a large number had heterosexual counterparts. Therefore, biology is at work, but there's more to it.

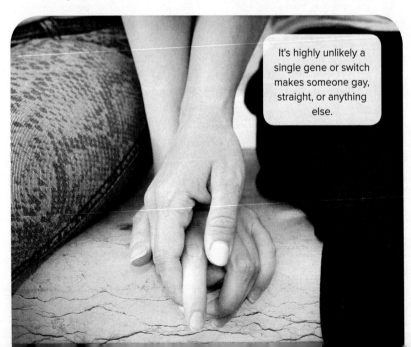

It's highly unlikely a single gene or switch makes someone gay, straight, or anything else.

The most fascinating ideas from this research relate to epigenetics. We have much more to learn before we'll truly understand our epigenetics. But this study of homosexual twins sees that our sexuality may also be influenced by sets of chemical markers that lie between, not within our genes.

The epigenome basically turns our genes on and off in response to a particular moment. Unlike our actual genome, the epigenome is constantly changing. Researchers explored the twin pairs' epigenomes, searching for patterns that could be predictive for sexuality.

In 9 different regions of the epigenome, that's exactly what they found. Their approach appears to be able to biologically identify homosexuality with a 70% accuracy rate.

That study and the hypothalamus studies all point to a common consensus: There are biological foundations of human sexuality. It's important to note that based on today's existing research, it's highly unlikely that there's a single gene or switch that makes someone gay, straight, or anything else. There's something much more complex going on, and we still have much to learn.

Suggested Reading

Harvey, *Almost a Man of Genius*.

Nelson, *Women in Antiquity*.

Peletz, *Gender Pluralism*.

Roughgarden, *Evolution's Rainbow*.

Questions to Consider

1. What specifically is the difference between biological sex, gender, and sexuality?

2. How do mainstream 21st-century ideas on gender and gender identity compare with the way people understood these concepts in Darwin's day?

3. Why are some people born neither fully male nor female?

Sex, Gender, and Sexuality

O ver the past several lectures we've been thinking about human diversity and cultural traditions, and we've seen how today's anthropologists continue to test and correct their understanding of human diversity. We started with cultural evolutionism but we've gotten a lot sharper since those early days of anthropology. And when it comes to the culture & biology question, we've done a complete 180-degree turn. Now, we generally understand biology and culture as distinct facets of our humanity. Well today, we're going to take a similar approach with sex and gender. It's time to use anthropology to see how biological sex, gender, and sexuality are unique threads of our humanity. And we'll start with an amazing true story from Salisbury, CT.

The year was 1843, some 76 years before women's suffrage. In Salisbury, officials were gearing up for a super competitive race when a young man named Levi Suydam registered to vote. Levi was keen on supporting the local Whig party candidate and was beyond excited that he was about to cast his first vote ever. And everything seemed in order because as a male property owner, Levi was entitled to vote. Period. Not so fast.

Doing everything they could to defeat the Whigs, the opposition in this small town didn't go down without a fight. In fact, when they heard of Levi's intent to vote, they protested. You see, the opposition party officials challenged the legality of Levi's vote because, in an era where females were not legally permitted to vote, they accused Levi of being more female than male. Seriously. And the argument intensified until finally, Dr. William James Barry was called upon to settle this matter once and for all. Determined to exercise his democratic duty, Levi consented to a medical examination to legally attest to his maleness, and thus his legal right to vote.

Dr. Barry's examination preserved Levi's right to vote. But things got a little more complicated a couple days later when it was confirmed that despite having male genitalia and a faint interest in sex with women, Levi also had a vaginal opening, and he menstruated. The rest of the story, as featured in Dr. Fausto-Sterling's 2000 book, *Sexing the Body*, is unclear. We don't know whether Levi's vote was counted or not, but we do know the Whigs, the party he supported, won that election by a single vote.

This story starts us on our way because, in it, we see a revealing example of how biological sex can be much more complicated than we may have been led to believe. Dr. Barry, for example, was conflating biological sex, gender, and sexuality. But, not only is our biological sex distinct from gender and sexuality, but we've come to find it isn't even a male/female binary. Like they say on facebook—it's complicated. Well, let's uncomplicate this right now.

So, here's a quick trivia question to get us started, What is the name of the Greek Goddess of Love? Aphrodite. And now for the bonus, who was her one-time lover, the messenger god of Olympus? That was Hermes. Now, as a result of the love affair between Aphrodite and Hermes, Aphrodite gave birth to a child, Hermaphroditus. Now, you can see the combination of the parents' names in Hermaphroditus. But more than that, you can also see the two principles they represent, Hermes, male. Aphrodite, female. In Hermaphroditus, these two principles are merged. In fact, both Greek myth and Greek art portray Hermaphroditus as two-sexed—having a female figure with male genitals. That's where our word hermaphrodite comes from. A hermaphrodite is an organism that has both male and female sex organs.

Now, as we'll see in a moment, hermaphrodites comprise a relatively small category within a larger group of people who, biologically, have both male and female sexual traits. We use the term intersex to describe this group, whose biology, not their sexuality, not their gender identity, whose biology is not exclusively female or male.

We've been taught that there are male and female people out there, but we don't hear as much about intersex people. Remember, we're not talking sexual orientation; we're talking strict physiology, here. And the fact is, some people are born intersex. What do you think the percentage rate is? Well, not everyone agrees on the exact number, but according to Anne Fausto-Sterling, the rate is 1.7 percent and that's a higher incidence than albinism.

Think of your high school or college. How many people were there? 100? 1000? More? Even in a group of 500 people, statistically, we'd expect that eight to ten of our peers would have been born intersex. So why don't we hear about this? Why don't we know more intersex people in our lives?

Well, as I said before, it's complicated. In the US, for example, our cultural and medical practices have classified intersex as a deformity rather than diversity. And as such, it's often the case that when intersex children are born, doctors and parents typically go into emergency mode. They make a decision. Male or female. And shockingly, the decision to go male or female is commonly made immediately based on the presence or absence of a functional penis. So while we might expect that around two percent of the people around us may be on the intersex spectrum, some of them, indeed most of them, will have had so-called corrective surgery.

On the one hand, it's easy to understand where a parent or doctor may be coming from here. Knowing that the world can be a cruel place for folks that don't fit in, parents consenting to corrective surgery are quite likely acting out of protective love for their new child. That said, studies have shown that people who received this sexual assignment surgery are at risk for gender identity issues later in life. Operating on the body can clear up biological ambiguity when it comes to sex organs, but those organs do not control one's sexuality or the gender identity. Rather than reinforcing a non-biological duality of male and female, maybe we should expand our understanding of other sexes?

So what we're seeing here is that we need to challenge the binary concept of biological sex, male and female. We need to broaden that concept to something that is more in line with biological reality. Simply put, we need to add intersex to the spectrum.

Now what I'd like to do is to drill down even further to show that even intersex itself isn't a simple concept. Because under the general term intersex there are various categories based on different biological characteristics.

It was biologist Joan Roughgarden wrote the book *Evolution's Rainbow*, in which we can see that intersex and non-binary sexes are more than common throughout the entire animal kingdom. And, without getting into the weeds here, Roughgarden discusses several types of intersexuality.

First, there's the hermaphrodite, and, despite the fact that this may be the most recognized or familiar form of intersex, it's actually the rarest of them all. One in about 85,000 to be exact. The hermaphrodite usually has some combination of testis and ovaries, and/or what they call ovotestis, which contain both ovarian and testicular tissue.

Then, there's the much more common hypospadias. Hypospadia is considered to be minor of the vent is off-center on the tip of the penis, and to one degree or another, this occurs in over 40 percent of males. But then, around one in 2000 males are born hypospadias major, which means that the vent is anywhere but the tip of the penis.

And even more common that hypospadias is chromosomal variation. This outcome happens about once in 1000 births. These varieties of intersex occur as the result of an additional chromosome so that instead of an XX female, or an XY chromosome male, we get XXY, XYY, triple X, or even double X-double Y.

There are other pathways to intersex, but let's go ahead and close up this first section, which has focused exclusively on biological sex because we can now explain that biological sex isn't really a binary male-female concept.

So let's make the move from our strict biological definition of sex to something more anthropological. Don't get me wrong, biology is a critical piece of anthropology, but unlike biology itself, anthropology garners its strength as a biology plus. So it is our duty as anthropologists to understand biological sex in terms of the physical body, but when we start talking about sexuality and gender, that's an entirely different story. Sexuality and gender are not strictly biological phenomena. And that's why cultural anthropology can help us see the distinct qualities of biological sex as opposed to one's sexuality and gender.

So in that spirit, let's move into cultural anthropology to see how some world cultures actually construct gender in a way that transcends the masculine-feminine binary.

Now, we've seen gender themes in our previous lectures, but when it comes to many of the foundational thinkers whose ideas shape anthropology to this day, people like Darwin, their proclamations on humanity as a single race, didn't technically include women. As we've seen elsewhere in this course, early anthropologists forced an evolution-inspired metaphor that ranked world cultures on the spectrum of savage to civilized. Social Darwinism, as this idea is sometimes referred to, extended well beyond culture grades. For example, as we'll see in a future lecture, some social Darwinists used the theory of evolution to describe different grades of religious belief and practice—from the so-called primitive religions to those that were ostensibly more advanced. But now, we're about to see how, when it came to the question of gender, the glow of evolutionary biology blinded even Darwin himself.

You see, Darwin and his contemporaries may have argued that we're ultimately all one race, *Homo sapiens*. But they were clear that half the human race wasn't built for thinking or anything else outside of the domestic sphere. And yes, you heard me right they were clear that women's role in human reproduction is so physically and mentally demanding that it actually limits their potential as scholars, doctors, and presidents. This isn't a glass ceiling, it was a biological one. And to understand just how deeply entrenched this attitude was, let me introduce you to one of my brilliant science heroes here. Her name is Clémence Royer. Royer was a French scientist who was hired by Darwin and actually eventually as the translator of his masterpiece, *On the Origins of Species*.

You see, as a scientist herself, Royer was the quintessential editor and translator. When she had a disagreement with Darwin, she let it be known sometimes in the form of intellectual discussion, and sometimes in the form of special footnotes for the reader. And that kind of transparent back-and-forth well, that's what built science as a trusted means to get progressively clearer on our big questions, like who are we and where do we come from?

So Darwin fired her. But, she continued her scientific career. And one of the reasons I first came to admire Royer was a remarkable and courageous speech she gave to the Société d'Anthropologie de Paris, in 1874. She was ahead of her times and most of anthropology with her succinct take on gender. Here's what she said,

> Up until now, science like law, made exclusively by men, has too often considered woman as an absolutely passive being, without instincts or passions or her own interests; as a purely plastic material capable of taking any form given her without resistance; a being without the inner resources to react against the education she receives or against the discipline to which she submits as part of law, custom or opinion. Woman is not made like this.

Royer's critique of European science in the late 19th century opened the door to a more comprehensive study of humanity—an anthropology that saw women in their fullness as human beings.

Of course, Royer was herself a European. But interestingly, even the non-Western people that early anthropologists studied could see a problem with this weird Western gender lens, or I should say lack thereof. Here's a British colonial anthropologist quoting an Asante man from West Africa, "The white man," he said, "has dealings with and only recognizes men; we supposed the European considered women of no account, and we know you do not recognize them as we have always done."

It's clear, anthropology reflected the societies that raised anthropologists in the first place. And for US anthropologists, that meant a society that didn't pass women's suffrage until 1920.

Undeniably, western gender traditions strongly separate men's roles from women's roles, thus developing a gender binary that is closely tied to the biology binary we discussed earlier. And when that binary is crossed or made more complex, the system breaks down like we saw in Salisbury, CT. But, as an empirical field of study, anthropology has tested and corrected these ideas, and cultural anthropologists, for one, have found and documented world cultures that don't have a strict male-female gender binary.

So, to get our cultural anthropology on, let's look beyond western conventions on gender. Let's see how several world cultures can help us expand our understanding of this thing called gender. First, let's go to the Zuni Nation of the American Southwest, where scholars like Will Roscoe have studied Native American cultures and the presence of two-spirit people. Practices and norms certainly differed between distinct Native American groups, but the presence of two-spirited people—or something similar—was actually quite common.

Two-spirited people are sometimes referred to as a third or fourth gender because they clearly contradict the male-female gender binary. Typically the two-spirited person is celebrated as a gifted member of society who may perform trades or skills that are associated with the so-called opposite sex.

Now, the literature is clear. Two-spirited people were not necessarily homosexual, but some certainly were. Additionally, these people were not on the down-low. They were not dressing or acting with the intent to deceive fellow community members, rather they were known and admired publicly for possessing both male and female qualities.

For example, among the Zuni, when a potential two-spirited child emerges, they had a ritual that served as a test to discern who this person truly was, regardless of biology. In the literature, there are a number of versions of these tests including one where a bow and and arrow is placed next to a basket, and the young child is asked to pick the one they like the most.

If a young girl, biologically speaking, picks up the bow and arrow, she is celebrated and accepted as a two-spirited person, and she will likely go on with life learning skills traditionally taught to biological males. Some great and celebrated Native American warriors, for example, were, in fact, two-spirited people, that an outsider might mistake for a masculine female.

Writings from explorers show two-spirited people across Native American nations, and they also reveal how European colonists brutally killed many of these people as deviants. Native cultures, on the other hand, respected and fully integrated two-spirited people into their concept of human gender diversity. Like us 21st-century anthropologists, the Native American tradition of the two-spirited people utterly shatters over simplified dyads like the male-female sex and gender binary. And it turns out Native Americans aren't the only ones who see more than two genders among their brothers and sisters.

In India, we have the hijras, and in the Balkans, we have what are called the sworn virgins. Then in the Pacific, we find the fakaleitis of Tonga and even the mahu of Hawaii. And there are others, but, what all these cultures and people have in common is a rather modern separation of biological sex, gender, and sexuality. One's genitalia, among the Zuni for example, did not automatically determine one's gender identity or even their sexuality. So, when we're talking about biological sex, we were strictly speaking about the bodies, stark physiology. But then we turned to gender, which shifted us away from biology and into identity. So now, let's bring it on home, and let's see how sexuality actually fits into the picture. After all, if gender isn't sexuality, and if biological sex doesn't determine sexuality, then what exactly is sexuality?

From an anthropologist's point of view, sexuality is a great research theme because it requires an interdisciplinary, four-field approach one that bridges biology, archaeology, language, and cross-cultural research. We've already used cultural anthropology to look into two-spirited people, so let's go back and start with biological anthropology and its intersection with neuroscience.

One of the big questions that we think about when we explore human sexuality is whether our sexuality is the product of nature or nurture. As for the roots of sexuality, neuro anthropologists tell us that sexuality does, in fact, have biological foundations. Multiple studies confirm statistically significant correlations between human sexuality and several biological indicators, including the nerve cells in our hypothalamus.

Now, the hypothalamus is part of our limbic system, which helps us regulate things like our behavior, our motivations, memory, smell, and even our emotional life. Located at the center of the brain, just above the stem is the hypothalamus, which produces hormones. Now, inside this hypothalamus, we find suprachiasmatic nuclei, what we refer to as SCN. And the SCN, which controls our circadian body rhythm, gets us closer to some of the biological roots of our sexuality.

Research by noted neurobiologist Dick Swaab and others reveal a relationship between SCN size and homosexuality. Simply put, the SCN of homosexual men are actually larger than the SCN in heterosexual men. They're about twice as large in terms of physical size and the total amount of neurons.

So let's try a mini thought experiment. If the SCN of homosexual men are around twice the size of their heterosexual counterparts, how will the SCN in homosexuals compare with heterosexual females? Will it be larger like the homosexual male SCN? Or maybe somewhere between male and female? Those are both good guesses, but they're wrong. Dead wrong. As it turns out, female SCN are much smaller than even heterosexual males, meaning women have smaller SCNs, and therefore fewer SCN neurons.

So that is one window into the neurobiology of sexuality. And similarly, Swaab and others like neurobiologist Simon LeVay also identified another compelling connection between our brains and our sexuality. They identified another nucleus that is understood as the sexually dimorphic nucleus. It's called INAH3, which is short for the third interstitial nucleus of the anterior hypothalamus. Like SCN, the INAH3 is significantly larger in males than females and even larger in heterosexual males than in homosexual males and heterosexual females. So, the INAH3 is smaller in people who are sexually attracted to males.

Last, I'd be remiss if I failed to mention one important fascinating biological connection between sexuality and biology. As I mentioned earlier, the hypothalamus helps us with our olfactory system. So scientists looked into how the hypothalamus responds to the smells of estrogen as found in female urine and testosterone from male sweat. The results were definitive. The hypothalamus of people who were sexually attracted to women, that's heterosexual men and homosexual women, responded to the estrogen. And, the hypothalamus of people attracted to men, including both homosexual men and heterosexual women, they responded to testosterone.

The SCN and ANAH-3, and even the olfactory studies we just discussed help open the door for the biological study of sexuality. And, there's more work to be done if we want a truly more comprehensive view of the biology of human sexuality. Nonetheless, pioneers like Swab and LaVay give us a terrific start.

Now, to cap off our micro-study of human sexuality, let's look at one more intriguing area of sexuality research. Twin studies. Returning to Roughgarden and her classic exploration of sexuality and sex, *Evolution's Rainbow*, we get a great birds-eye view of twin studies that have looked at sexuality. Collectively, these studies ask the central question we've been thinking about in the final third of our lecture, is there evidence for the biological foundations of sexuality?

The idea is simple. They compare identical versus fraternal twins. The thought is, that if homosexuality is in fact biologically based, identical twins who are both gay would be more common than two fraternal twins who are both gay. Makes sense. Well, it's true. When they tested this theory, they found it was accurate. Here are a few striking specifics.

A 2015 University of California study presented at the American Society of Human Genetics examined the genomes of 47 sets of identical twins. 37 of the sets had one homosexual twin and one heterosexual twin. The remaining 10 sets, just over 20 percent of the study population, all identified as homosexuals. And in two ways, that's quite revealing.

First, identical twins are exact genetic copies of each other. So, if sexuality is exclusively biological, aren't we going to see that, for identical twins, homosexuality comes in pairs? It turns out, they do. For over 20 percent of the twin sets, both twins were gay. And that rather high incidence indicates that there is something biological going on here. But, and here's the second piece. Not all identical twins mirror each other's sexuality. When we found twin sets with at least one gay twin, a large number of them had heterosexual counterparts. So biology is at work, but there's something more to it.

The most fascinating ideas coming from this research relate to epigenetics. We have a lot to learn about epigenetics before we're really going to understand them. But this study of homosexual twins sees that our sexuality may also be influenced by sets of chemical markers that lie between, not within our genes. What we know about our epigenome, is that it basically turns our genes on and off in response to a particular moment. Amazingly, unlike our actual genome, the epigenome is constantly changing.

So researchers explored the twin pairs' epigenomes, searching for patterns that could be predictive for sexuality. And in nine different regions of the epigenome, that's exactly what they find. Their approach appears to be able to biologically identify homosexuality with a 70 percent accuracy rate. Now that's a lot more accurate than just random guessing. Again, that study as well as the hypothalamus studies, all point to a common consensus. Without a doubt, there are biological foundations of human sexuality, but what's important to stress is that we're talking foundations with an "S," plural. Why? Because based on existing research, at this point it's highly unlikely that there's a single gene or switch that makes someone gay, straight, or anything else. Eye-color, sure, but not sexuality. There's something much more complex going on, and we still have a lot to learn.

So, what's our big take-home lesson today? Where exactly does our exploration of sexuality, gender, and biological sex leave us? Well, it appears that over the long arch of the human experience, homosexuality and femininity are no more pathological than heterosexuality and masculinity. They are all expressions of humanity.

Like gender and biological sex, diverse sexual identities are manifestations of the rich cultural and biological heritage we inherit and then pass on. And it's rather comforting to know that just about any cohabitation arrangement and sexual preferences among humans, well, we can find parallels in the wider animal kingdom. We're not alone. And we're not crazy. We're just upright, walking, and talking apes.

That is, walking and talking apes who understand and connect with each other through biological sex, gender, and sexuality, among other things. Distinct, yet certainly intertwined threads of our humanity, these three concepts help us see, and maybe even practice, much more inclusive expressions of human diversity. And, as we've seen so often before, when we explore the diversity of humankind, we also discover its unity. The unity of our human race.

LECTURE 17

Religion and Spirituality

nthropology is the study of humankind. If we want to understand humanity, there are few topics more important than religion. After all, religion is one of the ways we make sense of our lives, our universe, and even our morality. And while it is true humans have used religious differences to exact horrific violence and emotional pain upon each other, religion is also the source of unfathomable love and belonging in the form of selflessness and communion. This lecture takes an anthropological look at religion. The lecture traces its origins and examines how anthropologists have changed some of their thinking on religion.

Animal Grief

Ritual burials, feasts, cave art, and even primatology provide a fairly comprehensive glance at the origins and development of the human religious experience.

Modern apes, including nonhuman apes, express empathy and mourn the loss of others. Biological anthropologists like Barbara King argue that the empathy and mourning we observe in nonhuman apes gives us insight into the roots of the human religious experience.

One example is a famous gorilla named Koko. Koko is a special gorilla who is uniquely suited to bond with her human caretakers, as she has mastered American Sign Language. She showed grief when her companion gorilla passed in the year 2000. She had a similar reaction when she learned of the death of the comedian Robin Williams, with whom she had a bond.

The compassion, empathy, and mourning we see in apes like Koko give us great clues about the foundation, or early roots, of the human religious experience. Empathy and mourning are not religion in themselves. But over

generations and generations, they can evolve into what we recognize as modern world religions.

Parietal Art

Parietal art is early art found on cave walls or rocks. Dating back tens of thousands of years, the world's parietal art collectively documents our pre-modern religious life.

One example is a mysterious supernatural image of a therianthrope found in southern France at a site named Trois-Frères. Dating back to around 13,000 B.C.E., the therianthrope is half man, half animal. His finely detailed figure with muscular legs and antlers dominates the large cavern, known as the sanctuary, that houses it. It was likely used for shamanistic rituals and trances.

Archaeologists call this image the sorcerer. It's one of our earliest images of the pre-modern religious experience.

Parietal art is a great window into pre-modern religious life, and it shows that early *Homo sapiens* practiced shamanism. They contemplated alternative universes and what happens after death.

Burials

Archaeological evidence supports the theory that humans, and even Neanderthals, were thinking of the great beyond and doing purposeful burials as early as 90,000 years ago.

In Israel's Qafzeh cave, archaeologists found bones and fragments from over 24 burials. Nearby, on Mount Carmel, archaeologists discovered purposeful burials dating over 100,000 years ago.

The most famous Neanderthal burial is called the Old Man, and he was found purposely buried by others. Because he was old, frail, and essentially toothless, his remains show us that he had a small community who must have cared for him in his old age and upon death.

Modern World Religions

Around 3000 B.C.E., we begin to see archaic religious traditions and rituals emerging in Mesopotamia, then Egypt, China, and eventually in

Greek, Celtic, and Native American cultures.

Determining start dates for religions can be tricky. Some religions, like Judaism, can be considered to have a more specific date such as the year 1812 B.C.E. That was when Abraham made his covenant with God.

But other religions, like Hinduism, have a more nebulous starting point unless we pick an important date like the creation of the Upanishads text around 700 B.C.E.

Around 2500 years ago, humankind began to develop one religion after another. Jainism emerged around 600 B.C.E., with Zoroastrianism not long after that. Buddhism arrived right around 400 B.C.E.

Christianity emerged around the year 30 C.E. Islam came around 600 years later. More recently, Sikhism began in the 1400s, followed by the Baha'i and Mormon faiths in the 1800s. In the 20th century, we developed new religious traditions including Rastafarianism and the Jehovah's Witnesses.

By the time Edward Burnett Tylor and anthropology emerged in the late 1800s, the work of 2 scholars forged the foundation for what later becomes the anthropology of religion.

Émile Durkheim

The anthropology of religion essentially starts with the sociologist Émile Durkheim. In one of his most important works, *The Elementary Forms of The Religious Life*, Durkheim explains religion as a social phenomenon that hinders our selfish, individualistic proclivities, and instead promotes social cooperation.

For Durkheim, religious symbols used during religious rituals help reinforce this collective consciousness and cooperation. And as a result, individuals are dependent on society, as they are on God.

Durkheim points out that societies classify all things into 2 piles: the sacred or the profane. The sacred is the social, the ideal, and the divine; the profane is the personal, physical, and earthly realms. Ultimately, in any religion, rituals help promote and prescribe that which is sacred, and thus allow us

Buddhism arrived right around 400 B.C.E.

the opportunity to dwell within the sacred ourselves.

James Frazer

In 1890, James Frazer published another foundational text on the anthropology of religion: *The Golden Bough*. This immensely popular book was a compendium of accounts of world religious practices and knowledge.

Frazer's big idea was that archaic religions were basically fertility cults, and they provided opportunities to worship and make periodic offerings or sacrifices to a sacred king.

He also concluded that the great arc of the human religious experience takes humanity from an initial magical phase, through to religious belief, and then culminating in scientific thought.

Origins of Evolutionism

Frazer, like Tylor and many of his contemporaries, were cultural evolutionists who were keen on studying so-called primitive religions. That's because they were searching for clues about the origins of religion as a sociocultural phenomenon.

In their search for those origins, they went with cultural evolution or social Darwinism. To cultural evolutionists, the monotheism of the Abrahamic faiths was the most advanced form of religious belief. Other religions lagged behind at a lingering evolutionary pace.

According to these folks, advanced religions focus on texts, while primitive religions are oral traditions. In addition, advanced religions focus on the afterlife and are universal in scope. They apply across cultures, unlike primitive religions, which are more culturally specific.

Finally, advanced religions compartmentalize the sacred and profane, but in primitive religions, religion and daily life are organically intertwined and inseparable.

Tylor was in line with that school of thought on the evolution of religion. The racial context of cultural evolutionism tarnishes Tylor's legacy as one of the earliest anthropologists, but he is still widely regarded as a pioneer in the anthropology of religion.

Tylor assumed that the stages of humanity's material "advancement" also corresponded with parallel stages of our spiritual nature. Oddly, Tylor himself had interests outside the confines of conventional monotheism. He didn't write about this publicly, but he was into the Spiritualist movement, intrigued by claims of their evidence that the human personality continues after death.

Functionalism and Cultural Relativity

Tylor and cultural evolutionists weren't the only people investigating world religions. Some rising anthropology stars, like Franz Boas and Bronisław Malinowski, emerged around the turn of the 20th century. They refused to accept the pseudoscience of social Darwinism and the cultural evolutionists.

These men and their scores of students were a new and highly influential wave of cultural relativists. And they were curious about sociocultural functions of religion, very much in the Durkheim tradition.

In a 1925 essay on religion, Malinowski distinguishes the difference between magic and religion. Drawing on Frazer's magic-religion-science discussion, he says that magic is always utilitarian, and magic rituals are generally expected to get specific results, like healing a sick stomach.

Religion was different. Specifically, religious rituals usually focus on more ambiguous results. People don't take communion or fast for a month because they expect a priest cure their ills.

Malinowski and the cultural relativists did long-term participant-observation field research. Living day-to-day with their research communities, the cultural relativists ended up paying much less attention to the origins of religion. Instead, they sought to unveil the inner mindset of the people, including the practical and rational nature of their religious life and rituals. For them, the anthropology of religion was about deciphering the sociocultural function of rituals and religious life.

Cultural relativity in the anthropology of religion endured well into the 20th century, with students of Malinowski and Boas taking their questions to new regions and new people.

One Malinowski student who worked in Africa among pastoralists and farmers, Edward Evans-

Pritchard, focused on social systems. He looked into local religious practices of his East African study communities.

When he did, he broke away from the primitive/modern lens. Instead, he portrayed the rational and functional dimensions of religious practice. His main argument was that for the Azande and Nuer people, as well as any other group, rituals sustain and moderate social structures.

Interpretivist and Feminist Approaches

An emerging interpretivist critique of strict empiricist anthropology stemmed from the idea that regardless of how long an anthropologist does field research, he or she will never get a full and truly authentic picture.

These critiques ushered in yet another way for anthropologists to explore and understand religion. The new phase often focused inward rather than outward, and its holy grail was the search for meaning: symbolism and interpreting myths.

Two preeminent figures, Claude Levi-Strauss and Clifford Geertz, took this less empirical approach to understanding world religions and the religious experience in general.

Levi-Strauss explored the idea that universally, humans are wired to see and understand our world through binaries like raw versus cooked, wild verus civilized, and sacred versus profane. Religion, he wrote, works to mediate these binaries.

Clifford Geertz's anthropological definition of religion summarizes the way many anthropologists approach religion in their work today. Simply put, Geertz says that religion is how we make meaning. To paraphrase: Religion is a system of symbols, which motivate and emotionally resonate with people. It does this by articulating a general order of existence that gives meaning to our lives.

Beyond Levi-Strauss and Geertz, scholars continued to look at the structural and symbolic dimensions of religion. Researchers have developed even more ways to anthropologically cast even more light on humanity's diverse religious practices or beliefs, despite the structural unity of the human religion experience.

One example: Some feminist approaches look at religion as a social institution with gendered relations of power. Feminists have also enriched the anthropology of religion by examining female deities and goddess cultures in pre-Judeo-Christian times.

Suggested Reading

Frazer, *The Golden Bough*.

King, *Evolving God*.

Lowie, *Primitive Religion*.

Questions to Consider

1. What are some of the roots of the human religious experience, and when did it emerge in the evolution of our species?

2. How did Boas and Malinowski explain the existence of religion, and how did their ideas differ from what earlier theorists like E.B. Tylor and Morgan first said?

Religion and Spirituality

Anthropology is the study of humankind. And surely, if we want to understand humanity, there are few topics more important than religion. After all, religion is one of the ways we make sense of our lives, our universe, and even our morality, let alone our quotidian behavior. And while it is true humans have used religious differences to exact horrific violence and emotional pain upon each other, it's also the source of unfathomable love and belonging in the form of selflessness and communion.

So today, we're going to take an anthropological look at religion. We'll trace its origins and see how anthropologists have changed some of their thinking on this thing we call religion.

To begin our survey of religion and spirituality, let's travel first to Mali, where I do anthropology field work. Listen to this vignette, and I want you to think about what you might have done if you had been in my shoes.

Now, remember, when you're in the field as an anthropologist, you're not in a lab. You're in a community and that means that the fieldwork experience is riddled with moments that teach you just as much about yourself as the people you came to study. So, imagine you're an anthropologist. You've been in the field for a month, and you're finally getting the flow of the language. Every day, you've been working alongside your hosts, harvesting sorghum. But today is special. Today it's Ramadan, and the entire community is vibrantly dressed for the occasion. And, as a sociable and informal procession passes along the dirt path in front of your house, you rush to put on that new newly tailored outfit you made just for this day. Grabbing your camera, you run out the door to join your host family and everyone works their way to the edge of the village for a special outdoor prayer service.

You arrive to find all the males seated on prayer mats in neatly organized rows. Behind them the women and their mats. And taking up the very rear, it's the children, and they're seated in pairs on large palm leaves. And before you know it, the service begins with praying. Fortunately, you're seated on a mat with one of your new friends, Isa. He grabs your hand and brings you to your feet with a concerned yet reassuring smile. It's time to pray.

Now, remember, when I was in this situation, I came to it as a non-Muslim. But that shouldn't matter, I tried to assure myself. I was an anthropologist doing participant observation, and everyone understood that. They knew that my

M.O. was to live and participate in the daily life of this community for the next couple years. I worked when they worked. I slept when they slept. And we all ate and socialized together.

So, as my friend grabbed my hand and lifted me up to start the prayers, I started to watch and mirror Isa. But then panic set in. I got way too reflective. I simply couldn't determine whether or not I should be praying like this. I mean, despite being a non-Muslim, I've always been in the camp that, when it comes to religion, there are many roads to the summit, but only one destination.

My panic ensued because I didn't want to offend people, and I didn't know the best way to do that. One the one hand, people could interpret my non-believer praying as disrespectful. But then, would they be even more insulted if I sat off to the side and didn't participate at all? I freaked because, ultimately, I was already attached to this community, and I really didn't want to mess up what was a fairly smooth transition into my new village home.

So, if you were a non-Muslim and, if you were there in my shoes, what would you have done? Would you just go with it, and mirror Isa as you work out what to do? Or instead, do you feel so awkward about this entire situation that you make-up a polite excuse like you're not well? Maybe you just whisper the truth to Isa and ask him if it's OK? Well, as a cultural relativist who didn't feel that praying with my hosts violated my own spirituality, I went with the flow.

Now, if you would have made a different choice, that doesn't mean you're a terrible anthropologist. As the anthropologist, it's simply up to you to work out your boundaries, so you and your hosts know and agree, when participant observation is inappropriate. My response to this early fieldwork episode gives you a vicarious preview of how anthropologists think about religion. We study religion to study humans, not God and the great beyond. But, what is it that distinguishes anthropological approaches to religion as opposed to say religious studies, for example? Well, that's exactly what we're going to work out today. And, first on our list let's start with the origins of religion.

Ritual burials, feasts, cave art, and even primatology provide a fairly comprehensive glance at the origins and development of the human religious experience. Modern apes, including non-human apes, express empathy and mourn the loss of others. And biological anthropologists like Barbara King, argue that the empathy and mourning we observe in non-human apes can actually give us insight into the roots of the human religious experience.

Whenever I think of apes and their ability to express empathy, my thoughts go straight to a famous gorilla named Koko. Koko is a special gorilla who is uniquely suited to bond with her human caretakers and others. Why? Well, because Koko has mastered American Sign Language understands human speech. Dubbed an experiment in interspecies communication, Project Koko started in 1972 when a psychology student named Penny Patterson chose to teach this young gorilla sign language for her Ph.D. at Stanford. And, long story short, Koko's voice has revolutionized how we think about our ape relatives.

I'm sharing Koko's story because we're seeking out evidence that other apes do in fact experience a sense of loss upon the death of an acquaintance, not to mention empathy more broadly.

It was in the year 2000, when Koko's companion gorilla of some 24 years, Michael, passed away, and it emotionally destroyed her. Crying late into the night, she started requesting a night light, and during the day, she sat staring into space, chin down and bottom lip out. She even stopped purring when saw friends or received food. But then, less than a year later, the famous comedian Robin Williams visited Koko as part of this documentary film project, and wow, those two hit it off. Koko's caregivers remember that after months of despondent sorrow, Robin's visit changed everything. He had restored Koko's gleeful smile.

But, the story of Koko's empathy and capacity for mourning doesn't stop there. Sadly, the day Robin Williams passed, calls came pouring in at the Gorilla Foundation. And because Koko recognizes speech, she figured out something was terribly wrong. Dr. Patterson cried as she explained that Robin was gone, and Koko signed back an empathetic, "Cry, no." Patterson reported that Koko resumed the somber posture and quivering bottom lip. Years after Michael passed on, Koko had lost another dear friend.

Now, we've seen elephants, for example, express extreme grief upon the death of a child or family member, but what's remarkable about Koko is that she talks with us, using words about her subjective experience of loss. And that's an important first step for us, because, as biological anthropologist Barbara King argues, the compassion, empathy, and mourning we see in apes like Koko give us great clues about the early foundation, or early roots, of the human religious experience.

So we're not saying that empathy and mourning are religion, but what we are saying is that those experiences and emotions, over generations and generations, can evolve into what you and I recognize as modern world religions. To see the origins of what we recognize as religion. It's time we move into archaeology and parietal art.

In *Fields of Blood*, a book on the histories of world religions and their relationship with violence, former nun Karen Anderson explains that, "From the first, one of the major preoccupations of both religion and art—the two being inseparable—was to cultivate a sense of community—with nature, the animal world, and our fellow humans."

So let's look to some of the earliest art on record, in search of some evidence of religious activities or imagery. And indeed that's exactly what we find. Now, parietal art, as we call it, is the term we use to describe early art found on cave walls or rocks. And dating back tens of thousands of years, the world's parietal art collectively documents our pre-modern religious life.

One of my absolute favorite archaeological treasures is a mysterious supernatural image of a therianthrope found in southern France at a site named Trois-Frères. Dating back to around B.C.E. 13,000, the therianthrope is half man/half animal, and his finely detailed figure with muscular legs and antlers dominates the large cavern that houses it. Dubbed the sanctuary, it was likely used for shamanistic rituals and trances.

Archaeologists call this image the Sorcerer or the Horned God. And it's one of our earliest images of the pre-modern religious experience. Not far from the sanctuary, is another small cavern known as the Chapel of the Lioness. This image is engraved into the wall, and it depicts a lioness on an altar of sorts, and remarkably, the people who visited this image placed special offerings or objects into the wall below the lioness where archaeologists discovered all kinds of animal teeth, stone tools, and shells. Parietal art is a great window into the pre-modern religious life, and it shows that early *Homo sapiens* practiced shamanism, and they contemplated alternative universes as well as what happens after death.

The details of the parietal art at Trois Frères, are terrific because they show us some early precursors to modern world religions, and we can see evidence of rituals connecting people to a greater reality and to each other.

But, other archaeological evidence supports the theory that humans and even Neanderthal were thinking of the great beyond and doing purposeful burials as early as 90,000 years ago. That was in Israel's Qafzeh Cave where archaeologists found bones and fragments from over two dozen burials. And nearby, over on Mt. Carmel, we discovered purposeful burials dating to just about over 100,000 years ago. That's right around the time we transitioned from the archaic form of *Homo sapiens* into our present state as *Homo sapiens* modern.

So, maybe it's not so surprising that our now extinct Neanderthal cousins were also burying their dead. The most famous Neanderthal burial is called the Old Man, and he was found purposely buried by others. And knowing he was old, frail, and essentially toothless, his remains show us that he had a small community who must have cared for him in his old age, and upon death.

As humans continued burying their dead and seeking answers for the mysteries of life and death, we see compassion leading to purposeful burials and explorations into what lies beyond. And it's parietal art that gives us a glimpse of that some 10,000-plus years ago but then, along the arc of the human religious experience, we start seeing more and more archaeological evidence of ritual objects, such as the nearly 4,000-year-old figurines discovered in Peru at the Vichama site. Coming in at nine inches, the tallest of the three appears to be a priestess with a beaded necklace, long, dark hair, and a face painted white peppered with red dots. Archaeologists found the statues facing each other in a basket, and they believe the figures were actual ritual offerings for the construction of the building in which they were housed.

So with that glimpse into our pre-modern religious sensibilities, things like compassion, purposeful burials, mourning, shamanism, and even ritual objects, in some ways, the roots of the human religious experience can be found in archaeology. So let's our eyes to what we might consider full-fledged religion. When do the major modern world religions emerge?

Now, around B.C.E. 3,000 we begin to see archaic religious traditions and rituals emerging in Mesopotamia, and then Egypt, China, and eventually in Greek, Celtic, and Native American cultures too. Start dates for religions can be tricky because, like Christianity and Judaism, the history of the latter is in some ways the history of the former. Some religions like Judaism can be considered to have a more specific date such as when in the year B.C.E.1812 Abraham made his covenant with God. But other religions like Hinduism have a more nebulous starting point unless we pick an important date like the creation of the Upanishads text around B.C.E. 700.

So yes, we see Judaism emerge as one of the first world religions, and then, once we get to around two-and-a-half thousand years ago, something happens all around the world as we see humankind developing one new religion after another. We get Jainism around B.C.E. 600 with Zoroastrianism not long after that. And then we see Buddhism emerge right around B.C.E. 400. And after that, we have another burst of new world religions when we get Christianity around the year C.E. 30. Then Islam arrives around 600 years after that. More recently, we see Sikhism begin in the 1400s, followed by the Baha'i and Mormon faiths in the 1800s. And humanity hasn't stopped there. In the 20th century, we developed new religious traditions including Rastafarianism, Jehovah's Witness, and even the Saint John Coltrane Church in San Francisco.

In sum, up to around 1400, most of Europe was in the dark ages, while China, Africa, the Arab World, India, and parts of the Mediterranean thrived. That explains the early rise of world religions in the Middle East and Asia. They had the art, technology, scholars, and trade routes to both build and spread religion. But then, Europe rose from the 1400s on, as sailing technologies, the printing press, and the Protestant Reformation ushered in the Enlightenment. Thereafter, the Atlantic Slave Trade and the expansion of global empires only further fueled this rapid growth.

And that's when Anthropology comes in, connecting new corners of the world that hadn't yet touched. Anthropologists were making sense of a world full of religious rituals and practices among other things. So let's see what they found. Let's see how anthropologists have deconstructed and made sense of the world's religions.

By the time Edward Burnett Tylor and Anthropology emerge in the late 1800s, the work of two scholars forged the foundation for what later becomes the anthropology of religion. Now, borrowing from our sociological cousin Émile Durkheim, the anthropology of religion essentially starts with him. In one of his most important works, *The Elementary Forms of The Religious Life*, Durkheim explains that religion is a social phenomenon that hinders our selfish, individualistic proclivities, and instead, promotes social cooperation.

For Durkheim religious symbols used during religious rituals help reinforce this collective consciousness and cooperation. And as a result, individuals are dependent on society, as they are on God. And Durkheim points out that societies—and this is religion at work here—classify all things into two piles, there is either the sacred or the profane. The sacred is the social, the ideal, and the divine; whereas the profane is the personal, physical, and earthly realms.

So from this sociological pioneer, we begin to discover the mechanics of modern world religions. Ultimately, in any religion, rituals help promote and prescribe that which is sacred and thus allow us the opportunity to dwell within the sacred ourselves.

In 1890, James Frazer published a second foundational text on the anthropology of religion, *The Golden Bough: A Study in Magic and Religion*. This immensely popular book was a compendium of accounts of world religious practices from all around the world. And Frazer's big idea was that archaic religions were basically fertility cults, and they provided opportunities to worship and make periodic offerings or sacrifices to a sacred king. He also concluded that the great arc of the human religious experience takes humanity from an initial magical phase, through to the religious belief, and then culminating in scientific thought.

Frazer, like Tylor and many of his contemporaries were cultural evolutionists who were keen on studying so-called primitive religions. Why? Well, it's because they were searching for clues about the origins of religion as a sociocultural phenomenon. They wanted to explain when and how religion begins. And, in their search for those origins, they went with cultural evolution or social Darwinism. You see, they were sure that traditional religions in remote corners of Africa and the South Pacific were living windows into the primitive pre-religious lives of humankind. To these cultural evolutionists, the monotheism of the Abrahamic faiths was the most advanced form of religious belief, behind which other religions lagged at a lingering evolutionary pace. So according to these folks, what's the primary difference between primal religions versus these advanced world religions?

Well, advanced religions focus on texts, while primitive religions are oral traditions. In addition, advanced religions focus on the after-life and are quite universal in scope. They apply across cultures unlike primitive religions, which are more culturally specific. And last, advanced religions compartmentalize the sacred and profane but, with primitive religions, religion and daily life are organically intertwined and inseparable.

So, that's the prevailing thought on religion and religious diversity, or at least it was at the dawn of anthropology. And EB Tylor was right there with the evolution of religion. The racial context of cultural evolutionism tarnishes Tylor's legacy as one of the earliest anthropologists, but he is still widely regarded as a pioneer in the anthropology of religion. Tylor assumed that the stages of humanity's material advancement also corresponded with parallel stages of our spiritual nature. Oddly Tylor himself had interests outside the confines of conventional monotheism. He didn't write about this publicly, but he was really into the spiritualist movement, intrigued by claims of their evidence that the human personality actually continues after death.

So, Tylor represents for us the moment in anthropology when primitive religions were all the rage. We'll label this era the origins era, as evolution-oriented questions were at the heart of anthropological inquiry, especially in Tylor's day. But Tylor and cultural evolutionists weren't the only people investigating world religions. Some rising anthropology stars, like Franz Boas and Bronislaw Malinowski, emerged around the turn of the 20th century, and, as we've seen in earlier lectures, they refused to accept the pseudo-science of social Darwinism and the cultural evolutionists. These gentlemen and their scores of students were a new and highly influential wave of cultural relativists.

And they were curious about sociocultural functions of religion, very much in the Durkheim tradition.

In a 1925 essay on religion, Malinowski distinguishes the difference between magic and religion. Drawing on Frazer's magic-religion-science discussion magic, Malinowski says, is always utilitarian, and magic rituals are generally expected to get very specific results, like healing a sick stomach.

But religion, on the other hand, was different. Specifically, religious rituals usually focus on more ambiguous results. You don't take communion or fast for a month because you expect the priest himself to cure your stomach ache or whatever else ails you. It's more like the God moves in mysterious ways kind of thing.

Malinowski and the cultural relativists didn't just read up on comparative religions from their armchairs. No, these science-minded scholars left their desks and did long-term participant observation in field research. And living day to day with their research communities, the cultural relativists ended up paying much less attention to the origins of religion. Instead, they sought to unveil the inner mindset of exotic people, including the practical and rational nature of their religious life and rituals. For them, the anthropology of religion was about deciphering the sociocultural function of rituals and religious life.

Now, cultural relativity in the anthropology of religion endured well into the 20th century, with students of Malinowski and Boas taking their questions to new regions and new people. One Malinowski student who worked in Africa among pastoralists and farmers, Edward Evans-Pritchard focused on social systems so naturally, he looked into local religious practices of his East African study communities. When he did, he broke away from the primitive/modern lens. Instead, he portrayed the rational and functional dimensions of religious practice. His main argument was that for the Azande and Nuer people, as well as any other group, rituals sustain and moderate social structures.

Another Malinowski student, Alfred Radcliffe-Brown, collected field data that helped him argue that myths maintain and reinforce social structure and order. After all, especially among the Andaman Islanders that he studied with, people need group solidarity in order to survive, and it's religious rituals that generate this solidarity. Uniquely, Radcliffe-Brown analyzed myths using binaries like the sacred and the profane, and he inspired a whole new way to do the anthropology of religion.

An emerging interpretivist critique of strict empiricist anthropology stemmed from the idea that regardless of how long an anthropologist does field research, he or she will never get the real picture, a full and truly authentic picture. These critiques ushered in yet another way for anthropologists to explore and understand religion. The new phase often focused inward rather than outward, and its holy grail was the search for meaning. Symbolism. Interpreting myths.

Now, we don't have enough time to go deep into this interpretivist school, but two preeminent figures, Claude Levi-Strauss and Clifford Geertz, can help us quickly work out this less empirical approach to understanding world religions and the religious experience in general.

Like Radcliffe-Brown and Durkheim, Levi-Strauss explored the idea that universally, humans are wired to see and understand our world through binaries like raw and cooked, wild and civilized, let alone sacred and profane. And he says it's religion that works to mediate all of these binaries.

So, rather than an evolutionary lens to explore religion, Levi-Strauss asserted that so-called primitive and modern religions, well, structurally speaking, are the same. We're both structurally working out the ultimate binary, what is sacred, and what is profane? And putting aside the surface-level content of religions, Levi-Strauss focused on the structure of their myths and stories, among other things because that structure reveals the social, political, economic, and even the cosmological dimensions of humanity.

Similarly, Clifford Geertz also used a structural approach. In fact, his anthropological definition of religion summarizes the way many of us approach religion in our work today. Simply put, Geertz says that religion is how we make meaning. To paraphrase, religion is a system of symbols, which motivate and emotionally resonate with people And it does this by articulating a general order of existence one that gives meaning to our lives and to all existence.

Beyond Levi-Strauss and Geertz, scholars continued to look at the structural and symbolic dimensions of religion, and researchers have developed even more ways to anthropologically cast even more light on humanity's diverse religious practices or beliefs, despite the structural unity of human religion experience.

One brief example, feminist approaches look at religion as a social institution with gendered relations of power. Feminists have also enriched the anthropology of religion by examining female deities and goddess cultures in pre-Judeo-Christian times. I'd love to dive deeper into that, but it's time we wrap things up.

Now, today, we've seen, that there are plenty of reasonable ways to define and study religion even amongst just us anthropologists. Nonetheless, we can summarize the past century or so of the anthropology of religion into three major streams. There's the functional approach of Boas and Malinowski that looks at the roles religion plays in society. Then there's the analytic approach of Evans-Pritchard that asks how religion is manifested in society? And last, we get the more essentialist school of Levi-Strauss and Geertz, and they analyzed the basic nature of religion itself.

So to wrap this up, religion can be a difficult topic to discuss because of its primacy in our lives. But, with anthropology, we get traction on this slippery slope, because, our quest is not to improve our understanding of God. We're not out to prove or disprove the existence of anyone's God. This isn't metaphysics here. Instead, we study religion to understand humankind.

So the next time you may find yourself having difficulty relating to someone else's religious practices and beliefs, take a moment and try to connect with them anyway. Try out a little anthropology of religion without that fear of trespass that I encountered when I started my fieldwork in Mali. And remember, you don't have to abandon your own faith or lack thereof to embrace and understand our common humanity. All you need is just a little anthropology.

Art and Visual Anthropology

Whether we're aware of it or not, our cultural background impacts the way we see and understand the world. And nowhere is this fact more clearly on display than when we compare the visual arts of various cultures around the globe. This lecture does some of that comparative work. We'll start by asking: What is art? Then, we'll see how, in the 20th century, new perspectives emerged to challenge outmoded assessments of non-Western art. Next, we'll move to the work of visual anthropology and the history of ethnographic films to see how video technologies can both support and complicate anthropological missions.

What Is Art?

There is no clear line distinguishing art from non-art. For some, art is about technical prowess, while for others, it's about imagination. And then there are others still, who might focus on message or level of abstraction.

The question of what art is has only grown thornier as anthropologists have expanded our cultural horizons. Over the 20th century, Western anthropologists and art historians grappled with the way they should categorize non-Western art.

When anthropologists and others first examined the art of the ancient Aztecs or the art of Africa more generally, they labeled it "primitive" art. But is this is a helpful way to look at non-Western art? Consider the *ciwara* sculpture that is a unifying symbol among the Bamana people of West Africa.

The *ciwara* depicts an antelope, which is carved out of wood and features intricate details in the horns, face, body, and mane. Some brilliant *ciwara* pieces were made recently by sculptors who live in Bamako, the capital city of Mali. These artists are modern, not primitive, and so are their sculptures.

But there was a time when anthropology identified African art and the *ciwara* as primitive art, whether it was produced 500 years or 1 day ago. Early anthropologists like Edward Burnett Tylor viewed Africans as primitive humans, so that's how they categorized African art and cultures.

The Mid-20th Century and Art in Anthropology

Things changed around the middle of the 20th century, when new generations of anthropologists staked new territory on the interpretivist or symbolic side of the discipline. Rather than testing and correcting theories built on materialist, empirical data, these anthropologists embraced the immeasurable interpretive dimensions of the human experience.

With an eye for interpretation, symbolic anthropologists like Claude Levi-Strauss and Victor Turner embraced artistic expressions as treasure troves of cultural insight and data. In his classic book, *The Savage Mind*, Levi-Strauss described artists as hybrids that are part scientist and part craftsman—or what he calls a bricoleur. The craftsman builds a material object, but that object is an object of knowledge. It contains an entire worldview.

In an ever more connected world, people in the mid-20th century were more exposed to non-Western art and cultures than those who lived just a few decades before. And this art captured the imagination of the West. People considered an African mask a look into Africa and Africans themselves.

Pablo Picasso developed his cubist perspective in large part through his appreciation and emulation of African masks and sculpture. African masks plainly have ears, eyes, and a mouth. They clearly represent a person, or perhaps an antelope. But they aren't like any real-life person or antelope; for instance, they might have elongated noses or sharp points for ears.

Viewed from the side, or in different light, these masks transform. It's almost like the carver

African masks

created the mask using more than 1 set of eyes. That's essentially the inspiration for Picasso's cubism: All those strange shapes and forms are multiple perspectives, visualized at once.

The Late 20ᵗʰ Century

The extinction of the term *primitive art* occurred as anthropologists continued to expand the ways they looked at art. New research themes emerged to look at the sociocultural context of art, the politics of representation, and even the repatriation of art to places like Mali, Egypt, and Greece.

The sociologist Howard Becker was a leading figure in the movement to understand art as a sociocultural process, not just as objects produced by artists. In one of his most celebrated texts, *Art Worlds*, he reframes art as a cooperative network of everyone who participates in the production of art. There's the artist of course, but there are also the people who make and sell art supplies, dealers, critics, consumers, installation experts, business managers, and so on.

Recording an event may change the behavior of the people taking part in the event.

With social Darwinism subsumed by cultural relativity, anthropologists found it no longer tenable to classify non-Western art as primitive. The *ciwara* sculpture isn't more primitive than the Mona Lisa; it's just from a different art world.

Visual Anthropology

Visual anthropology is an inclusive specialization that examines humanity by studying performance, media, and media technologies. Perhaps the most well-known aspect of visual anthropology is ethnographic photography and film—doing anthropology with cameras.

Since the early days of anthropology, researchers have used still and moving images to document and analyze culture. And the evolution of ethnographic film, aided by remarkable developments in sound and camera technology, inspired anthropologists to incorporate progressively more and more of the emic perspectives of their research populations.

Film captures body language and the environment surrounding whatever is being filmed—whether that's a ritual, a performance, or any other aspect of daily life. These are details we don't get from interviews, surveys, and other anthropological methods.

One caveat: The presence of a camera can change a person's behavior. For example, someone might put more effort into their appearance if they know a camera is going to be around. This is called cinestance.

Nanook of the North

Nanook of North, from 1922, was the first major anthropology film, and it helped pioneer the documentary film genre. The filmmaker, Robert Flaherty, filmed episodes from the daily lives of Arctic people, and audiences loved this rare and close-up glance into Inuit life.

Modern critics, however, have critiqued Flaherty for leaving out all signs of modernity. Despite the presence of his equipment, not to mention modern conveniences at the local trading post, Flaherty chose to show only scenes that portrayed an illusion of the pristine and so-called primitive state of Inuit people living off the frigid land.

This was before the era of documentary film, so it's not surprising that Flaherty violated a few of the cardinal rules that would later define the genre. For example, the scenes and people he shares with the audience aren't entirely authentic. He cast the film.

The main star, Nanook, is indeed a hunter, and we see him harpooning a seal, wrestling with a walrus, and building remarkable igloos. But his loving wives and adorable children weren't his own. Flaherty staged the family and their activities to capture Inuit life. But he exaggerated the remoteness and the pre-modern qualities of their daily lives.

Flaherty was open about this, and his staged scenes mesmerized audiences nonetheless. It seems that Flaherty avoided the cinestance problem by recruiting actors to film scenes of daily activities and family life in an igloo.

The Ax Fight

Flaherty's approach to visually recording and presenting Inuit life inspired countless documentary filmmakers, among many

anthropologists. One instance: Napoleon Chagnon and his film partner Timothy Asch left a camera on during a fight they witnessed among a band of South American forest dwellers.

They edited the footage into an anthropological film titled *The Ax Fight*, which is now shown in hundreds of anthropology classes every year. Unlike Flaherty, Chagnon and Asch didn't stage an ax fight with local actors. Instead, the camera just sits there, drawn on a confusing scene.

The year was 1971, and Chagnon and Asch had just arrived in their Amazonian host village. A fight broke out right before their eyes. An enduring tension in the community boiled over when visitors from a neighboring community asked to be fed despite their refusal to help out in their hosts' gardens.

Then, a male visitor beats a woman who has refused to give him food. Distraught, she screams and is comforted by family. Her brother and husband settle the dispute, first with clubs, and then with axes and machetes.

However, only 1 man is hurt, and there isn't much fighting because people intervene, placing themselves between the 2 groups.

Chagnon and Asch captured about 10 minutes of this half-hour episode. For their final film, they show the footage 4 times in a row.

- In the 1st version, they show an unedited version with sound. Some comments by the filmmakers as they were filming are audible, but there is no narration or explanation of what the viewer is seeing.
- In the 2nd version, Chagnon's voice explains what's happening, including his own original confusion about what was going on.
- In the 3rd version, the filmmakers diagram the lineages and families involved in the fight. This presentation makes it clear that the viewer is witnessing 1 episode of an enduring conflict.
- Finally, Changnon and Asch present an edited version of the fight, which transparently illustrates how the filmmakers' own cultural point of view is at work when they're making ethnographic films.

Their approach recognized that ethnographic filmmakers reveal culture at work on both sides of the lens. They dealt with that challenge through transparency. From raw footage and confusion to an edited episode that finally makes sense, Chagnon and Asch bring the viewer into their process before sharing their final version of events.

Jean Rouch

The ethnographic filmmaker Jean Rouch took the Chagnon and Asch approach to another level. Rouch did some amazing ethnographic film projects with the Dogon cliff dwellers in northern Mali. And he never lets the viewer forget that they're seeing a constructed scene. Specifically, he himself steps into the lens with his subjects, and he even shows film and equipment in his shots.

For the viewer and the participant, Rouch felt that cinestance was unavoidable. He decided to accept it and make it part of the story. His approach was a logical step toward what is now called participatory video and photography. Participatory visual approaches are quite common in anthropology today.

These ethnographic encounters place photography and video equipment into the hands of research populations themselves, who then visually reveal their attitudes and worldview with images instead of words.

Suggested Reading

Becker, *Art Worlds*.

Errington, *The Death of Authentic Primitive Art*.

Henley, *The Adventure of the Real*.

Marion and Crowder, *Visual Research*.

Questions to Consider

1. What is visual anthropology, and what kinds of work do visual anthropologists do?

2. How do social Darwinist ideas spill over into the way we categorize non-Western art?

3. What are some classic anthropology films, and how do they differ in terms of the way their filmmakers chose to document or present non-Western cultures?

Art and Visual Anthropology

W hether we're aware of it or not, our cultural background impacts the way we see and understand the world. And nowhere is this fact more clearly on display than when we compare the visual arts of various cultures around the globe. Today we're going to do some of that comparative work. We'll start by asking, What is art anyway? And then we'll see how, in the 20th century, new perspectives emerged to challenge the outmoded assessments of non-western art. Once we've done that, we'll move to the work of visual anthropology and the history of ethnographic films, like *Nanook of the North*, to see how video technologies can both support and complicate our anthropological mission to document world cultures.

So, let's start with the big question. What exactly is art? To get at that question, I show various images to my students, and I ask them to rate particular works of art on my handy dandy art-o-meter. Basically, they take in an image, and then, they score it on a scale from 1 to 5. 5 being the gold standard. Pure art.

For example, the majority always rates Van Gogh's *The Starry Night* a perfect 5/5 on the art meter. But then, when I show them another classic, like Gustave Courbet's *The Artist's Studio*, most students appreciate his sharp technical precision as a painter, but there are always students who give Courbet less than 5 out of 5 because to them it's not quite as arty as Van Gogh.

They typically explain that Courbet's sharp realism is basically a straight-forward, objective photograph. They rate him lower because they are missing the vibrant and imaginative twist that they see in Van Gogh's *Starry Night*. Ultimately, my students do agree that some works are definitely art just not as pure, not 5 out of 5 art. Andy Warhol's soup cans fall into this less than pure category of art. Same thing goes when I show them tourist art from West Africa, like a miniature bicycle sculpted out of wire and old pesticide cans. But, to wrap up the exercise, I drop the big one A half completed paint-by-numbers picture of a snowy barn or a sailboat. And without fail, this painting gets the lowest rating of all. Some students simply can't imagine giving anything but a 0 to a paint-by-numbers painting.

At any rate, one of the lessons that usually comes out of the art-o-meter exercise is that there's really not a clear line distinguishing art from non-art. For those who penalize paint by numbers art, for them, art has a critical, technical skills threshold that has to be met. If you have to paint by numbers, this group would say, you don't have the technical chops to rank among the greats.

But then, the folks who downgraded Courbet's realism because it was too representational, too much like a photograph, well for them art has a definitive creativity threshold. So Courbet may have the technical chops, but somehow the way he sees the world is less imaginative, less creative than the way other artists do. At least that's what some critics would argue.

So, the truth is, there just isn't a single definition of art. For some, it's about technical prowess, while for others, it's about imagination. And then there are still different people, who might focus on message or level of abstraction. The definition of art, ultimately, is like beauty, it's in the eye of the beholder.

The question "What is art?" has only grown thornier as anthropologists have expanded our cultural horizons, particularly, over the 20th century, as we learned more and more about cultural diversity. Western anthropologists and art historians grappled with the way they should categorize non-western art. Walking down the grand marble floors in our art museums, we Westerners have long felt at home with the familiar schools there's the Greco-Roman wing, the Renaissance collection, the Neoclassical rooms, the Impressionists But what about non-western art?

Well, when anthropologists and others first examined the art of the Ancient Aztecs or even the art of Africa more generally, they labeled it primitive art. But is this is a helpful way to look at non-Western art? I mean just consider the ciwara sculpture that is a unifying symbol among the Bamana people of West Africa. The ciwara depicts an antelope, which is carved out of wood, and it features intricate details in the horns, face, body, and mane. And, I've seen brilliant ciwara pieces that were made less than a decade ago by sculptors who live in Bamako, the capital city of Mali. These artists are modern, not primitive and so are their sculptures. But there was a time when anthropology identified African art and the ciwara as primitive, whether or not it was produced 500 years ago or yesterday morning. Primitive.

As we've seen in other lectures, this problematic way of categorizing other cultures can be traced to the industrial revolution, when Westerners began to view progress as the dominant narrative of humanity. Our ultimate purpose. And that's when anthropology first got its footing as an academic discipline. So, it's not surprising that early anthropologists like Edward Burnett Tylor incorporated this progress narrative into their work. People like Tylor viewed Africans as primitive humans, so that's how they categorized African art and cultures as well.

Actually, art itself wasn't a major specialization in the early years of anthropology. In fact, when it came to art, early anthropology was a split house. On the one hand, we had the archaeologists, who unearth and interpret art from all across the world. In these early days, they examined artistic artifacts by studying their form and meaning. Then, on the other hand, sociocultural anthropologists, at least in those early days, they tended to shy away from studying form. They were really looking more at the functional dimensions of art objects.

But things changed around the middle of the 20th century when new generations of anthropologists staked new territory on the interpretivist or symbolic side of our discipline. Rather than testing and correcting theories built on materialist, empirical data, these anthropologists embraced the immeasurable interpretive dimensions of the human experience. And that included an appreciation for different modes of artistic expression. So, with an eye for interpretation, symbolic anthropologists like Claude Levi-Strauss and Victor Turner, embraced artistic expressions as treasure troves of cultural insight and data.

In his classic book, *The Savage Mind*, Levi-Strauss described artists as hybrid creatures that are one part scientist and one part craftsman—or what he calls a bricoleur. The craftsman builds a material object, say a ciwara sculpture, but that object, he says, is an object of knowledge. It contains an entire worldview—or as the famous Russian painter, Wassily Kandinsky called it, "A whole lifetime of fears, doubts, hopes, and joy."

In an ever more connected world, people in the 20th century were more exposed to non-western art and cultures than those who lived just a few decades before. And this art captured the imagination of the West, which was decades away from YouTube and Google. People looked into African masks to look into Africa and Africans themselves. And, in the art world, primitive art eventually emerged as an exciting trend when visionary artists like Pablo Picasso, for example, developed his cubist perspective, in large part, through his appreciation and emulation of African masks and sculptures.

If you look at African masks, for example, you can plainly see ears, eyes, and maybe even a mouth You can tell that it represents a person or perhaps an antelope. But it isn't like any person or antelope that you've ever seen. From elongated noses, sharp geometric points for ears, and for the ciwara antelope, the head is bigger than the entire body. And then, if you look at that same mask from the side, or in a different light, it transforms. It looks completely different than it did. It's almost as if the carver created that mask using more than one set of eyes, more than a singular point of view. Plural vision, if you will.

Well, that's essentially the inspiration for Picasso's cubism. All those strange shapes and forms are multiple perspectives, visualized at once. Duchamp's *Nude Descending a Staircase* is another great example. In it we see a figure descending down a set of stairs, but in one glance we see the figure at all moments of her descent, from start to finish. Simultaneously.

Now, we'll leave the official definition of cubism for art historians, but from our anthropological approach, we can appreciate connections between the emergence of cubism and the rise of primitive art in the West. And, once more, we need to ask, Given that connection—given the influence of so-called primitive forms on Western art is primitive the word we should be using? Well, that question got answered in the late 20th century, when primitive art, as a museum-approved genre, came of age, and then shortly after promptly went extinct.

This extinction of the term primitive art occurred as anthropologists continued to expand the ways we were looking at art. Adding to the symbolic and interpretivist work of the mid-century, new research themes emerged to look at the sociocultural context of art, the politics of representation, and even the repatriation of art to places like Mali, Egypt, and Greece.

Sociologist Howard Becker was a leading figure in this movement to understand art as a sociocultural process, not just as objects produced by artists. In one of his most celebrated texts, *Art Worlds*, he reframes art as a cooperative network of everyone who participates in the production of it. There's the artist of course, but there are also the people who make and sell art supplies, there are the dealers, there are critics, there are consumers, installation experts, business managers, and the list goes on.

With social Darwinism subsumed by cultural relativity, anthropologists found it no longer tenable to classify non-western art as primitive. The ciwara sculpture isn't more primitive than the Mona Lisa, it's just from a different world. As such, the idea of primitive art fell out of favor, and in its place, anthropologists and others now use geographic categories to describe what we once called primitive art.

So at the Met in New York, for example, the Rockefeller Wing now houses a collection called the Arts of Africa, Oceania, and the Americas. It has thousands of artifacts dating from B.C.E. 3000 right through to the present. Art is culturally, temporally, and geographically bound, but it's not primitive. And that's a major change since the origins of anthropology and the study of non-western art. Through cultural relativity and symbolic and interpretivist methods, anthropologists discovered new ways to think about the human condition.

Specifically, they created visual anthropology, an inclusive specialization that examines humanity, by studying performance, media and media technologies. Basically, they look at any visual representations from art to urban murals, and just so much more. Perhaps the most well-known aspect of visual anthropology, at least for the general public, is ethnographic photography and film—doing anthropology with cameras. So as I promised at the outset, I'm going to pivot now from looking at the broader world of the visual arts to looking specifically at the visual work of anthropologists themselves.

Since the early days of anthropology, researchers have used still and moving images to document and analyze culture. And in the evolution of ethnographic film, aided by remarkable developments in sound and camera technology, inspired anthropologists to incorporate progressively more and more of the emic perspectives of their research populations. They were bringing a core lesson from early anthropologists like Bronislaw Malinowski who trained his students to record ethnographic utterances. To record the words and thoughts of your study community using, at least at first, their own exact words.

Many saw additional advantages to the camera. Even a grainy early film of a Zuni harvest festival dance would be far superior to a written account recorded by a visitor. After all, unlike other ways of recording culture, film captures body language and the environment surrounding whatever it is that's being filmed— whether that's a ritual, a performance, or really any aspect of daily life for that matter. These are details we don't get from interviews, surveys, and other anthropological methods. There's something special about getting data on film. Well maybe. I mean, sure, film can be an important tool in our anthropological kit. But let me add just one caveat.

People often work on the assumption that a picture, whether it's still or moving, provides purely objective information. With a video camera, we can not only interview our research subjects, but we can record their actual responses right down to their posture and the pitch of their voice. We get these voices & images directly from the study community pure, unadulterated data. Wrong. Yeah, there was a time when we thought that cameras don't lie, but in a Photoshop world, now we know much better.

Take some of my Malian photography. I love taking photos of children. But you know, those photos don't provide an unfiltered perspective. Let me explain. So, I'll be walking around, doing something in the village, when I the corner and come across an endearing scene of a young mother bathing a child in a bright pink plastic tub. They're laughing and splashing, and with the big mango tree over there and a goat right behind them, the picture is perfect.

So I do as I'm supposed to. I greet them both and then ask, may I please take your picture? Sure, they say, but and this happens every single time, the mother goes running off into her hut, she changes into one of her best outfits, as then she wraps a beautiful scarf around her head.

My ideal representation of the daily beauty and pace of my host community, this mother bathing her child in the shirt and skirt I always see her in, well, that isn't the representation she wants me to put on film. If we were simply chatting or maybe even recording our voices, the moment would have proceeded without the urgent costume change. But kind of like working at home in your pajamas. When it's time to Skype your boss, you'd better change your shirt and check your hair. We all do this.

The last time I was filmed teaching, I put on a coat and tie. My students almost didn't recognize me. But the cameras were coming, I told them. I had to sharpen up. So that video shows real students in a real class. And it was the real lecture too. But I modified my performance and my appearance in a way that I wouldn't have on any other day. The camera's presence changed my entire approach.

We call that cinestance. The second a camera emerges, bang cinestance. We act different than we would otherwise. As if the whole world was watching, we tend to tighten things up for the camera. And, as you can imagine, that's a challenge to visual anthropology. If we're using cameras to observe and record the way people live and think in their day to day lives, won't bringing a camera ruin everything? Won't cinestance just kick right in, giving us an altered reality one with nicer than normal clothes, for example?

Well, now that we've had our caveat about the assumed objectivity of film, let's close by examining how anthropologists have learned to deal with cinestance. We're going to review a few pivotal anthropology film classics that document the development of ethnographic film as a genre. And we'll see how our anthropological quest to accurately portray our research communities inspired more participatory and transparent approaches that eventually have us stepping in front of the lens, and sometimes even handing over our cameras.

Our brief ethnographic film festival begins with *Nanook of North*. This 1922 film is the first major anthropology film, and it actually helped pioneer the documentary film genre. The filmmaker, Robert Flaherty, filmed episodes from the daily lives of Arctic people, and audiences loved this rare and close-up glance into Inuit life. Modern critics, however, have critiqued Flaherty for leaving out all signs of modernity. Focusing the lens only on the activities and scenes that emphasized the pre-modern living conditions of the Inuit. Despite the presence of his equipment, not to mention modern conveniences to be found at the local trading post, Flaherty chose to show only scenes that portrayed an illusion of the pristine and so-called primitive state of Inuit people living off the frigid land.

Now, this was before the era of documentary film, so it's not surprising that Flaherty violated a few of the cardinal rules that would later define the genre. For example, the scenes and people he shares with the audience aren't as authentic as you might expect. You see, he actually casted this film. The main star, Nanook, is indeed a hunter, and we see him harpooning a seal, wrestling with a walrus, and building remarkable igloos—including a small one just for his puppies. But his loving wives and adorable children weren't his own. Flaherty staged the family and their activities in order to capture Inuit life. But he exaggerated the remoteness and the pre-modern qualities of those daily lives.

But Flaherty was open about this, and his staged scenes mesmerized audiences nonetheless. Seated in their velvety theater seats, his audience snacked on popcorn as Nanook and his family delivered an enthralling and visually stunning peek into the stark way of life they could hardly imagine.

So *Nanook of the North* remains a classic, despite the fact that it is anything but cinema verite. It seems that Flaherty avoided the cinestance problem by recruiting actors to film scenes of daily activities and family life in an igloo. And while it may be disappointing to learn that Rainbow, a beautiful 4-month old baby, isn't really Nanook's child, one still gets the feeling that we're seeing something genuine on that screen.

There's a magical moment in the film where Nanook's children tumble with puppies on a soft, furry animal skin. Even knowing that was a staged scene, the images feel authentic. The tender parenting and the arduous physicality of daily life combine in a way that the viewer recognizes a common humanity in people that early anthropologists first classified as primitive.

Flaherty's approach to visually recording and presenting Inuit life inspired countless documentary filmmakers, among them dozens and dozens of anthropologists. One of my former teachers, Napoleon Chagnon, and his film partner Timothy Asch left a camera on during a rather violent fight they witnessed among a band of South American forest dwellers. They edited that footage into an anthropological film titled *The Ax Fight*, which is now shown in hundreds of anthropology classes every year. Unlike Flaherty, Chagnon and Asch didn't stage an ax fight with local actors. Instead, the camera just sits there, drawn on a confusing scene, accompanied by some terribly discomforting wails.

You see, it was 1971, and Chagnon and Asch had just arrived in their Amazonian host village. And to their surprise, a fight broke out right before their eyes. We learn that an enduring tension in the community boiled over when visitors from a neighboring community asked to be fed despite their refusal to help out in their hosts' gardens. Then, one of the visitors, a man, beats a woman who has refused to give him food from her garden. Distraught, she screams and is then comforted by family, while her brother and husband settle the dispute, first with clubs, and then with axes and machetes. But, it's not as gory as it may sound. Only one man is hurt, and there isn't much fighting because people intervene, placing themselves between these two groups.

The gravity of the episode neutralized the cinestance factor, but the filmmakers did something else to deliver an authentic representation of this ax fight. Chagnon and Asch captured about 10 minutes of this half-hour episode, and for their final film, they show the footage four times in a row. First, they show an unedited version with sound. You can hear some comments by the filmmakers, as they were filming, but there is no narration or explanation of what we're seeing. Then in the second version, Chagnon's voice explains what's happening including his own original confusion about what was going on.

In the third version, the film-makers diagram the lineages and families involved in the fight, and this presentation makes it clear that we are witnessing one episode of a long-enduring conflict. Then, last, Chagnon and Asch present an edited version of the fight, which transparently illustrates how the filmmakers' own cultural point of view is at work when we're making ethnographic films. In a sense, they're showing their work. They give you the raw material, they then share some background context, and then they show an edited version that flows as a film while presenting credible scholarship in a visual form.

Their approach recognized that ethnographic filmmakers reveal culture at work on both sides of the lens. And, they dealt with that challenge through transparency. From raw footage and confusion to an edited episode that finally makes sense, Chagnon and Asch bring the viewer into their process before sharing the final version of what really went down.

Ethnographic filmmaker Jean Rouch took the Chagnon and Asch approach to a whole new level. Flaherty, remember, carefully kept his lens away from any hint of modernity—including his own equipment and supplies. We get so enticed by this remote and foreign world, that we forget there were a camera and filmmaker directing all the shots.

Rouch did some amazing ethnographic film projects with the Dogon cliff dwellers in Northern Mali. And he never lets you forget we're seeing a constructed scene. Specifically, he himself steps into the lens with his subjects, and he even shows film and audio equipment in his shots.

On Saturday Night Live, sometimes the boom mic messes up the frame and we see that thing poke right into the top of the frame. Well, Jean Rouch wanted that mic seen. For the viewer and the participant, Rouch felt that we can't avoid cinestance. So if you can't beat it, join it. Accept it and make it part of the story.

His approach was a logical first step to what we now call participatory video and photography. Participatory visual approaches are now quite common in anthropology today. These ethnographic encounters place photography and video equipment into the hands of research populations themselves, who then visually reveal their attitudes and worldview with images instead of words.

I did this once with my Malian friends. I gave them cameras to document our visit to an agricultural research station. Seeing the station through their eyes helped me see what mattered to them. They took pictures of some interesting types of sorghum that they hadn't seen before. There were a few pictures in the lab and of people, but these farmers showed me that when it comes to the agricultural research station, it was the sorghum trials that most interested them.

In this lecture, we've seen how evolution-inspired ideas of cultural progress initially seeped into the anthropological view of so-called primitive art. And this is consistent with what we've seen in other lectures. From religion to family structure, to human sexuality, and even artistic expression, or just about any other facet of human diversity, anthropology was initially rooted in social Darwinism. We saw non-western people as primitive, as a sharp binary opposite to the modernity of the West. But because our discipline is a scientific one, we've always tested and corrected what we learn and think. And today we've once more discovered that social Darwinism is refuted by empirical data.

This correction has had an impact even upon the way that we collect and interpret information on film, as the lens through which anthropologists view cultural diversity has changed with the times. And that change—that adjustment of our lens—has brought the common elements of the human condition more sharply into focus.

Conflict and Reconciliation across Cultures

For the rest of this course, we'll see how anthropological perspectives contribute ideas and insight on some of the biggest challenges facing humankind. From war and conflict to criminal justice and health, anthropology has exciting contributions to share. This lecture focuses on violence and conflict resolution. In addition to uncovering the roots of violence, anthropology has important lessons to bring to the table when it comes to providing cross-cultural insight on conflict resolution.

Three Myths

Anthropology, with its 4-field approach, can clear up 3 major myths about humanity and our so-called violent nature. Myth 1 is that there has always been warfare. Myth 2 is that we are biologically predisposed to violence. Myth 3 is that warfare is universal.

Regarding myth 1: War is a recent development in our human history. *Homo sapiens* have been around for 200,000 years. But the archaeological record indicates that war, as we understand it today, only emerged some 10,000 years ago, around the advent of agriculture.

- Before agriculture, the survival of small hunter-gatherer groups required a level of cooperation that made conflict counterproductive for all parties. That's not to say that hunter-gatherers never fought, but these fights (or forager conflicts) rarely erupted into all-out war.

- Because non-sedentary people can't accumulate property the way Westerners do today, they cultivated communal lifestyles. In fact, we can observe this lifestyle,

even in contemporary hunter-gatherer societies.

- For example, anthropologist Marjorie Shostak lived among the San people of the Kalahari, and she recorded the life history of a remarkable woman named Nisa. Nisa's community taught her that feeding all bellies takes precedence over individual accumulation or surplus. Without a surplus or troves of private property, the incentive for war isn't eliminated, but it is greatly reduced.

- However, the agricultural revolution changed all that. Sedentary living and agricultural surplus sparked the inevitable rise of urban centers and empires. And that's where the archaeological record shows us a sharp increase in large-scale, organized violence or war.

- The 20th century, according to historian Eric Hobsbawm, was the most murderous in recorded history. Warfare took the lives of some 187 million people in the 20th century.

- At first, 20th-century warfare was essentially limited to inter-state conflict and civil wars. But now, in the 21st century, armed conflict is no longer in the exclusive hands of governments and states. In fact, 21st-century warfare permeates civilian lives.

- At the turn of the 20th century, only 5% of combat casualties, on average, were civilians. By the end of World War II, however, the tide turned. Around two-thirds of World War II casualties were civilian deaths.

- That figure has risen today to approximately 90%. That is to say, 90% of those who die in 21st-century war are civilians. Ultimately, warfare as we know it today is truly a modern phenomenon.

Now for myth 2, the idea that we are biologically predisposed to violence. Are we programmed for conflict? In some ways, it appears so. But there might be other biological factors at work that show us that we're not predestined for violence and war.

- In their book *Demonic Males*, Richard Wrangham and Dale Peterson show that chimpanzees, like some humans, beat, rape, and kill other chimpanzees.

- But this predilection of our chimpanzee cousins is not an inevitable dimension of our own biological being.

Around two-thirds of World War II casualties were civilian deaths.

- Wrangham and Peterson remind us to look at bonobo apes too.
- Unlike the violence we see in chimpanzees and gorillas, bonobos are relatively peaceful. Bonobo males do occasionally become aggressive and violent, but they rarely kill or rape. Why? Tight-knit bands of female bonobos gang up on and attack male counterparts who act up.
- Regarding humans: Our propensity for conflict resolution may be every bit as powerful as any inclination to fight. The anthropologist Carolyn Nordstrom documented this phenomenon among humans in her civil war ethnography: *Mozambique: A Different Kind of War Story.*
- Nordstrom looked deeper into unfathomable violence and aggression of 15 years of war. She described soldiers, wartime profiteers, and ordinary people. Remarkably, she revealed a propensity for cooperation among ordinary people in the face of extraordinary violence.
- Ordinary people created ways to actively socialize combatants back into a life of nonviolence. They established a network of healers, and they even kidnapped and reconditioned soldiers toward nonviolence.
- It's apparent that there are some biological mechanisms at work in the history of human warfare. But we're equally or perhaps more inclined toward making peace.

Now for the 3rd myth: the idea that war is universal. There are people in the human family tree who make peace seem as inevitable as war.

- The Amish, for example, refuse to fight in wars. They don't even take disputes to court unless all internal efforts to resolve a conflict have failed. One of the core Amish beliefs is the doctrine of non-resistance.
- Similarly, India's Jain religion sees the path to peace as our ultimate purpose. Ascetic Jain monks even go to the extreme of sweeping a path as they walk, to avoid stepping on small unsuspecting insects. Jains practice peace by fostering goodwill toward

The Amish refuse to fight in wars.

others, based on the unity and sanctity of all life.

- Another example: The San of the Kalahari region are super-sharers. They're one of the oldest indigenous populations on earth, tracing their cultural history back some 20,000 years. They are known as peaceful people who discourage fighting, aggression, and even competition.

Conflict Resolution: Mali

This lecture will now look at some cross-cultural examples of conflict resolution strategies to see how people transcend violence and aggression elsewhere in the world.

Once a thriving global empire, today the Republic of Mali suffers from extreme poverty and a complex security crisis that may take decades to repair. But thousands of rural Malian villages show 2 terrific daily practices that foster peace over conflict.

One example is joking cousins. Take the example of the Sangare and Doumbia families. These families have exchanged friendly insults and laughs for generations. Every family name in

Mali has a joking cousin alliance. Anthropologists interpret this cultural phenomenon as a safety valve preventing enduring social conflict.

Another safety valve that promotes unity over conflict is the village meeting. The meeting protocol itself is an act of consensus building: If someone's making a request, that request is passed up and down the line of village elders, repeated until everyone has had the opportunity to add their voice to the discussion.

Conflict Resolution: Liberia

In West Africa in Liberia, the Kpelle people use a conflict resolution practice that they call the Kpelle moot. The Kpelle are more interested in reconciliation rather than retribution-style justice. When an offense threatens the social harmony within a Kpelle village, the entire community comes together to hold a moot.

The moot forum takes place in someone's home rather than a courthouse. The complainant who calls for the moot selects a family member to serve as a mediator, but there's no singular judge or jury. In the moot, everyone chimes in.

The conflicting parties directly question and challenge each other. The mediator and the audience all join in. By the end of this long discussion, the person deemed at fault publicly apologizes and presents a small gift or two to the aggrieved. The aggrieved, in turn, offers a smaller gift to help repair the relationship.

Anthropologists attribute this form of reconciliatory justice as unique for small-scale societies. In a small, rural, cooperative community, people simply can't afford enduring conflict. It undermines the social harmony required to make a living off the land. The Kpelle moot vets justice while maintaining the social network that ensures the village's bellies will remain full.

Anthropology in the Military

In Afghanistan, the US military has used anthropology to better understand the sociocultural dimensions of Afghan communities. Anthropology in the military is not new, but not everyone agrees that this is a good idea.

In the 1960s, the US government launched Project Camelot, based at American University. Project Camelot was a counterinsurgency study led by anthropologists and other social and behavioral scientists.

The logic was that US national security was threatened by insurgencies throughout the world, particularly in neighboring Latin America. The orchestrators of Project Camelot brought in some anthropology because insurgencies signaled a breakdown of social order, and the military wanted to better understand the social and cultural processes that lead to these transformative events.

Shortly after collaborating Latin American scholars discovered the true purpose and funding source, the military shut down the program. But evidence shows that similar work continued nonetheless.

Prominent anthropologists like Margaret Mead and Marshall Sahlins were so affronted by this military coopting of anthropology, particularly in Latin American and in Southeast Asia, that they helped pass a stern resolution by the American Anthropological Association (AAA).

Simply put, they made it clear that anthropologists should not participate in what they labeled "clandestine" intelligence work.

In 2007, the American Anthropology Association once again spoke up about the militarization of anthropology. Specifically, they were reacting to a new military project operating in Afghanistan. It was called the Human Terrain System (HTS).

The 2007 statement was clear that the HTS violated the AAA code of ethics, which mandates that anthropologists do no harm to research subjects. Unlike previous militarized applications of anthropology, the HTS teams put anthropologists in actual warzone combat units. They could carry weapons alongside other, more conventionally trained soldiers.

The goal, however, was for the anthropologists to help soldiers and local populations communicate and understand each other. One example would be the Jirga. The Jirga, a community's council of elders, was a setting in which community members build consensus with guidance and discussion of Muslim values and teachings.

Anthropologists helped the US military recognize the importance of these civic conversations. They brought combat units and communities together. They won friends through understanding and collaborative social projects like building schools, or helping the elders deal with problems ranging from food scarcity to fears of being attacked.

The primary rationale for deploying HTS teams in Iraq and Afghani war zones was to curtail casualties by winning the hearts and minds of local communities. And to an extent, they did. But the program quickly faltered after critiques of mismanagement and dubious behavior along with a serious challenge by professional anthropologists across the nation.

In reaction to the HTS program, the AAA made it clear they were concerned by the fact that anthropologists were working in combat units. Specifically, critics questioned the voluntary subject's ability to "decline" participation in HTS anthropology work, owing to the presence of weapons.

Similarly, if any of the information warzone anthropologists produce

is used by the military for the purpose of designing combat campaigns and identifying potential targets, then the participating anthropologists are in clear violation of their code of ethics. Their work directly alters the winds of war, and this can undermine the work of all anthropologists abroad. The last thing a field researcher needs is to be accused of being a meddling intelligence officer.

At its peak, the HTS employed some 500 people, who served in 30 teams deployed in both Iraq and Afghanistan. The cost: $725 million. But once the majority of US troops were pulled out of these theaters, the HTS program was officially closed in 2014.

However, it's likely we'll see new forms of anthropology in combat zones and elsewhere in the military. Despite the controversy, the generic idea of a military with cultural knowledge of the people they engage with has become essential.

Suggested Reading

Little and Smith, *Mayas in Postwar Guatemala*.

Nordstrom, *A Different Kind of War Story*.

Peterson and Wrangham, *Demonic Males*.

Price, *Weaponizing Anthropology*.

Questions to Consider

1. What are two specific examples of how people across cultures work together to reduce conflict?

2. Are humans biologically predisposed toward war and violence?

Conflict and Reconciliation across Cultures

Today, and for the rest of this course, we'll see how anthropological perspectives contribute ideas and insight on some of the biggest challenges facing humankind. From war and conflict to medicine, criminal justice, and health, anthropology has exciting contributions to share. And today, we're going to think about violence and conflict resolution. These are crucial themes for anthropologists because, the United Nations and the US military and intelligence agencies all agree our increasingly dynamic climate, along with growing populations, will lead to increased conflict throughout the world.

In the modern age, war and violence appear as an unavoidable aspect of the human experience. In fact, over the past 5000-6000 years or so, we've seen over 14,000 major wars. And millions upon millions of lives have been lost in these conflicts. But, as an anthropologist, I know, that these grim statistics are not the whole story. Because, in addition to uncovering the roots of war, anthropology has important lessons to bring to the table when it comes to providing cross-cultural insight on conflict resolution.

So today, we're going to see how anthropology can be a peacemaker, or at least a conflict manager. But first, let's return to the history of human violence to clear up three big myths about humans and their proclivity toward violence and war.

Anthropology, with its four-field approach, can clear up three major myths about humanity and our so-called violent nature. One there has always been warfare. Two we are biologically predisposed to violence. And, three Warfare is universal. Let's think about each one of those myths.

First, have humans always been at war? Or is this form of organized violence more of a recent phenomenon? Well, as it turns out, war is a recent development in our human history. It's not really a long standing practice. *Homo sapiens*, as we have learned, have been around for 200,000 years. But the archaeological record indicates that war, at least as we understand it today, only emerged some 10,000 years ago, right around the advent of agriculture. Now, it would be an oversimplification to state that our foraging past was an era of peace and cooperation, but to some degree, that's some useful hyperbole. Let me explain As we learned in previous lectures, prior to our transition to agriculture we fed ourselves by foraging and hunting. And this hunter-gatherer way of life shaped our social worlds. Specifically, it kept our communities small, and it leveled out inequality.

We know that hunter-gatherer communities were much smaller than the farming-based societies that emerged within the past 10,000 years. Archaeologists and others note that bands of foragers typically maxed out at around 100-150 people. And, in a society where everyone knows each other, if there's a slacker in our midst, we'll know it. You can't slide by in the shadows of a band of foragers. Furthermore, like a college student always on the move between dorm, home, and summer jobs, foragers were nomadic. They were always on the run.

They followed the food on seasonal migrations. And as non-sedentary people, they couldn't afford to invest in personal property or surplus. In a way, they had to restrict their earthly possessions and food stores, to what they could carry. Compared with our overstuffed pantries today, foragers didn't have a place to store a few cans of soup for that rainy day. When they wanted soup, they started from scratch.

So, the survival of these small groups required a level of cooperation that made conflict counterproductive for all parties. Now, that's not to say that hunter-gatherers were peaceniks who never fought, but, we are saying that these fights or forager conflict rarely erupted into all-out war. Because non-sedentary people can't afford to accumulate property the way we Westerners do today. They cultivated more communal lifestyles. In fact, we can observe this lifestyle, even in contemporary hunter-gatherer societies.

For example, anthropologist Marjorie Shostak lived among the San people of the Kalahari, and she recorded the life of a remarkable woman named Nisa. When she recounts Nisa's childhood, she describes the emotional dimensions of growing up in this kind of sharing economy. She remembers, for example, how proud and excited she was when her father returned after a successful hunt. But, knowing that that bounty would be shared with the community, she said it always stung just a little bit. It's like any kid learning how to effectively socialize, Nisa's community taught her that feeding all bellies takes precedence over individual accumulation or surplus. And without that surplus or troves of private property, the incentive for war, well, it isn't eliminated, but it is greatly reduced.

However, the agricultural revolution changed all of that. Sedentary living and agricultural surplus sparked the inevitable rise of urban centers and empires. And that's where the archaeological record shows us a sharp increase in large-scale, organized violence or war. The 20th century alone, according to Eric Hobsbawm—was the most murderous in recorded history. He reminds us warfare took the lives of some 187 million people in the 20th century, alone.

At first, 20th-century warfare was essentially limited to inter-state conflict and civil wars. But now, in the 21st century, armed conflict is no longer in the exclusive hands of governments and states. In fact, 21st-century warfare permeates our civilian lives. And that's a major change. At the of the 20th century, only five percent of combat casualties, on average, were civilians. By the end of World War II, however, we see the tide has ed. Around 2/3's of WWII casualties were civilian deaths. And, believe it or not, that figure has risen today it's around 90 percent. That is to say, 90 percent of those who die in 21st-century war are civilians. Ultimately, warfare, as we know it today, is a truly modern phenomenon, and the pervasive armed conflict we see in our lives is fundamentally different from previous wars.

So war is a rather recent phenomenon when we consider the long arc of *Homo sapiens* history. So, let's turn to our second myth—the idea that we are biologically predisposed to violence. Famous scholars like Hobbes and Darwin are not alone in thinking that we humans are hard-wired toward aggression and violence. In fact, many researchers have traced this tendency to our primate cousins.

Primatologist Richard Wrangham, for example, has studied what he calls coalition killing among non-human species, and he finds that other animals like chimpanzees do participate in coalition killings. These killings involve bands of chimps collectively carrying out violent attacks on rival groups— which sounds a lot like warfare, doesn't it? Wrangham explains that coalition killings are an expression of dominance. But is that the full story? Are we programmed for conflict? Well, in some ways it appears so. Yet, there might be other biological factors at work that show us that we're not predestined for violence and war.

In his book *Demonic Males*, Wrangham shows that chimpanzees, like some humans, beat, rape, and kill other chimpanzees. We've even observed them drinking the blood of their victims. But let's not take this predilection of our chimpanzee cousins as an inevitable dimension of our own biological being. Why? Well, Wrangham reminds us to look at bonobo apes too. He shows us that chimpanzees and bonobos have much different lives, despite their shared genetics.

Unlike the violence we see in chimpanzees and gorillas, bonobos are the free-loving hippies of the ape family. Bonobo males do occasionally become aggressive and violent, but they rarely kill or rape. Why is that? Well, Wrangham explains that tight-knit bands of female bonobos actually gang up on and attack male counterparts who act up. Ultimately, Wrangham prescribes a little bonobo living as a remedy for our 21st-century conflict. He argues that if humans emulated bonobos in terms of increasing the political power and voice of women, our species can suppress excess male aggression.

Another primatologist, Frans DeWaal, has found that, among chimpanzees, attraction between opponents actually increases after a fight. Apparently, make-up sex isn't necessarily a human invention. And our propensity for conflict resolution may be every bit as powerful as any inclination to fight. Anthropologist Carolyn Nordstrom documented this phenomenon among humans in her civil war ethnography, *Mozambique, A Different Kind of War*. In her timely study, Nordstrom looked deeper into unfathomable violence and aggression of 15 years of war. She connects us with soldiers, with wartime profiteers, and even ordinary people. And remarkably, she revealed the unrelenting and pervasive propensity for cooperation among ordinary people in the face of extraordinary violence.

In an astonishing refusal to accept a life of war and violence, ordinary people created ways to actively socialize combatants back into a life of non-violence. They actually established a network of healers, and, they even resorted to kidnapping soldiers. They kidnapped and reconditioned them toward non-violence. So, it's apparent that there are some biological mechanisms at work in the history of human warfare. But, we've also discovered that biologically and culturally, we're equally, or perhaps more inclined toward making peace, not war.

And now for our third and final war myth. The idea that war is universal. Even in an era where gun violence impacts our daily lives, we have brothers and sisters on our human family tree who make peace seem as inevitable as war. Take the Amish, for example. I grew up in Ohio, not far from a thriving Amish community. And the Amish refuse to fight in wars. They don't even take disputes to court unless all internal efforts to resolve a conflict have failed. One of the core Amish beliefs is the doctrine of non-resistance. As Christians, they walk the walk when it comes to the Sermon on the Mount—they the other cheek.

Similarly, in India, we find the Jain religion, which sees the path to peace as our ultimate purpose. In the words of Lord Mahavira, "One weapon will always be stronger than the other, but the path of Ahimsa, or peace, remains unsurpassed." Ascetic Jain monks even go to the extreme of sweeping a path as they walk, to avoid stepping on small unsuspecting insects. All Jains practice peace by fostering goodwill toward others, based on the unity and sanctity of all life. This worldview has endured since the B.C.E. sixth century. And like the Amish, Jains may live in a world of armed conflict, but they don't participate.

A final group from Southern Africa can get us an even closer look at a human society that makes peace, not war. This is the San of the Kalahari region. They are super-sharers. They're one of the oldest indigenous populations on earth, tracing back their cultural history some 20,000 years. Now, with only a few dozen remaining villages, most of this once hunter-gatherer population has transitioned into sedentary farming lifestyles. But, nonetheless, their peaceful foraging legacy remains. The San are known as peaceful people who discourage fighting and aggression, and even competition.

Anthropologists explain the peaceful lives of the San, in part, by looking at the lives of their children. You see, in this cooperative, transient lifestyle, there are fewer children around, and these kids don't see much aggression growing up. Typically, San bands only have around a dozen plus children of all ages, at any given time. And, with such a small crew, competitive games and team sports are less feasible than cooperative activities.

Remarkably, the San have no living memory of fighting over territory. No living memory of war. Richard Lee studied the San for over three years, and in all, he recorded only 58 actual arguments, of which 34 became physical fights. Put another way, in three years, this tight-knit crew sees a physical fight less than once a month, and a single argument only once every three weeks. And that's not per person. That's the whole community.

So if we look to our past—and even to groups, like the Amish and the San, we learn that war itself is A, a relatively recent phenomenon; B, not biologically inevitable, and C, it isn't universal. If anything, we've seen that humans are wired for conflict resolution, not just coalition killing. Even non-human primates demonstrate their inclination toward reconciliation.

So for this final segment of our lecture, let's take a closer look at this idea of humans as peacemakers Let's see some cross-cultural examples of conflict resolution strategies to see how people transcend violence and aggression elsewhere in the world.

Once a thriving global empire, today the Republic of Mali suffers from extreme poverty and a terribly complex security crisis that will probably take decades to repair. Nonetheless, I want to bring you back to Mali, not to discuss the on-going conflict there, but because thousands and thousands of rural Malian villages show us two terrific daily practices that, again, foster peace over conflict.

First, joking cousins. In Mali, my name is Solo Sangare. Solo is the name of my adoptive Malian father, and Sangare, that's our shared family name. And every time I go to our neighboring village for the weekly market day, my tea and sugar guy, Cemoko Doumbia always greets me with a lively insult, usually something about flatulence. And, he's insulted if I don't engage in a short round of exchanging barbs. I'm kind known for slinging insults about eating donkey meat, but you'd probably have to be there to find any of this humorous.

You see, it's not that we hate each other. In fact, that's the exact opposite of what's really going on here. We're terrific friends. But, he's a Doumbia, traditionally a blacksmith's name. And me, I'm a Sangare, we're herders. And we Sangare's and Doumbia's have long been joking cousins. These two families have exchanged friendly insults and laughs for generations. But, god help you if you try and join in without being a Sangare or Doumbia. Every family name in Mali has a joking cousin alliance. And anthropologists interpret this cultural phenomenon as a safety valve that prevents enduring conflict.

And more broadly, there's another safety valve that promotes unity over conflict. And this is one of my favorite parts of daily life in the village. It's the village meeting. Malians do meetings that can include very heated exchanges, but, when it comes time for a meeting, there is a protocol that diminishes the chance, that things will get really out of hand. In fact, the meeting protocol itself is an act of consensus building.

These meetings always start slow, as the elders walk to the chief's hut. Everyone brings a goat skin mat to sit on in a loose oval pattern, with all parties facing each other. And, when I did my first village meeting to ask permission to live and study with this community, this is exactly how it went down. The elders were seated, surrounded by on-lookers, who often chime in when appropriate. Then my host brother, Burama, addressed the chief and elders, explaining that I'd like to speak to them. And even though every one of us heard that message, the junior most elder thanked Burama. And then, right in front of us, he proceeded to repeat exactly what Burama had said. But he said it to the next elder in order of age.

Some elders add a couple lines, and some take a few away. But, by the time the last elder gives this redundant message directly to the chief, everyone has had the opportunity to add their voice to the discussion. And if you're into quick meetings, stay the heck away from the village because now, the chief will initiate the message chain again, explaining that I may continue with my actual question. And once that message is passed down, out loud, through each elder, the final elder, the youngest one, gives the message back to my brother Burama, who in turn, gives it to me. But I heard that message. Each and every time. All the way from the Chief to Burama, and back. This process continued, up and down the line, until we settled our business. When I asked the chief about the meeting protocol, he explained, that it allows everyone to be heard, and, it brings everyone together in the spirit of compromise or, in moments of conflict, reconciliation.

Leaving Mali, let's stay in West Africa and head to Liberia, where we'll meet the Kpelle people and a conflict resolution practice that they call the Kpelle Moot, or in the local language, *berei mu meni saa.*

The Kpelle are hard-working, rural rice and peanut farmers. And their small, close-knit communities have a great method for resolving conflict. Unlike the US justice system that deploys court hearings to mete out justice through punishment like fines or jail time, the Kpelle, are more interested in reconciliation rather than retribution-style justice. When an offense threatens the social harmony within a Kpelle village, the entire community comes together to hold a moot.

The moot forum takes place in someone's home rather than a courthouse. You see, the complainant who calls for the moot selects a family member to serve as a mediator, but really, there's no singular judge or jury. In the moot, everyone chimes in. The conflicting parties directly question and challenge each other the mediator and the audience all join in and, by the end of this long discussion, the person deemed at fault publicly apologizes, and presents a small gift or two to the aggrieved, who, in offers a smaller gift to help repair the relationship. Then the guilty party picks up the tab and provides enough rum or beer so everyone can share a drink.

Similar mediation rituals are found elsewhere in Africa. And anthropologists attribute this form of reconciliatory justice as unique for small-scale societies. You see, in a small, rural, cooperative community, people simply can't afford enduring conflict. It undermines the social harmony required to make a living off the land. So, rather than developing a justice system that locks up people who are desperately needed for fetching water or farming rice, the Kpelle moot vets justice while maintaining the social network that ensures the village's bellies will remain full.

So let's keep with our cross-cultural case studies of conflict resolution, but for our last example, we'll head to Afghanistan to see how the US military has used anthropology to better understand the sociocultural dimensions of

Afghan communities. Now, anthropology in the military is not a new thing but, not everyone agrees that this is a good idea.

Back in the 1960s, the US government launched Project Camelot, based at American University. Project Camelot was a counterinsurgency study led by anthropologists and other social and behavioral scientists. The big idea was, to better predict and perhaps influence social movements in other countries. The logic was that US national security was threatened by insurgencies throughout the world, particularly in neighboring Latin America. So, the orchestrators of Project Camelot brought in some anthropology because insurgencies signaled a breakdown of social order, and the military wanted to better understand the social and cultural processes that lead to these transformative events.

Shortly after the true purpose and funding source was discovered by collaborating Latin American scholars, the military shut down this program. But, evidence shows that similar work continued nevertheless. Prominent anthropologists like Margaret Mead and Marshall Sahlins were so affronted by this military co-opting of anthropology, particularly in Latin American and in Southeast Asia, that they helped pass a stern resolution by the American Anthropological Association. Simply put, they made it clear that anthropologists should not participate in what they labeled clandestine intelligence work.

And that's why, in 2007, the American Anthropological Association, once again, spoke up about the militarization of anthropology. Specifically, they were reacting to a new military project operating in Afghanistan. It was called the Human Terrain System. The 2007 statement was clear that this HTS violated the AAA code of ethics, which mandates that we do no harm to our research subjects.

Now, one thing that especially raised ethical concerns was that, unlike previous militarized applications of anthropology, the Human Terrain teams put anthropologists in actual warzone combat units. They could carry weapons alongside other, more conventionally trained soldiers. The goal, however, was for the anthropologists to help soldiers and local populations communicate and understand each other. One example would be the Jirga. The Jirga, a community's council of elders, was a setting in which community members build consensus with guidance and discussion of Muslim values and teachings.

Anthropologists helped the US military recognize the importance of these civic conversations, and they brought combat units and communities together, winning friends through understanding and collaborative social projects like building schools, or helping the elders deal with problems ranging from food security to fears of being attacked.

The short of it is that the primary rationale for deploying Human Terrain Teams in Iraq and Afghani war zones was to curtail casualties by winning the hearts and minds of local communities. And to an extent, they did. But the program quickly faltered after critiques of mismanagement and dubious behavior compounded with a serious challenge by professional anthropologists across the nation.

In reaction to the Human Terrain System Program, the American Anthropological Association made it clear they weren't averse to the idea of reducing casualties, but they were genuinely concerned by the fact that anthropologists working in combat units clearly violated our disciplinary code of ethics. Specifically, critics questioned the voluntary subject's ability to decline participation in the team's anthropology work. After all, if a guy with a gun came to my door and asked if I'd agree to join him for an interview, I'm thinking that gun is going to work better than a pretty please. I'd say yes, even if it was against my wishes. Maybe, deep down, I'm worried that anti-US forces will see me talking with these soldiers, and when they leave, my family could be killed or tortured.

Similarly, if any of the information warzone anthropologists produce is used by the military for the purpose of designing combat campaigns and identifying potential targets, then the participating anthropologists are in clear violation of our code of ethics. Their work directly alters the winds of war, and this can undermine the work of all anthropologists abroad. The last thing a field researcher needs is to be accused of being a meddling intelligence officer.

At its peak, the HTS had employed some 500 people, who served in 30 teams deployed in both Iraq and Afghanistan at a cost of over $725 million. But once the majority of US troops were pulled out of these theaters, the HTS program was officially closed in 2014. But it's likely that we'll see new forms of anthropology in combat zones and elsewhere in the military because, despite the controversy, the generic idea of a military with cultural knowledge of the people they engage with has become essential.

And I got to tell you I've found that over time, more and more of my students side with the Human Terrain System concept, arguing that the potential to save lives makes the compromise worthwhile. Others, however, do challenge this position in the spirit of our anthropological code of ethics. Regardless, it's always a thoughtful and important discussion.

So, with that, let's wrap up our anthropological exploration of conflict and reconciliation. Deep in our bones, we typically interpret as scary and threatening, that which is unfamiliar and unrecognizable. And those feelings can quickly escalate into a maelstrom of violent behavior and, of course, even war. But, through biology, language, and culture, anthropologists help us render the unfamiliar familiar, or, as Ruth Benedict once said, to make the world safe for human differences.

Today, we saw how humankind has a wealth of cultural and even biological resources that we can use to transform conflict into reconciliation and harmony. And, in a conflict-ridden world where 90 percent of wartime deaths are civilian casualties, we'd do well to build on what we've learned about the anthropology of war, in order to develop an anthropology of conflict resolution too. Let's spread the word.

Forensics and Legal
Anthropology

A nthropology is not just an intellectual exercise that satisfies
our curiosity about human cultures, past and present. It's also a
discipline that can give us practical insight into some of the biggest
challenges we face as our planet's 21st-century caretakers. This lecture
expands on that theme by looking at the fascinating topics of forensics
and legal anthropology. Forensic anthropologists essentially analyze and
identify unknown human remains, among other things. They do so for legal,
criminal justice, and humanitarian purposes.

A Hypothetical Discovery

By studying skeletal remains, forensic anthropologists can determine a victim's age, sex, ancestry, health, and more. Perhaps most importantly, they can determine identity and cause of death.

Let's walk through the basic process that a forensic anthropologist would go through to determine a victim's biological profile and cause of death. Imagine we're in New Mexico. A construction crew on the Turquoise Trail in Cerrillos just unearthed a rather grim discovery: a battered box containing an assortment of what appears to be human bones.

Once the bones are in hand, it's off to the lab. We inventory what we have as we lay them out on a table in their respective anatomical positions. Specifically, we determine that these are indeed human bones.

Next, we work out whether we have 1 or more individuals here. For our hypothetical remains, the bone ratios point toward there being just 1 individual.

Finally, we figure out whether we have a male or a female here. The pelvis is the best indicator for biological sex for 3 reasons:

- The pelvic brim—the hollowed-out ring inside the pelvis—is heart shaped in males but more circular in females.
- The sacrum is the tip of the tailbone. Here, a forensic anthropologist would look to see if it points more inward or outward. If the sacrum is pointing inward, the specimen is a male.
- The pubic angle or arch is the outside angle created by the bottom front of the pelvis. If the pubic arch is less than 90 degrees wide, that indicates a male. Arches greater than 90 degrees are indicative of a female pelvis.

Now for our hypothetical discovery: Let's say we examine the remains and discern a heart-shaped pelvis with an inward-pointing sacrum and a pubic arch under 90 degrees. The conclusion: We're looking at a male.

Alternatively, we can usually work out biological sex simply by examining the skull. Just like the pelvis, there's sexual dimorphism in human skulls:

- Male skulls tend to have square-shaped chins. A female skull will have a more rounded, almost pointed chin.
- The back of the jaw is also an indicator: On males, we'll see more of a 90-degree angle, versus the female's wider Angle.
- Another feature that can help discern the biological sex of human remains is the orbital margin—or eye sockets. If we look to the top edge of the socket, we see the supraorbital margin or

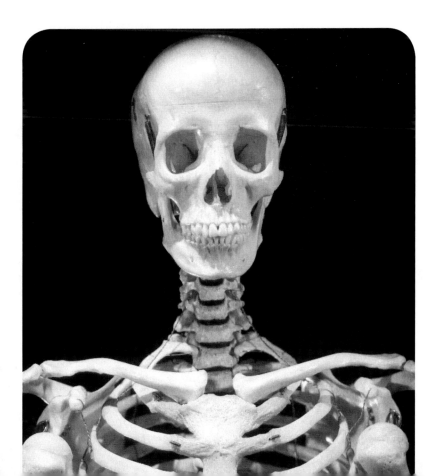

edge. In females, this edge is sharper. In males, we see more rounded upper edges, with larger brow ridges and a more sloped forehead. By comparing a specimen with established comparative measures, we can accurately determine biological sex most of the time.

Returning to our hypothetical victim: The chin, jaw, and orbital margins all confirm what we concluded from the pelvis. We have a male.

Aging the Remains

But what was his age at death? Here again, the skull can help us. Skulls have sutures that join the bones of the skull, and as we age, they begin to fuse. Going straight down the middle of the skull is the sagittal suture, which fuses between the ages of 29–35. On our hypothetical skull, the sagittal suture is about halfway fused. We can conclude that the male was in his early 30s.

How long has he been dead? To figure this out, we "read" insects. Some insects arrive right away and get right to the corpse. Insects

like blowflies or maggots can be present for a few days to a year or more. Seeing where blowflies are in their development cycle can help us get an accurate date of death down to the day.

For our hypothetical human remains, those "early" insects are long gone. These bones are at least a couple years old, and probably older. We'll have to send them out for some further testing.

Forensic Anthropology

Forensic anthropology has a comprehensive protocol and toolkit to work out the details we need to assemble a profile for unidentified human remains. In our hypothetical lab, we got a great start with just a few quick procedures: We definitely have a male in his early 30s who died at least a couple years ago.

Forensic specialists can bring new tools: forensic archaeology, DNA analysis, chemical dating techniques, and so on. There are even dental, pollen, geological, and isotope analysis. Isotope analysis is a fascinating technique that can even be used to reconstruct the diet of prehistoric peoples.

Forensic anthropologists work in a wide variety of occupations and settings. Many forensic anthropologists work as professors and academic researchers, but most of them work outside of academia in government agencies, the armed forces, medical examiner's offices, and more.

Following the catastrophic loss of life in tragedies like the 9/11 terrorist attacks, forensic anthropology has taken on a new humanitarian dimension. Forensic anthropologists have increasingly been called on when it comes time for identifying victims in mass-casualty events.

We can also see the growing importance of forensic anthropology through the lens of the US Congress. In the 1990s, 2 important pieces of federal legislation were passed—and they both demonstrate the broad social role that forensic anthropologists are playing outside of the academy.

- One is the Native American Graves Protection and Repatriation Act, or NAGPRA. Passed in 1990, NAGPRA

requires that the remains of any Native Americans, once discovered, be turned over to the appropriate contemporary Native American nation. This protection covers gravesites and artifacts, not just human remains.

- In 1996, the Aviation Disaster Family Assistance Act was signed into law. This act basically formalized a federal infrastructure aimed specifically at dealing with the aftermath of mass casualties. This infrastructure supports and dispatches DMORT units, or Disaster Mortuary Operational Response Teams.

- These teams include forensic anthropologists and a host of other experts. There are funeral directors, medical examiners, pathologists, medical records experts, fingerprint specialists, forensic dentists, X-ray technicians, and more.

- Following a string of airline accidents, the first DMORT team was launched in the 1990s. The role of forensic anthropology was integrated into the DMORT system, specifically for identifying remains in humanitarian service to the deceased and their families.

Anthropologist Snapshot: Mary Manhein

Mary Manhein is a gifted storyteller and an honored member of the prestigious American Academy of Forensic Sciences. She is the archetype of the forensic anthropologist. In her career, she handled over 1000 forensic cases, and was called upon by law enforcement agencies from all over the US.

Notably, she was a critical force in establishing the Louisiana Repository for Unidentified and Missing Persons Information Program. This program brings closure to case after case of unidentified and missing persons in Louisiana.

Manhein has been an active DMORT team member for years. Her notable cases have ranged from Hurricane Katrina to serial killers. One of her toughest cases was the recovery of 7 astronauts following the *Columbia* space shuttle disaster in 2003. The account of this surreal experience is a standout chapter in her book, *Trail of Bones*.

Anthropologist Snapshot: Gillian Fowler

Another major figure in this field is Gillian Fowler. Fowler took up forensic anthropology after hearing horrific stories told by Guatemalan refugees in Mexico. The refugees had fled a civil war that claimed some 200,000 lives.

Fowler happened to be teaching English in Mexico when these stories compelled her to get a degree in forensic archaeology and eventually to head to Guatemala. There, she spent 6 years exhuming mass graves.

She was determined to help families find loved ones who were taken away and never seen again. Their experience of loss without closure had gone on for over 2 decades. But Fowler's efforts helped change that.

The healing nature of this work was evident in Fowler's lab where victim's families actually visited their skeletal remains. Fowler's account mentions family members laying flowers, lighting candles, praying, and crying.

The candles and flowers in Fowler's lab are clear signs that the power

The Aviation Disaster Family Assistance Act formalized a federal infrastructure to deal with the aftermath of mass casualties.

of forensic anthropology extends beyond legal and criminal cases. Forensic anthropology has a critical, deeply effective, humanitarian purpose. Furthermore, the data that forensic anthropology produces has been used in criminal courts to successfully prosecute many of the leaders responsible for mass killings.

Suggested Reading

Clarke, *Fictions of Justice*.

Ferllini, *Silent Witness*.

Manhein, *The Bone Lady*.

Steadman, *Hard Evidence*.

Questions to Consider

1. How is it that forensic anthropologists can actually examine unidentified human remains, even if all that remains is a skull and/or pelvis?

2. How and why has the work of forensic anthropologists integrated a new "humanitarian" focus in the 21st and late 20th centuries?

Forensics and Legal Anthropology

As we began to see in our last lecture, anthropology is not just an intellectual exercise that satisfies our curiosity about human cultures, past and present. It's also a discipline that can give us practical insight into some of the biggest challenges we face as our planet's 21st-century caretakers. Today we're going to continue that theme by looking at the fascinating topics of forensic and legal anthropology.

What can anthropology contribute to criminal justice and the legal system? Well, when you hear news stories about the tragic discovery of human remains, for example, they always include a pretty good description of the deceased. They may even have an actual name. But how do they do that?

You see, there's people behind the scenes who work that all out, and, a lot of them are anthropologists. Of course, they are. We've seen how biological anthropologists retrieve and identify hominin bones from millions of years ago. Why not do the same exact thing with more recent bones? We'll get to legal anthropology later in the lecture, but let's start with forensics.

Linguistically the origins of forensics is *forensis*, which we translate as of or before the forum. The forum was where Roman citizens once presented criminal cases in public. True, modern Romans have swapped the forum for courtrooms. Yet, modern forensics, in a tip of the hat to Ancient Roman justice, is still about good evidence, and a well-built case.

Now, when it comes to anthropology, forensic anthropologists essentially analyze and identify unknown human remains, among other things and they do so for legal, criminal justice, and now humanitarian purposes. Remarkably, by studying skeletal remains, forensic anthropologists determine a victim's age, sex, ancestry, health, and a lot more. And perhaps most importantly, they determine identity and cause of death—two critical pieces of information in legal, criminal, and humanitarian contexts.

So, how about a hypothetical discovery? Let's walk through the basic process that a forensic anthropologist would go through to determine a victim's biological profile and cause of death. Even if all we have are barely recognizable human remains, forensic anthropologists can still work out the mystery.

Imagine we're living in New Mexico, teaching forensics and biological anthropology there. And since we started this teaching gig, we've been on the call list at the coroner's office. You see, she regularly calls us in to help figure out what's going on when someone discovers unidentified human remains.

And today, while at school, we got the call. Turns out, a construction crew out on the Turquoise Trail in Cerrillos just unearthed a rather grim discovery—an assortment of what appears to be human bones. When we get there, we discover, that we have is a good piece of a skull, most of the pelvis, and a few hand and arm fragments. And so be it. That's all we need. I mean, more would be great, but this will do it. So, with bones in hand, it's off to the lab and let's work this mystery out.

Now, once we've recovered the perceived victim's remains, we inventory what we have as we lay them out on a table in their respective anatomical positions. Specifically, we determine that, indeed, these are human bones. Next, we work out whether we have one or more individuals here. And if there are multiple people, we have to figure out how many there are.

For our hypothetical remains, the bone ratios, which is the number and types of bones we find, well they're pointing pretty clearly in the direction of just one individual. There's zero indication that we have any commingled remains. That said, if you wanted to double check, we could always bring in some DNA analysis. But again, things look good. This is almost certainly one person. Now, let's figure out whether we have a male or a female. Let's work out the biological profile. And we're fortunate to have a nearly complete pelvis because it's the pelvis that's the best indicator for biological sex. Why is that?

Well, when forensic anthropologists look at a pelvis, they can see certain characteristics that distinguish a male from a female. Let's look at three. First, there's the general shape of the human pelvic brim. That's the hollowed out ring inside the pelvis. If the pelvic brim is heart shaped, we've got a male. A more circular pelvis is going to indicate it could a female.

Similarly, we can look at the sacrum. The sacrum, is the tip of the tailbone, at the rear of the pelvis. Here, a forensic anthropologist would look to see if it points more inward or outward. If the sacrum is pointing inward, again, we're looking at a male. But what if the remains were so badly destroyed that we didn't have the sacrum. What if we just had a front fragment of the pelvis?

Well, we also can look at the pubic angle or arch. This is the outside angle created by the bottom front of the pelvis. If the pubic arch is less than 90 degrees wide, say in the 50 to 80-degree range, that's a classic sign that indicates a male. And arches greater than 90 degrees are indicative of a female pelvis. And, we can figure out what's going on with these differences if we consider their implications on birthing a child. The wider, more rounded female pelvis has evolved to facilitate childbirth.

Now, putting all of this together for our hypothetical discovery on the Turquoise Trail, let's say we examine the remains, and we discern a heart-shaped pelvis, with an inward-pointing sacrum and a pubic arch under 90 degrees. What's our conclusion? We're looking at a male.

But get this. We can usually work out biological sex simply by examining the skull. Just like the pelvis, there's sexual dimorphism in human skulls. Check it out.

Again, we can look at three areas to work this out. First the chin. With our male skull, we'll see a square shaped chin. A female skull will have a more rounded, almost pointed chin. Or, there's the back of the jaw. Here, on the male, we'll see more or a 90-degree angle versus the female's wider angle. And last, another great skull feature that helps us discern the biological sex of human remains is the orbital margin—our eye sockets. If we look to the top edge of the socket, we see the supraorbital margin or edge. In females, this edge is sharper, and in males, we see more rounded upper edges, with larger brow ridges, and a more sloped forehead. Then, by comparing our specimen with established comparative measures, most of the time, we can accurately determine biological sex.

So, returning to our hypothetical victim, let's say that we examine the skull and find that the chin, jaw and orbital margins all confirm what we concluded from the pelvis, there's no doubt we have a male here. But how old is this guy? What was his age at death?

Here again, the skull can help us. You see, when we're born, we've got these sutures or seams on our skull. And, without them, childbirth simply wouldn't work. The sutures join the bones of the skull, and as we age they begin to fuse.

And right at the top, going straight down the middle, we have the sagittal suture. This is a great way to read the age of our victim, because, for all of us, our sagittal suture fuses between the ages of 29-35. Sure, there's mild variation here, but we can use this technique with great certainty.

So, if we look at our hypothetical skull, and we see his sagittal suture is about half-fused, then we can conclude that this guy was probably in his early 30s.

Information like age and sex helps develop a biological profile for unidentified remains, but one last analysis we need to do, is to work out how long it's been since our 30-something, male victim met his end. How long has he been dead? Well, we have a great way to do this, but it's not for the light-stomached. Simply put, we use insects. Actually, it's more like we read the insects.

You see, some insects arrive right away and get right to the corpse. Insects like blowflies or maggots can be present for a few days or even up to a year or more. Blowflies, amazingly, when we see where they are in their development cycle can help us get an accurate date of death down to the day. Or at least really, really close to that. For our hypothetical human remains, those early insects are long gone. These bones are at least a couple years old, and maybe even a little older. We're going to have to send them out for some further testing.

So, it's clear. Forensic anthropology has a comprehensive protocol and toolkit to work out the details we need to assemble a profile for unidentified human remains. In our hypothetical lab, we got a great start with just a few quick procedures. We definitely have a male, he's in his early 30s and he died at least a couple years ago. It wasn't a recent death. And as we need to work out more, we can collaborate with other forensic specialists who bring new tools for our forensic toolkit.

There's Forensic archaeology, DNA analysis, chemical dating techniques. There's even dental, pollen, geological and isotope analysis. Isotope analysis is a fascinating technique that can help us determine a lot of things. It can even be used to reconstruct the diet of prehistoric peoples.

Like many professional anthropologists, forensic anthropologists work in a wide variety of occupations and settings. As you can imagine, many forensic anthropologists work as professors and academic researchers, but actually, most of them work outside of academia in government agencies, the armed forces, medical examiner's offices, and so many more places. One increasingly important and necessary application of forensic anthropology puts these men and women to work in non-governmental organizations and war crimes tribunals.

The United Nations reports that into the 21st century, we can expect increased instability arising from human conflict and natural disasters. And we're going to need forensic anthropologists to help work us through this.

For example, following the catastrophic loss of life in tragedies like the 9/11 terrorist attacks, forensic anthropology has taken on a new humanitarian dimension. Because of the skills and methods we've already talked about, forensic anthropologists have increasingly been called on when it comes time for identifying victims in mass-casualty events.

We can also see the growing importance of forensic anthropology through the lens of the US Congress. In the 1990s two important pieces of federal legislation were passed—and they both demonstrate the broad social role that forensic anthropologists are now playing outside of the academy. The first of these two legislative acts is the Native American Graves Protection & Repatriation Act, otherwise known as NAGPRA. Passed in 1990, NAGPRA requires that the remains of any Native Americans, once discovered, must be turned over to the appropriate contemporary Native American nation. And this protection covers gravesites and artifacts, not just human remains.

Then in 1996, the Aviation Disaster Family Assistance Act was signed into law. This act basically formalized a federal infrastructure aimed specifically at dealing with the aftermath of mass casualties. This infrastructure supports and dispatches DMORT teams otherwise known as Disaster Mortuary Operational Response Teams.

These teams include forensic anthropologists and a whole host of other experts. There are funeral directors, medical examiners, pathologists, medical records experts, fingerprint specialists, forensic dentists, x-ray technicians, and the list goes on.

For logistical purposes, the US is divided into ten regions, each of which has a dedicated DMORT team that is ready for deployment anytime there's a possibility of mass casualties. And once human remains are actually recovered, the DMORT team process begins in earnest. It's actually a two-part system, and it's all brought together with a remarkable computer program called VIP. The Victim Identification Profile. This program can integrate nearly a thousand different data types.

On one end of the system, families and others help the team collect and input all kinds of antemortem or pre-death data. That can be things like photographs, ex-rays, DNA, dental records, whatever they can get. Then, on the other side of things, the forensics experts recover the physical remains.

And as they inventory and analyze these remains, they enter that postmortem data into the VIP program. Following a string of airline accidents, the first DMORT team was launched in the 1990s. And from the beginning, the role of forensic anthropology was central. And more importantly, it was integrated into the DMORT system, explicitly for identifying remains in humanitarian service to the deceased and their families.

OK. Let's take a quick pause here. We've just learned that forensic anthropologists contribute their knowledge of human biology to one, assist in criminal and legal processes, and two, more recently, they've become critical responders after natural disasters and human catastrophes. And before that, we took a peek to see how forensic anthropologists actually create profiles of victims by analyzing their remains.

So now, let's go deeper. Let's dig in, and see some real forensic anthropology in action. I want to look at snapshots of forensic anthropologists, who I think are nothing short of superheroes. First, the Bone Lady and I'm not talking about that Cleveland Browns superfan with dog bones in her hair. No. I'm talking about Mary Manhein. A gifted storyteller, and an honored member of the prestigious American Academy of Forensic Sciences. Manhein is the archetype of the forensic anthropologist. In her career, she handled over 1,000 forensic cases and was called upon by law enforcement agencies from all over the United States.

Notably, she was a critical force in establishing the Louisiana Repository for Unidentified and Missing Persons Information Program. This program brings closure to case after case of unidentified and missing persons in Louisiana. In fact, the program helped find and reconnect over half of the thousands of missing people cases resulting from regional hurricanes since Katrina. Hearing that, I'm sure you won't be surprised to learn that Mary Manhein has been an active DMORT team member for years. Her notable cases have ranged from Hurricane Katrina to serial killers. And actually, come to think of it maybe you've seen her work on television. For a while, she did some clay facial reconstructions of unidentified persons for the Fox TV show Missing. And, you know what, if her work still doesn't ring a bell, I know you've heard of one of her toughest career cases. The recovery of seven astronauts following the Space Shuttle Columbia disaster in 2003.

As part of her regional DMORT team, Manhein was called up once it was clear that the Columbia had exploded over Hemphill, Texas. Oddly enough, her father's hometown, Hemphill leaped to the occasion, as Manhein remembers thousands of people helping the cause. There were advance teams all across the region. And when they identified human bones or other remains, Manhein and others rushed to the scene. The riveting account of this surreal experience is one of my favorite chapters of her book, *Trail of Bones*.

So that's one forensic anthropology giant. Let's go to Guatemala where I want you to meet another major figure in this field—Gillian Fowler. Starting off as a secondary geography teacher, Fowler took up forensic anthropology after hearing the horrific stories told by Guatemalan refugees in Mexico. The refugees had fled a civil war that claimed some 200,000 lives. Fowler just happened to be teaching English in Mexico, when these stories compelled her to get a degree in forensic archaeology and eventually to head to Guatemala. And it was there that she spent six years exhuming mass graves. A quick read from her published field journals provides a powerful dose of the grimness, and, the inevitable emotional toll of this important work. Fowler writes

> I pick up a skull and begin to gently rub away some of the dirt. This young man was executed with a single gunshot to the back of the head, a common form of execution in this area strewn with clandestine graves. I begin to lift the ribcage where I find signs of another impact wound; he was shot twice from behind. A quick check of the rest of the skeleton gives me an idea of his age at death, less than 21. Next to him lies the skeleton of his father. What must their last thoughts have been before the bullets struck?

She continues

> I glance up. The woman beside me is peering into the grave, watching my every move with a look not of despair or fear but of strength and courage. She is their wife and mother, the only member of her family to survive the horrors that were inflicted upon so many some 25 years ago.

The moment Fowler arrived in Guatemala, she found her calling. She was determined to help families find loved ones who were taken away and never seen again. Their experience had gone on and on for well over two decades. This is loss without closure. But Fowler's efforts helped change that. The healing nature of this work was evident in her lab where victim's families actually visited their skeletal remains. Fowler's account mentions family members laying flowers, lighting candles, praying, and, of course, crying. The gist of it is that putting to rest a loved one that was taken through a horrific tragedy like the Guatemalan civil war doesn't bring the people we miss back, but it does bring them home.

The candles and flowers in Fowler's lab are clear signs that the power of forensic anthropology extends well beyond legal and criminal cases. Forensic anthropology has a critical, deeply effective, humanitarian purpose as well. But it's important to note that the work of Fowler and other like-minded researchers does more than help with closure following atrocities like a civil war. The data that they produce has been used in criminal courts to successfully prosecute many of the leaders responsible for mass killings, all the way from soldiers up to generals.

This role of anthropology in the courts provides us with an opportunity to think more broadly about the way that anthropologists study justice across cultures. Because in fact, there's a deep tradition of what we call legal anthropology, and is dates back to the some of the founders of modern anthropology. For example, in his 1926 book, *Crime and Custom in Savage Society*, one of the early builders of the anthropological method, Bronislaw Malinowski, examined non-western cultures, and, among other things, their versions of law and order.

As we learned in our previous discussions on race, culture, and religion, social Darwinism had a profound influence upon early anthropologists and their ideas about the history and diversity of human cultures. This same perspective affected their interpretation of the world's legal systems. As early as the 1860s, the jurist Henry Sumner Maine used legal rights to classify cultures. Specifically, he distinguished between societies in which legal rights are based on contractual agreement versus class or social status. But it's classic legal anthropology who gives us a clearer view of this ranking system.

E. Adamson Hoebel studied the legal systems of a number of world cultures. He checked out the Cheyenne, the Comanche, the Shoshone, the Pueblo, and even Pakistani legal systems. His seminal 1954 text, *The Law of Primitive Man*, outlined a comprehensive scheme of human legal systems. Hoebel ranked each on their complexity and, believe it or not, their perfection.

His complexity vector was right in line with earlier anthropological approaches that looked at evolutionary models of culture, from primitive or simple to complex. But the perfection vector was rather new. Without going too far into this one, he's basically looking at another evolutionary arc. This one looked at the social realities of law versus law in pursuit of ideal perfection.

Today, Legal Anthropology is thriving, and it's as interdisciplinary as ever. And like forensic anthropologists, legal anthropologists are citizen scholars because, in addition to teaching and research, their applied work extends far beyond the academy. So let's close with one final case study to get a taste of some amazing legal anthropology work.

Kamari Maxine Clarke at Yale University uses cultural anthropology to examine international criminal courts and Africa. After a comprehensive analysis, Clarke concludes that indictments and prosecuting leaders are important, but that those efforts fail to confront the foundational causes of some crimes. So what does work? What might work? Well here's one idea from Africa.

In a 2013 New York Times Op-Ed titled "Treat Greed in Africa as a War Crime," Clarke mentions a summit meeting involving leaders of the African Union. And it was in that meeting, she notes, that AU leaders proposed expanding the power of the African Court on Human and Peoples' Rights. Specifically, AU leaders called for expanding corporate criminal liability for the illicit exploitation of natural resources, and for trafficking in hazardous waste. This idea might not align with pure entrepreneurial vim and vigor, but according to Clarke, it could go a long way towards turning around the ecological degradation that undermines, not only the future of Africa but the future of our world.

And now, Clarke's second idea, the big one, is that the post-colonial complexities of cross-cultural justice and human rights are incompatible with the International Criminal Court based in the Hague. In fact, AU leaders openly rallied its members to refuse cooperation with the ICC's plan to arrest the president of Sudan back in 2009.

In her book, *Laws and Societies in Global Contexts*, legal anthropologist Eve Darian-Smith, calls for a more global and inclusive approach when it comes to international cooperation on transnational issues. Such an approach is exactly what African Union leaders appear to be pining for. International collaboration that respects local and regional perspectives, even where they may contradict prevailing ideas and policy.

And with that, my anthropology friends, we've come to the end of another lecture. Today, we've applied anthropology in legal, criminal justice, and humanitarian contexts. And we saw that anthropology is not an ivory-tower discipline. Rather, it offers hands-on expertise that helps us meet real-world 21st-century challenges.

As we've said before, anthropologists are bridge builders. Whether we're addressing human rights in Africa or the ravages of a hurricane in Louisiana, we take an interdisciplinary approach to problem-solving. And that interdisciplinary mindset well, that's the strength that supports our anthropological mission to connect different ways of thinking about humankind and the world we live in.

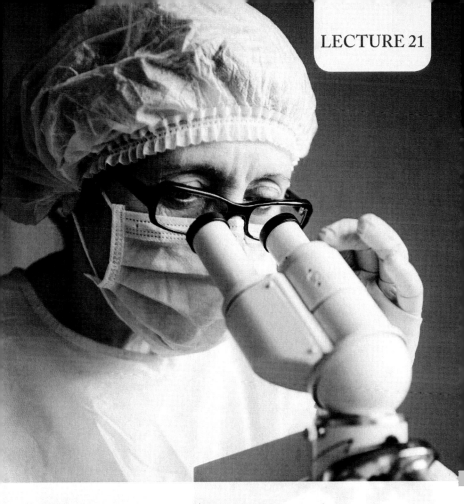

Medical Anthropology

This lecture discusses one of the largest specialization areas in contemporary anthropology: medical anthropology. Contemporary medical anthropology is a definitive example of applied anthropology. From an interdisciplinary, 4-field approach, medical anthropologists chart the depths of the biology-culture nexus of human health and healing. The central ethos of medical anthropology is inspired by the idea that human health (good and bad) is much more than biology.

The Nocebo Effect and Medical Anthropology

Researchers who study the nocebo effect—damage caused by negative thoughts—essentially look at the various ways that our thoughts can harm us. A recent study showed the sheer power of thought on human heath.

In this nocebo study, patients were given sugar water and told that it would induce vomiting. A full 80% vomited upon drinking that sugar water. The power of the brain and what we think can definitely impact our biological health. Therefore, we'd do well to understand what's going on with this body-mind connection.

That's one of the reasons medical anthropology emerged in the first place. Doctors and other health professionals, long aware of the placebo and nocebo effect, increasingly recognized that illness and health aren't exclusively biological.

Because anthropologists are equally at home with biology and culture, anthropology emerged as a great way to think about medicine and health beyond, yet inclusive of, our biology.

Malaria

About half the world's population is vulnerable to malaria. Hundreds of thousands of people die of malaria every year. It's a disease that not only kills people, but also drains households of the critical labor they need to keep bellies full throughout the year.

Malaria is caused by certain blood parasites that are carried and transmitted by mosquitos, so having malaria is definitely a biological event. But it's also much more.

Malaria used to be a major public health issue in the United States. Prior to the mid-20th century, malaria crippled the nation's labor force and military. For example, when Franklin D. Roosevelt signed the Tennessee Valley Authority into law, malaria affected almost a third of the Tennessee Valley population.

In World War II, the US lost somewhere around 60,000 soldiers to malaria. In the 1940s, the military enacted the Malaria Control in War Areas program. A few years later, that was transformed into the Communicable Disease Center. Today, it is the CDC, or the Centers for Disease Control and Prevention.

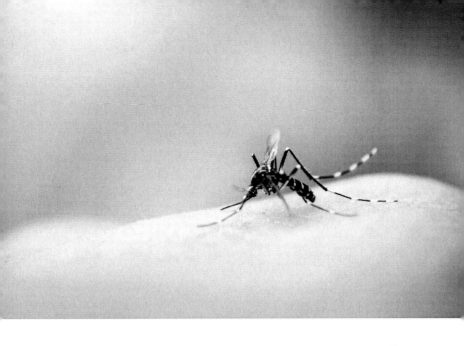

The initial campaign against malaria was nothing short of remarkable. With the widespread application of DDT, the CDC partnered with state and local health agencies in 13 states. And starting in 1947, they commenced spraying millions of homes, and with stunning results. In less than 5 years, the CDC completed its job. Malaria and reports of malaria infections were eliminated from the US.

Most Americans will never experience malaria in their lifetimes. But the eradication of the disease in the US was not a question of human biology at work. Without the collective action and organized campaigns of the 1940s, malaria would still be endemic in the US. Rather than biological adaptations or vaccines, the economy and political history of the US wiped malaria out.

"Pocahontas Goes to the Clinic"

A great example of medical anthropology in action is a piece by Cheryl Mattingly called "Pocahontas Goes to the Clinic." Mattingly took her medical anthropology to a clinic. Her work showed that biomedicine is most effective when it's paired with cultural understanding and competence. In her study, she revealed how cultural misunderstanding undermines the quest for positive health outcomes.

One of her examples involves sickle cell anemia patients, particularly black teenage males. Mattingly documented how black teen males with sickle cell anemia are often mislabeled as "med-seekers."

With regular hospital visits for pain treatment, the sociocultural context of being a black teenager in the US carries the additional burden of being labeled a troublesome med-seeker, when in fact these kids need medical treatment.

Another angle from Mattingly's study featured Spiderman. Mattingly described a boy who was neutralized by fear of medical personnel and equipment. Doctors needed to test the boy's mobility to properly diagnose and treat him, but he was nonresponsive.

Mattingly explained that as soon as the attending physician found a common cultural frame through which to relate to this boy, everything changed. They assuaged the boy's fears and turned him into a compliant patient by talking about Spiderman. When the medical procedures were

translated into superhero tests aimed at helping this young patient and his crime-fighting ambitions, the boy leaped into action.

Adequate biomedical treatment was made possible only after the boy jumped out of his shy demeanor. And that happened only because the physician built a cross-cultural bridge of understanding. Mattingly shows us with her vignettes that, along with technical prowess, we also need health professionals with strong cross-cultural skills.

Illness versus Disease

In medical anthropology, illness is defined as a cultural experience. For example, consider someone who is a diabetic. The illness is not diabetes. Illness is the person's subjective experience as a diabetic. Arthur Kleinmann, a foundational figure in medical anthropology, tells us that illness is an individual's "culture-bound understanding of the event."

It's the business of medical anthropology to separate the subjective experience of illness from the biological phenomenon of disease. Treating a disease is not the same as treating an illness. Disease requires a physician to fix what's going on in the body. But illness requires someone to fix what's going on in the patient's head—the subjective experience. Disease is not culture bound, but illness is.

As an example, take diabetes in the US versus Mali. Being a rural Malian diabetic is quite different from having this condition in the US. Namely, in rural Mali, one may not even be diagnosed as a diabetic. One might not even get tested for diabetes because their illness— their subjective health experience— could very likely be attributed to some other condition.

Even in the event of a diagnosis for a Malian, the treatment plan presents a number of complications. Many families have to scrape to buy a single dose of antibiotics, let alone a recurring expense for insulin injections. Even if the medicine, supplies, and test kits were made affordable, there may be no refrigeration or electricity for the insulin.

Health

Anthropologists need a precise yet inclusive definition of health. The World Health Organization (WHO) defines health as "complete

physical, mental, and social well-being." And they add that it involves "the capability to function in the face of changing circumstances."

The WHO also strives to promote the highest possible level of health so that people can live a "socially and economically productive life."

The WHO's definition of health clearly goes beyond mere biology. And this anthropological perspective is not coincidence. Medical anthropologists play a host of roles in the WHO and in related institutions throughout the world.

Case Study: Medical School

The rest of this lecture takes a look at 2 classic medical anthropology case studies. First up is a 1993 study by Byron J. Good and Mary-Jo DelVecchio Good. The Goods studied the peculiar culture of first-year students at Harvard Medical School.

One section of this study described the close and rather tender relationships that many medical students develop with their first cadavers and sometimes their surviving families. After spending

hours and hours with a cadaver, the students realize that the deceased's selfless gift is what opened the path to becoming a practicing physician.

Another observation from the study is that as medical students go through their first year, the experience literally changes the way they see the body. By breaking down the body into its constituent systems and components, the lab and lectures eventually render the complexity of the human body into a machine with working parts.

Case Study: *Flexible Bodies*

A second classic medical anthropology case study comes from anthropologist Emily Martin. In her 1994 book, *Flexible Bodies*, Martin delves deep into the collective American psyche to learn about how non-physicians understand the immune system.

Martin did a comprehensive field study at 3 sites. She worked at an immunology research lab. She volunteered at a house for HIV-positive individuals. And she also worked as an AIDS activist. In all, she did some 200 interviews, along with untold pages of field notes from her day-to-day observations.

Martin shows that the metaphors we use reveal a lot about how we understand our immune system. Specifically, she showed that in the US, militarism is a common metaphor for medicine and treating disease. We *fight* cavities and *battle* cancer.

But ultimately, Emily Martin's study shows that in the US, people are shifting from militaristic, fortress-based metaphors for the body and health. Instead, from the age of AIDS onward, she sees our metaphors focusing more on dynamism and flexibility.

As they entered the 21st century, Martin says, Americans were no longer focusing on the idea of the immune system as a battle-ready army of white blood cells. Rather, they were coming to see and value adaptability in the immune system. This new focus on flexibility has broader cultural ramifications, including openness to new ways of treating disease and maintaining our health.

Suggested Reading

Brown and Closser, *Understanding and Applying Medical Anthropology*.

Packard, *The Making of a Tropical Disease*.

Wilkinson and Kleinmann, *A Passion for Society*.

Questions to Consider

1. What's the difference between illness and disease, and why do medical anthropologists underscore these differences?

2. In the hospital, in what ways might cultural differences or misunderstandings result in making you sicker?

3. How do anthropologists contribute to working out global health challenges?

Medical Anthropology

T oday, we're going to talk about one of the largest specialization areas in contemporary anthropology, Medical Anthropology. And, I'll warn you the following lecture may seriously impact your health. And I'm hoping in a good way. You see, once you start thinking about health from an anthropology point of view, things like illness, germs, and hospitals will never look the same again.

Take magical death, for example. Do you ever worry about that? It's kind of scary, and I assure you that it's real. Basically, it's a condition in which your thoughts can kill. Seriously. It's called the nocebo effect. The word *nocebo* comes from a Latin word meaning to harm, and the nocebo effect is the opposite of the placebo effect—which may be more familiar to you. Researchers who study the nocebo effect essentially look at the various ways that our thoughts—just our thoughts—can harm us. And it turns out, that the power of negative thoughts can not only make you sick, it can actually kill you. This is the phenomenon we call magical death.

Take the 18th-century story from a Vienna medical school. It was here that a crew of students pulled what became a tragic prank. You see, they bound up a bothersome peer, and blindfolded him. As he struggled to break free, the group explained that they were going to cut off his head.

So, with his head on the chopping block, pranksters grabbed a wet cloth and dropped it on the poor guy's neck. The moment it hit his skin, the young man died on the spot. His brain, the power of his expectations and thoughts ostensibly transformed that wet cloth into a razor sharp blade. And there are similar stories like this across cultural traditions. The nocebo effect is real. So let's start with that to begin our exploration into medical anthropology.

Here's a more recent and scientifically documented study that shows the sheer power of thought on human health. In this nocebo study patients were given sugar water and told that it would induce vomiting. What do you think happened? How many of these people actually threw up? Was it 10 percent? maybe as high as 25 percent? Would you believe that a full 80 percent, a clear majority, vomited upon drinking that sugar water? The power of the brain and what we think can definitely impact our biological health. So we'd do well to understand what's going on with this body-mind connection.

That's one of the reasons medical anthropology emerged in the first place. Doctors and other health professionals, long aware of the placebo and nocebo effect, increasingly recognized that illness and health aren't exclusively biological. And because anthropologists are equally at home with biology and culture, anthropology emerged as a great way to think about medicine and health beyond, yet inclusive of, our biology.

So the big picture, the central ethos of medical anthropology is inspired by one general idea. That is, that human health good and bad is much more than biology. But what do we do with that? Well, take malaria for example About half the world's population is vulnerable to suffer from malaria. Despite great progress and new medications in recent years, hundreds of thousands of people die of malaria every year. And millions and millions more suffer through it. And as someone who's had it over a half dozen times, I can tell you, malaria can be horrible.

Every year in my project communities in Mali, we do our best, but we witness malaria deaths, particularly in elderly and infant populations. It's a disease that not only kills people, but it also drains households of the critical labor they need to keep bellies full throughout the year. And getting malaria, experiencing it, treating it, are all shaped by sociocultural, political, and environmental factors. It's bigger than sheer biology.

Malaria is caused by certain blood parasites that are carried and transmitted by mosquitos. So, having malaria is definitely a biological event, let's not forget that. But, it's so much more than that. Let's work this out with a brief history of why only half the world's population has to worry about this disease. Malaria, you may be surprised to learn used to be a major public health issue right here in the United States. Historical sources and government data are clear about that. Prior to the mid-20th century, malaria crippled the nation's labor force and military.

Right here in North America, our workforce was regularly hampered by this pervasive disease. When Franklin D. Roosevelt signed the Tennessee Valley Authority into law, malaria affected almost a third of the Tennessee River valley population. It was like the common cold, except, this one knocks you out of the workforce for longer in fact, as we've heard, it can kill you. Similarly, with the world wars and the collection of US territories around the world, the US military also reeled from malaria.

In World War II alone, sources estimate that we lost somewhere around 60,000 US soldiers who died not from bullets, but from malaria. In the 1940s, the military enacted the Malaria Control in War Areas program, which a few years later was transformed into the Communicable Disease Center, or what we now call the CDC or the Center for Disease Control. The short of it was, that the US brought home its efforts to combat malaria. The CDC brought malaria eradication not only to US military bases in the south but to the general public as well. Ultimately these efforts established the CDC as a national malaria eradication agency.

Despite the prevalence of malaria in the world today, the US did wipe out malaria on its own soil. If you contract malaria as an American citizen today, it's because you picked it up somewhere else. The initial campaign was nothing short of remarkable. With the widespread application of DDT, the CDC partnered with state and local health agencies in 13 different states. And starting in 1947, they commenced spraying millions of homes with stunning results. In less than five years, the CDC completed its job. Malaria and reports of malaria infections were eliminated from our shores. Following this success, the CDC expanded its mission to incorporate both domestic and international programs. And they now work on all diseases, including everything from hepatitis to Ebola.

So, most Americans will never experience malaria in their lifetimes. But the point of all this is that the eradication of the disease in the US was not a question of human biology at work. Without the collective action and organized campaigns of the 1940s, malaria would still be endemic in the US. Rather than biological adaptations or vaccines, it's the economy and political history of the United States that wiped malaria out.

But it's not only one's exposure to malaria that is bigger than biology. It's also the malaria experience and treatment as well. It all transcends sheer biology. That is to say, that if you catch malaria today, there are great medicines out there, but socio-economics make it so that the majority of Africans who get malaria have trouble accessing these drugs. So, from how you get a disease, to how you treat it, and how you experience and understand an ailment in medicine and health, there's way more than biology at work.

Anthropologists, as we've described them in just about every one of our lectures, are bridge builders. They connect people and cultures across cultural, linguistic, and just about any other human divide. And in the case of medical anthropologists, they build their bridges to connect biology and culture. One of my favorite examples of medical anthropology in action is a curiously titled piece by Cheryl Mattingly called *Pocahontas Goes to the Clinic*.

Rather than seeking exotic cultures from the other side of the globe, Mattingly takes her medical anthropology to the clinic. To the modern hospital. Just like the ones near your hometown. And anyone who's spent any amount of time in a hospital knows that Mattingly is spot on when she describes it as a cultural border zone. I mean, whether we're talking about patients, families, doctors, nurses, or anyone else in that zone, we see all kinds of human diversity in terms of language, religion, ethnicity, gender, age, economics, and so much more.

When I was hospitalized following a trip to Ghana, West Africa, I remember being astonished to wake up at George Washington University Hospital only to find a Ghanaian nurse attending to me. It turned out, she was from the town I had left just a few days before in Africa. The clinic is indeed a border zone, and that's terrific, but there's something potentially menacing amongst all that diversity. Namely, we're talking about the potential for misunderstandings. Misunderstandings that could mean life and death. Or at the very least, discomfort versus excruciating pain.

Well, Mattingly went to her field site, the clinic, the way I went to my partner villages in West Africa. And her work shows us that biomedicine is most effective when it's paired with cultural understanding and competence. In her study, she reveals how cultural misunderstanding undermines the quest for positive health outcomes.

One of her examples involves sickle cell anemia patients particularly black teen males. Mattingly documented how black teen males with sickle cell anemia are often mislabeled as med-seekers. Sickle cell is a painful inherited blood disease, that leads to sudden attacks of very severe pain. It comes on without warning. And, the pain is so disabling that hospitalization becomes a regular aspect of living with this condition. What scares most parents of sickle cell children, is that in addition to crippling pain, this condition can cause major organ damage including brain damage. With regular hospital visits for pain treatment, the sociocultural context of being a black teenager in the US carries the additional burden of being labeled a troublesome med-seeker, when in fact, these kids need medical treatment.

One of the best vignettes from Mattingly's study actually featured Spiderman, not Pocahontas. Think about it, depending on your cultural, religious, or even linguistic background, talking with doctors can be unnerving. What if you were an elder born to a religion and culture that made it inappropriate to discuss sexual or even digestive issues with younger people, or people of the opposite biological sex? Without compromising my spiritual life, how do I speak to that kind, young, female doctor about what's really going on with my body? I can use euphemisms perhaps, but that's only going to cloud the picture that my doctor needs to see. Mattingly shows how an anthropologically-minded physician can work to understand and identify these precarious, yet fairly easy to resolve, situations.

This doesn't have to involve far-away cultures; it might involve understanding differences between adult culture and children's culture. In one case, for example, Mattingly describes a boy, who for whatever reasons, was neutralized by fear of medical personnel and equipment. And that happens to all of us. Many of us get super nervous in the clinic, even when we share a language and cultural background with our providers.

Health professionals who administer blood pressure tests explain this white coat syndrome when patients are shocked to see their blood pressure numbers in the clinic are significantly higher than when they just did the tests at home. For many of us, simply the presence of a white coat and a medical test is enough to increase our blood pressure.

Now, Mattingly explains, doctors needed to test the boy's mobility to properly diagnose and treat him. But he was non-responsive. Well, Mattingly goes on and says that as soon as the attending physician found a common cultural frame through which to relate to this boy, everything changed. The short of it is that they assuaged the boy's fears and turned him into a super compliant patient by talking some Spiderman. When the medical procedures were translated into superhero tests aimed at helping young spiderman and his crime fighting ambitions, the boy leaped into action.

It sounds like a simple solution once we hear this after the fact, but, it's critical to note that adequate biomedical treatment was made possible only after the boy jumped out of his shy demeanor. And that happened only because the physician built a cross-cultural bridge of understanding.

What Mattingly shows us with a few of these vignettes is that, equal to technical prowess, we also need health professionals with strong cross-cultural skills. And that fact is not lost on the medical profession at large. Standardized testing for nursing certification now includes anthropological content. The nursing program at my university pushes students into anthropology courses, and they've even made anthropology a major component for our health studies minor. Anthropology clearly has a mission in the world of medicine. So now that we have a better idea of this mission, let's take a closer look at how medical anthropologists break down the phenomena of sickness and health.

How does our view of illness and health change when we see them as anthropologists? Well, linguistically we find that the word ill derives from an old Norse word meaning wicked, or morally evil. These connotations of wickedness indicate that prior to our modern understanding of pathology, we attributed illness to evil or malevolent forces, perhaps supernatural forces. But that old sense of the term is far from how medical anthropologists understand illness today. Simply put, in medical anthropology we define illness as a cultural experience. For example, I am a diabetic. But illness is not diabetes. Illness is my subjective experience as a diabetic.

Arthur Kleinmann, a foundational figure in medical anthropology, tells us that illness is an individual's culture-bound understanding of the event and as a physician and anthropologist, he should know. From his dual post at Harvard as Professor of Medical Anthropology and Professor of Psychiatry, Kleinmann has inspired scores of students, some of whom, have become medical anthropology giants in their own right. But in terms of medical anthropology, we don't conflate disease and illness. It's our business to separate the subjective experience of illness from the biological phenomenon of disease. But why?

Well, treating a disease is not the same as treating an illness. For the disease, I need my physician to fix what's going on in my body but, I'll also need someone to help me fix what's going on in my head. My subjective experience. One may think the disease/illness distinction is pedantic, but if your illness, your subjective symptoms, don't fit into the objective disease framework taught in medical schools, then you can't receive adequate treatment.

Disease, according to Kleinmann, is the biological event, regardless of the cultural context. So disease is not culture bound, but illness is. And that's why our health requires more than prescriptions and surgery. We tend to focus so much on the disease, we often neglect the person. And that's significant because we seldom pay enough attention to the vastly different ways we humans subjectively experience critical health events like disease.

Take diabetes in the US versus Mali. It's the same disease. It's all about blood glucose levels, period. However, being a rural Malian diabetic is quite different from having this condition in the US Namely, in rural Mali, you may not even be diagnosed as a diabetic. You might not even get tested for diabetes because your illness, your subjective health experience, could be very likely attributed to some other condition. But even if you do get diagnosed, the kind of treatment we'd want for our children—insulin, blood testing, special diet, etcetera—in the village, that treatment plan presents a number of complications.

First, there's economics. Many families in my host community have to scrape to buy a single dose of antibiotics, let alone a recurring expense for insulin injections. But even if the medicine, supplies, and test kits were made affordable, there's no refrigeration or electricity for that insulin. So while diabetes, as a biological event, is the same disease no matter where you go nevertheless, the experience of having this condition may differ from one person to the next, and from one culture to the next.

So illness, culture-bound; disease, not culture-bound. But what about health? Sometimes it's useful to define things by describing what they are not but, that's not the case for health. To understand health as simply the absence of sickness, that doesn't do health and healing justice.

The old English root for health is more on the mark. It refers to wholeness and being whole. But, like illness, being whole doesn't look the same for everyone everywhere. Across cultures, wholeness—or health—includes multiple dimensions; there's physical psychological, emotional, and spiritual health. And we measure our health in different ways. Sometimes we think of health as feeling good, or sometimes it's simply the absence of disease. But other times, we feel healthy as a consequence of living in harmony with our own subjective morality.

That's all a bit too loose for our practical purposes because as anthropologists, we need a precise yet inclusive definition of health. And we don't have to look far to find one. The WHO, the World Health Organization, defines health as "complete physical, mental, and social well-being." And they add that it involves the capability to function in the face of changing circumstances. The WHO also strives to promote the highest possible level of health so that people can live a socially and economically productive life.

The WHO's definition of health clearly goes beyond mere biology. And this anthropological perspective is no coincidence. Medical anthropologists play a host of roles in the WHO, and in related institutions throughout the world. For example, my former mentor Peter Brown, at Emory University, serves on the WHO advisory committee on malaria. And speaking of actual medical anthropologists and what they do, in this final segment of this lecture, let's take a quick look at two classic medical anthropology case studies.

One of my favorite medical anthropology studies is a 1993 study by Byron J. Good and Mary-Jo DelVecchio-Good. Like anthropologists bustling off to a remote fishing village in the Pacific, the Goods studied the peculiar culture of first-year students at Harvard Medical School. They used anthropology to peer into the minds of actual medical students as med school transforms their understanding of the human body from mystery to mechanics.

One of the more remarkable sections of this study described the close and rather tender relationships that many medical students develop with their first cadavers, and sometimes even their surviving families. After spending hours and hours with a cadaver, the students realize that the person who provided it, the deceased, their selfless gift is what opened the path to becoming a practicing physician. And the magnitude of this gift is not lost on these medical students.

Another great observation from the Goods' study is that as medical students go through their first year or so in med school, the experience literally changes the way the aspiring physicians see the body. By breaking down the body into its constituent systems and parts, the lab and lectures eventually render the complexity of the human body into a machine with working parts. For anyone who has ever contemplated or even finished the arduous med school path, this study is a terrific inside glance into how doctors are made.

A second classic medical anthropology case study comes to us from anthropologist Emily Martin. In her 1994 book, *Flexible Bodies*, Martin delves deep into the collective American psyche to learn about how us non-physicians understand the immune system. Martin did a comprehensive field study, rooted in the anthropological tradition. She did participate observation at three sites. She worked at an immunology lab. She volunteered at a house for HIV-positive individuals. And she also worked as an AIDS activist. In all, she did some 200 interviews, along with untold pages of field notes from her day-to-day observations. Again, her primary question was, how do regular people understand the immune system?

From the days of polio to the age of AIDS, Martin shows us that the metaphors we casually throw around reveal a lot about how we understand our immune system. Specifically, she showed that in the United States, militarism is a common metaphor for medicine and treating disease.

We fight cavities and battle cancer. My favorite commercial growing up featured this nasty crew of cavity creeps that's what they were called, and the only way to keep them from destroying your teeth with microscopic sledgehammers was to call the Crest team to fight a minty battle and brush them right away. It was basically a 30-second cartoon version of that awesome 1966 Hollywood gem, *Fantastic Voyage*. If you've seen this one, you're probably smiling a bit. In this sci-fi adventure, a group of American scientists jumps into a nuclear submarine, which is then zapped and shrunk down to the size of a blood cell. The submarine team then gets injected into the scientist's body, where they evade antibodies and other dangers, on their mission to save the scientist's life.

Ultimately, Emily Martin's study shows that in the US, we're shifting from militaristic, fortress-based metaphors for the body and health. Instead, from the Age of AIDS onward, she sees our metaphor changing into one of dynamism and flexibility. As we entered the 21st century, Martin says, Americans were no longer focusing on the idea of an immune system as a battle-ready army of white blood cells. Rather, we're coming to see and value adaptability in our immune system. And this conceptual shift, this new focus on flexibility, has broader cultural ramifications, including an openness to new ways of treating disease and maintaining our health.

Pioneering field researcher, Audrey Richards, presaged the value of medical anthropology with her nutritional studies that merged qualitative and quantitative field data in the early 1900s. And by the late 20th century, as we can see in the work of Martin and the Goods medical anthropology had come of age. Early scholars focused on the idea of ethnoscience or folk medicine. And among those early medical anthropologists was a consensus that biomedicine otherwise known as Western medicine was essential to boost so-called lesser developing nations. And that global project provided fertile ground from which medical anthropology blossomed.

Later in the 20th century, medical anthropology followed the currents we saw in cultural anthropology, with new interests in psychology, meaning, and symbolism. And as the century pushed on, so did the idea that non-western medicinal practices that is those outside conventional biomedicine were worth understanding from within, rather than through a western or scientific lens. And ever since, Medical Anthropology has become one of the largest and most prolific sub-sections of the AAA.

Contemporary medical anthropology is a definitive example of applied anthropology. From an interdisciplinary, four-field approach, medical anthropologists chart the depths of the biology-culture nexus of human health and healing. And it's exciting to imagine how medical anthropology will continue to flourish as new technologies and discoveries reveal more and more about how to foster our physical, emotional, and spiritual health.

Anthropology and Economic Development

I nternational development has been a cornerstone of anthropological inquiry for many decades. Development anthropology, as a specialization, puts anthropologists on the front lines of international development. Development anthropologists bring their research to bear on socioeconomic, cultural, ecological, and technical problems around the world. They collaborate with knowledge experts and communities in search of practical solutions to human problems. This lecture takes a close look at development anthropology.

Robert Redfield

When it comes to the theoretical origins of development anthropology, the best place to start is probably with Robert Redfield. Redfield was instrumental in shaping a subfield known as peasant studies, a term that has evolved into the anthropology of development.

His research approach reinforced the 4-field tradition that Franz Boas laid as the foundation of anthropology. He blended cultural studies, biology, linguistics, and archaeology to collect data on poor rural farming societies.

Redfield's career was launched by his work with Mexican immigrants in Chicago. Ruminating over the fact that these immigrants had lives in both the US and Mexico, his focus turned to what life is truly like in rural Mexican societies.

He went to Mexico in 1926 and 1928 with hopes of finding the perfect community to study a poor rural farming society. He found that community in Tepoztlán, a small agricultural community of family farmers.

It was here that Redfield developed 2 of his core concepts: the idea of the peasantry and the idea of the little community.

Redfield says the peasantry is a category of pre-urban human society. Peasant society has a few requisite traits.

- First, peasants eat what they grow. They are farmers.
- Second, peasants don't treat land as capital or commodity. Instead they follow traditional land-tenure systems, which can vary across cultures. In one Malian village, for example, families maintain their farming territory over the generations, only so long as they continue to cultivate them.
- Third, a peasant society is rather powerless against the greater forces that surround it. Finally, the peasantry lives in rural territory that is linked to the rest of the world through regional market towns.

Redfield created the idea of the little community to help us grasp what life is like among the peasantry.

According to the work of Robert Redfield, a peasant society is rather powerless against the greater forces that surround it.

The little community, says Redfield, represents an "earlier" form of human society. It is small and culturally homogenous. And daily life in the little community is structured by age- and sex-based groups.

Another characteristic of the little community is that things largely stay the same. One generation follows the next, living and working the same basic lives. A final trait: The little community is self-sufficient as an organic whole. In other words, the little community takes care of its people from birth to death.

Redfield taught that the little community was an essential unit of analysis for development anthropology around the world. He promoted participant observation as essential. For him, embedded, participant observation was the only way to get at a community's essence.

Lewis and Mintz

Redfield inspired new questions for a new generation of anthropologists. One of these scholars was Oscar Lewis. Lewis revisited Tepoztlán, and in 1951, he published a revised account. His primary departure from Redfield was that, for him, the little community was not a complete picture.

Per Lewis, the little community wasn't the primordially harmonious society that Redfield portrayed. And it wasn't a self-sustaining community. Instead, it was tied to a complex network of regional and global forces.

Another anthropologist who built on Redfield's work was Sidney Mintz. His methodological innovation was to use anthropology to study commodities instead of communities.

And this innovation has become a central thread in contemporary development anthropology. By examining coffee and sugar, for example, anthropological inquiry can now reveal the ties that Lewis identified as connecting little communities with the rest of the world. Mintz's groundbreaking history of sugar, *Sweetness and Power*, has become a foundational text across fields like the social sciences, history, ethnic studies, and international business.

The Mission

Development anthropology's mission has 3 components:

1. Development anthropologists must embrace their role as agents of social change.
2. Development anthropologists should apply their unique, 4-field approach and their specializations to produce empirical and comprehensive development studies.
3. Based on empirical field studies, development anthropologists should produce collective theories of development.

Theories of development are not the exclusive domain of anthropologists. But most development anthropologists work within a common framework of ideas called development theory.

Development theory is dynamic, which means we can trace the evolution of certain key concepts within the overarching framework. One of the most influential—and unfortunate—of these concepts is the idea of the so-called third world.

An alternate term is *majority world*. After all, the third world's population is clearly the majority. They're not some distant, disconnected population.

The pervasive, deceptive term *third world* dates back to author Alfred Sauvy, who coined the term *tiers monde*, or third world, in 1952. From his Cold War perspective, Sauvy identified the third world as unallied regions that had yet to fully realize the capitalist model of the United States (the first world), nor the socialist vision of the Soviet Union (the second world). Instead, these countries and their populations were territorial battlegrounds for the Cold War.

Today's Order

The global economic order we're familiar with in the 21st century has shallow roots that go back to 1944 in Bretton Woods, New Hampshire.

Emerging as a victor from both World Wars, the US had escaped the monstrous carnage of war-torn Europe and Asia. As such, its relative power rose as the US worked with the international community to organize and finance the rebuilding of the world and the modern global economy.

At the Bretton Woods Conference, representatives from allied nations brokered the development for what became the International Monetary Fund and the World Bank. The idea was that the global economy would shift from the gold standard to the US dollar. Those countries with surplus would contribute funds that would be provided to countries that were struggling with deficits.

The trouble is the strings attached to taking money from the World Bank. For starters, such money is a loan, and loans have interest. In addition, countries have to follow these rules:

1. A country must agree to reduce existing budgets by cutting subsidies and social services across the board.
2. A country must eliminate trade and investment restrictions to spur exports and industry.
3. A country must devaluate the local currency.

During the boom years of 1945–1969, this approach appeared to have some success. But since then, things haven't been so rosy. Global poverty endures, and arguably the system that emerged out of Bretton Woods undermines peace in our time.

So what happened? Anthropologists came to realize their principles of international development weren't as universal and objective as once believed.

Changes

Development theory changed over the course of the 20th century and into the 21st century. One way to examine this is through the modernization school. The modernization school asserts an evolutionary schema that places all the world economies on a continuum from traditional to modern. This theory is attributed to Walt Rostow, who articulated the 5 stages of economic growth.

The first of Rostow's 5 stages is the traditional, collective economy, which corresponds well with Redfield's idea of the little community.

The way out of this "primitive" state is to enter the stage Rostow refers to as the preconditions for takeoff. In this early phase, a society works out solutions for economic growth, but they're unsustainable.

Then, with some careful planning and external investment, a country can move up to what Rostow calls the takeoff stage, where resistance to steady growth is finally overcome.

For the modernization school, a "takeoff" economy is defined by the 10% rule. That means that the investment and savings rate rises to around 10%. As industry grows, more and more income is reinvested.

Once an emerging economy achieves take off, it evolves into to the drive to maturity. Rostow notes that this is the stage that economic growth finally becomes automatic.

That automatic, supposedly inevitable growth leads to the pinnacle of economic development: the age of high mass consumption. Ultimately, modernization theory is all about getting traditional farming societies to become high-consumption societies.

That's right in line with 19th-century anthropology, an era dominated by the widespread presumption that Western cultures were the biological and cultural pinnacle of human evolution. Anthropologists eventually grew critical of this presumption, and they updated their anthropological perspective of development.

Critiques of modernization began to emerge in the 1960s and '70s. Scholar Andre Gunder Frank was one of several theorists who represented the post-modernization phase of development theory: dependency theory.

Frank and dependency theorists pointed out the inconsistencies of modernization theory. They presented the world economy as an exploitative network of metropolis-satellite relationships. And the cards are stacked against so-called third world countries. They can't feasibly replicate the immeasurably costly arc that established the modern global economy. For example, Africa cannot enslave a foreign population to build a continental infrastructure for a new millennium.

Underdevelopment

Another brand of dependency theory is known as underdevelopment. This variation on dependency theory asserts that modernization actually produces under-developed societies.

Scholar and activist Walter Rodney wrote a book entitled *How Europe Underdeveloped Africa*, in which he eviscerated modernization theory. Rodney examined the socioeconomic, historical, and cultural dynamics of the developing world, and he documented how development programs in the modernization tradition actually produced underdevelopment.

He provided evidence for the colonial and postcolonial decline in status of women in developing countries. He also pointed out that conventional development schemes of the West invested in places like Africa for the purpose of resource extraction, which produced pockets of development. There are schools, roads, and health clinics in the third world, but they're not universal.

Development Today

How does development theory align with development practice in the 21st century? Development anthropologists may debate the specifics of development initiatives, but Redfield's instruction to embrace their role as agents for social change binds them.

The UN monitors real-world indicators like school enrollment numbers for girls.

Development anthropologists are serving in government and nongovernment positions focused on articulating and achieving a global standard for international development.

Today, this standard is largely reflected in the Millennium Development Goals (MDGs), which were adopted by the United Nations in 2000. These are a set of 8 goals that aim to reduce and then eliminate extreme poverty across the world in 1 generation.

The UN is measuring progress on 8 vectors. For example, the third MDG promotes gender equality. To measure country-by-country progress on this goal, the UN monitors real-world indicators like school enrollment numbers for girls and women's representation in national parliaments.

With tangible measurements like this over time, maybe it's not such a far-fetched idea that we might eventually eradicate extreme poverty and hunger across the globe.

Suggested Reading

Mintz, *Sweetness and Power*.

Nolan, *Development Anthropology*.

Redfield, *The Little Community and Peasant Society and Culture*.

Rodney, *How Europe Underdeveloped Africa*.

Questions to Consider

1. What is the role of the anthropologist in international development? How has that role changed (or has it changed) since Robert Redfield and the early days of development anthropology?

2. What is the theory of underdevelopment, and why does it take an anthropological perspective to see it?

Anthropology and Economic Development

H ave you ever thought about joining the Peace Corps? As a former volunteer, I enthusiastically encourage you to do so. Every volunteer has a unique experience, and it's definitely not easy, but for me, my time in Mali as an agricultural volunteer opened my eyes and mind to a future I wanted, yet had never known. It inspired me to become an anthropologist, to create a small non-profit called African Sky, and to find creative and useful ways to sustain my relationships with people and communities in Mali. And my story isn't unique. So many returned volunteers end up as anthropologists, and if not, their hands-on, embedded participation with their host communities makes them honorary anthropologists at the very least.

That hands-on experience, that training, explains why you see so many returned Peace Corps volunteers working in the Department of State and in NGOs and think tanks throughout the world. Put simply, their Peace Corps experience prepares them to make valuable contributions in the area of international development.

International development has been a cornerstone of anthropological inquiry for many decades. And in fact, development anthropology, as a specialization, puts anthropologists on the front lines of international development. Development anthropologists bring their research to bear on socio-economic, cultural, ecological, and technical problems around the world, and they collaborate with knowledge experts and communities all over the world, in search of practical solutions to human problems.

Today, we'll take a close look at development anthropology. First, I'll share some of my field research from Africa, then, we'll look to a pioneer of development anthropology to give us some theoretical grounding. And last we'll take a look at how anthropologists have thought about international development since WWII.

As an anthropologist, I've spent the past couple of decades working with farming communities in Mali, West Africa. And we've done some terrific projects together, but one of my all-time favorites thus far was a cross-generational women's that summit we organized in back in 2011. It all started around five to six years earlier when a Peace Corps volunteer in Mali suggested that I travel north to meet someone he deemed to be my kindred spirit in terms of community-based research and initiatives.

I ran to meet this man, Tamba, in a town called Markala. And I quickly discovered that when it came to assisting people in this community, Tamba was a tireless champion. The first day I met him, we hired a moped taxi to tour four groups he was mentoring. There was a local self-advocacy group for people with developmental and/or physical disabilities, and there was an amazing women's group led by a retired female physician. Then there was a bountiful garden managed by a man who gave orphans opportunities to learn how to garden. Then, as I sat eating a delicious plate of rice and peanut sauce with my newfound friend, two women from yet a fourth group approached us with a small folder and a proposal.

The group they represented included just over a dozen local women of varying levels of literacy and numeracy. And by the time they approached Tamba and I, they were as prepared as any group I've ever worked with in Mali. It turns out that the women had joined together to propose a soap-making venture. They had the predicted expenses, output, and income. They were intent on saving money by making soap for their households, but they also wanted to make some money by producing surplus soap to sell to friends and neighbors. All they needed was a little help launching their venture.

I loved this idea. I didn't have their full requested budget, but I invested on the spot. And I promised them that I would return in about six months to hear about their progress. My NGO and I weren't teaching them to make soap. We were deeply embedded in Malian communities, and through our social networks and anthropological depth; we positioned ourselves to support a remarkably well-defined local initiative. But that's not all. We invested in these women because they had something to teach us. The moment they placed that first proposal and budget in my hands, I was focused on learning how to replicate their terrific plan across Mali. So after six months, I returned to honor my promise, that I would invest a second round of funding. And my single prerequisite was that I wanted to hear how things went, including their figures and results.

Now, the group proudly showed their notebook as well as the tin box they used to store common funds. They didn't make a huge profit, but they did make money. I congratulated them and pulled out the money I had set aside to double down on our pay-it-forward social investment. But then the celebratory tone sharply turned to concern and confusion. They asked me if I understood the notebook they used for their records and accounting. I insisted I did. So why then, they asked me, was I offering more money? They told me to give it to somebody else because they made back their initial investment with profit. This was something I'd never heard before—they actually didn't want additional funds. Their model worked right away, and they were building it themselves.

A year later, the women changed sectors and went into peanut butter production. You heard that right. They bought large sacks of locally harvested peanuts, and a couple times a month they spent the afternoon together, roasting peanuts over an open flame using a contraption that they had commissioned from a local blacksmith. While socializing, they then crushed the roasted peanuts into the most absolutely delicious peanut butter that I have ever tasted. The women explained, that after a few discussions, they were convinced that making peanut butter, not soap, offered the best return on investment. Remember, not all these women were literate, but the project helped spread critical skills among group members.

These women continue to make and sell high quality, nutritionally charged peanut butter among neighbors and other locals. And they've demonstrated their methods for countless visitors, both Malians and Americans. And this is just one development success story I could tell you about. For example, there was another larger group in Markala that focused on training young, married women as apprentices. Young women who never got the opportunity to attend or complete primary school, were brought in and taught income generating skills ranging from making and selling dried foods, to tailoring and designing clothing, and even running small restaurants, and more. After visiting these two women's groups over several years, I had a flash. I reached out to everyone I knew, including Tamba and regional Peace Corps volunteers and we organized a cross-generational women's summit.

We brought together women of all ages from a half dozen communities across Mali. Many reported that they've never been more than a few kilometers from home. They stayed with local host families, and over the course of three days, we facilitated sessions that, one at a time, gave each visiting delegation an opportunity to lead a session. Some shared ways to make a non-toxic mosquito repellent that was perfect for infants. Others gave advice on running a successful women's organization, including how to do dues, officers, and mission statements. And of course, the Markala women's groups shared their stories and advice too.

The summit was so popular and well attended that one, a local radio station transmitted and recorded the sessions. And two, we received messages from the groups that when they returned to their communities, there were a lot of men who wanted in on the next one. This summit was a terrific, locally salient and locally controlled development initiative. African Sky and I didn't send in our experts to teach and organize the summit. We facilitated and funded a multi-day event in which women of all ages congregated to teach each other what they've been doing to make better lives for themselves and their families. Through an embedded, anthropological approach, people who might otherwise be labeled as project people in need, well, they became true partners and leaders in the way all of us think and do development. Together.

OK, now let's step back from the field for just a moment and look at the theoretical origins of development anthropology. Probably the best place to start is with Robert Redfield. Redfield was instrumental in shaping a subfield known as peasant studies, a term that has itself evolved into the anthropology of development. His research approach reinforced the four-field tradition that Franz Boas laid as the foundation of Anthropology. He blended cultural studies, biology, linguistics, and archaeology to collect data on poor rural farming societies.

Redfield's career was launched by his work with Mexican immigrants in Chicago. Ruminating over the fact that these immigrants had lives in both the US and Mexico, his focus turned to what life is truly like in rural Mexican societies. He went to Mexico in 1926 and 1928 with hopes of finding the perfect community to study a poor rural farming society. And he found that community in a place called Tepoztlán. Now, Tepoztlán was a small agricultural community of family farmers. And it was here that Redfield developed two of his core concepts. There is the idea of the peasantry and the idea of the little community. Let's take a look at each concept.

So what is Redfield's peasantry? Well, he offers what he describes as a loose definition. He says the peasantry is a category or a type of pre-urban human society. And peasant society has a few requisite traits. First, peasants eat what they grow, they're farmers. Second, peasants don't treat land as capital or commodity. Instead, they follow traditional land tenure systems, which, can vary across cultures. In my primary Malian host village, families maintain their farming territory over the generations, only so long as they continue to cultivate them. A third characteristic of Redfield's peasant society is that it is rather powerless against the greater forces that surround it. He called this phenomenon the guidance of the peasantry from above. Last, this peasantry lives in a rural territory that is linked to the rest of the world through regional market towns.

Having identified the central features of peasant society, Redfield created the term the little community to help us grasp what life is like among the peasantry. The little community, says Redfield, represents an earlier form of human society. It is small and culturally homogenous. And daily life in the little community is structured by age and sex-based groups.

Two other characteristics of the little community are that things kind of stay the same. One generation follows the next, living and working the same basic lives. And last, the little community is self-sufficient as an organic whole. Or, in other words, the little community takes care of its people from birth to death.

Redfield based these concepts on his observations in rural Mexico. But he taught that the little community was an essential unit of analysis for development anthropology around the world. In other words, his definition of the little community is dynamic enough to allow for meaningful, and universal, cross-cultural analysis. You can compare little communities across Mexico, Mali, and Madagascar for that matter. Despite the surface differences, Redfield challenged us to see their common condition. And last, Redfield promoted participant observation as essential. For him, embedded, participant observation was the only way to get at a community's whole, at their essence.

But even beyond his important emphasis on participant observation, Redfield made terrific use of the interdisciplinarity of anthropology as a four-field discipline. And he continued to build upon the methods and ideas of his predecessors, inspiring new questions for a new generation of anthropologists, scholars who continue, through anthropology, to test and correct the views of their predecessors.

One of these scholars was Oscar Lewis. Lewis revisited Tepoztlán, and in 1951, he published a revised account of this so-called little community. His primary departure from Redfield was that, for him, the little community, both as Tepoztlán, and as a universal unit of analysis, wasn't a complete picture. First, this little community wasn't the primordially harmonious society that Redfield portrayed. And second, It wasn't an intuitive, self-sustaining community. Instead, it was tied to a complex network of regional and global forces. Like Mexican immigrants sending remittances home, the little community is anything but an insular isolated place. And as such, the endemic poverty of the little community is a product of a system-wide failure. An economic system and history steeped in extreme inequality.

Another anthropologist who built on Redfield's work was Sidney Mintz. His methodological innovation was to use anthropology to study commodities instead of communities. And this innovation has become a central thread in contemporary development anthropology. By examining coffee and sugar, for example, anthropological inquiry can now reveal the ties that Lewis identified as connecting little communities with the rest of the world. Mintz's groundbreaking history of sugar, *Sweetness and Power*, has become a foundational text across the social sciences, history, ethnic studies, international business, and more.

From the work of Redfield, Lewis, and Mintz, we can begin to articulate a mission for development anthropology. And this mission has three components. One, development anthropologists must embrace their role as agents of social change. I model this ethos today when I teach anthropology as a tool for social transformation. Two, development anthropologists should apply our unique, four-field approach and our specializations to produce empirical and comprehensive development studies. Three, based on empirical field studies, development anthropologists should produce collective theories of development.

Now, theories of development are not the exclusive domain of anthropologists. That said, most development anthropologists work within a common framework of ideas. And we call this framework development theory.

Development theory is dynamic, not static—which means we can trace the evolution of certain key concepts within our overarching framework. And one of the most influential—and frankly, unfortunate—of these concepts is the idea of the so-called third world. I mean, where exactly is that? When did this third world even start?

I love how my mentor Peter Brown used the term majority world rather than third world. After all, the third world, the developing world, those living in poverty, are clearly the majority. They're not some distant, disconnected, third-degree population. They are literally, the majority world. But then how did we get this pervasive and insidiously deceptive term in the first place? Well, we go back to author Alfred Sauvy who coined the term Tiers Monde, or third world in 1952.

From his Cold War perspective, Sauvy identified the third world as unallied regions that had yet to fully realize the capitalist model of the United States the first world, nor the socialist vision of the Soviet Union the second world. Instead, these countries and their populations were territorial battlegrounds for the Cold War.

Let me explain. One of my most important African Studies teachers, Kofi Buenor Hadjor once worked directly for Kwame Nkrumah, the first president of decolonized Ghana. Hadjor explained that Nkrumah was masterful at playing the chess game, balancing US and Soviet interests toward the advantage of his own country. And by remaining unclaimed and unequally integrated into the first and second world economies, Nkrumah attempted to find Ghana its own unique and viable place in the global economy.

But, as Hadjor always taught, In the Cold War, you either picked a lane, the first or second world, or you got run over. And indeed, without getting into the thick of it, Nkrumah's Ghana fell. But how could that be? How could a global economy, one that has seen thousands of years of powerful empires across Africa and beyond, how could the world order shift so decidedly? Well, the global economic order we're familiar with here in the first half of the 21st century has pretty shallow roots. Roots that go back to 1944 in Bretton Woods, New Hampshire.

Emerging as a victor in two world wars, the US had escaped the monstrous carnage of civil infrastructure of war-torn Europe and Asia. And as such, its relative power rose as the US worked with the international community to organize and finance the rebuilding of the world and the modern global economy.

So at the Bretton Woods Conference, representatives from allied nations brokered the development for what became the International Monetary Fund and the World Bank. Basically, the idea was that first, the global economy would shift from the gold standard to the USdollar. Secondly, those countries with surplus would contribute funds that would be provided to countries that were struggling with deficits. One groovy world.

The trouble is, just like a payday loan, there were strings attached if you took money from the World Bank. First, it's a loan. You're paying it back, and with interest. And to do that, your country must agree to a Structural Adjustment Program. Or in other words, to get this loan you have to follow the rules. One, you've got to agree to reduce existing budgets by cutting subsidies and social services across the board. Second, your country must eliminate trade and investment restrictions, to spur exports and industry. And last, your country must devalue the local currency. That happened in Mali just before I arrived in 1994. Overnight, the value of local currency dropped in half. Why? Well, it brings in the outside money. Think of it this way: Local resources are now half price on the global market, and imports into your country now cost twice what they did yesterday.

During the boom years of 1945–1969, this approach appeared to have some success. But since then things haven't been so rosy. Global poverty endures, and arguably the system that emerged out of Bretton Woods is undermining peace in our time. So what happened? Well, if you ask an anthropologist, we figured out that our principles of international development weren't as universal and objective as we once believed.

To flesh that statement out, let's look at how development theory changed over the course of the 20th century and into the 21st. And let's start with the modernization school. We attribute this theory to Walt Rostow, who articulated the five stages of Economic Growth. The modernization school asserts an evolutionary scheme that places all the world economies on a continuum from traditional to modern.

The first of Rostow's five stages is the traditional, collective economy, which corresponds well with Redfield's idea of the Little Community. And the way out of this primitive state is to enter the stage Rostow refers to as the preconditions for take-off. In this early phase a society works out solutions for economic growth, but they're unsustainable. Then, with some careful planning and external investment, a country can move up to what Rostow calls the take-off stage, where resistance to steady growth is finally overcome. In fact, it's what we come to expect. Economic growth becomes sustainable.

For the modernization school, a take-off economy is defined by the 10 percent rule. That means that the investment and savings rate rises to around 10 percent. And as industry grows, more and more income is reinvested. The idea is that as incomes rise, people buy more food, homes, swimming pools, and all that stuff that makes us happy. And to do that we'll need more and more food producers, house builders, and pool makers. So once an emerging economy achieves take-off, it evolves into to the penultimate stage, the drive to maturity. Rostow notes that this is the stage that economic growth finally becomes automatic. It sticks. And, it's that automatic, supposedly inevitable growth, which leads to the pinnacle of economic development, the age of high mass consumption.

Ultimately, modernization theory is all about getting traditional farming societies, like the ones I work with in Mali, to become high consumption societies. Basically, we're talking about getting Amazon Prime and iPhones to the furthest corners of Timbuktu. Development is about creating more consumers. And that's right in line with 19th-century anthropology, an era dominated by the widespread presumption that western cultures were the biological and cultural pinnacle of human evolution. Anthropologists eventually grew critical of this presumption. And they updated our anthropological perspective of development. But why?

Well, voices from the majority world—the third world—they were unequivocal. Why in the world would developing countries want to follow the West into the suffering and violence of the West's 20th century? Why follow the path that led to two world wars? Or consider the pervasive fear and conflict of the Cold War era. World Wars and nuclear proliferation aside, in the US alone, there's Vietnam, racial violence, and decades of civil unrest punctuated by the senseless assassinations of JFK, Martin Luther King, RFK, Malcolm X, and so many more.

In response to all of this, critiques of modernization began to emerge in the 1960s and 1970s. Scholar Andre Gunder Frank was one of several theorists who represented the post-modernization phase of development theory, what we'll call dependency theory. Frank and dependency theorists pointed out the inconsistencies of modernization theory. They presented the world economy as an exploitative network of metropolis-satellite relationships. And the cards are stacked against so-called third world countries. They can't feasibly replicate the immeasurably costly arc that established the modern global economy. Africa can't enslave a foreign population to build a continental infrastructure for a new millennium.

Another brand of dependency theory is known as underdevelopment. This variation on dependency theory asserts that modernization actually produces under-developed societies. Scholar and activist Walter Rodney spoke for the developing or majority world. And he wrote a book entitled *How Europe Underdeveloped Africa*, in which he eviscerated modernization theory.

Specifically, Rodney examined the socio-economic, historical, and cultural dynamics of the developing world, and he documented how development programs in the modernization tradition actually produced underdevelopment. He provided evidence for the colonial and post-colonial decline in the status of women in developing countries. He also pointed out that conventional development schemes of the west invested in places like Africa for the purpose of resource extraction, which produced pockets of development. There are schools, roads, and health clinics in the third world, but they're not universal, they're in pockets. Benefits for the few, but at the cost of the many.

So where are we today? How does development theory align with development practice in the 21st century? Development anthropologists may debate the specifics of development initiatives, but Redfield's instruction to embrace our role as agents for social change binds us all. Into the 21st century, development anthropologists are serving in government and non-government positions focused on articulating and achieving a global standard for international development.

Today, this standard is largely reflected in the Millennium Development Goals, which were adopted by the United Nations in 2000. These MDGs are a set of eight goals that aim to reduce and then eliminate extreme poverty across the world, and in one generation, no less. It may be easy to doubt generational goals like the eradication of extreme poverty and hunger. But the UN is measuring our progress on eight vectors, to plot the trajectory of progress in very specific ways.

For example, the third MDG promotes gender equality. And to measure country-by-country progress on this goal, the UN monitors real-world indicators like school enrollment numbers for girls and women's representation in national parliaments. And with tangible measurements like this over time, maybe it's not such a far-fetched idea that we might eventually eradicate extreme poverty and hunger across the globe. After all, we've already worked through millennia of war, not to mention the Atlantic Slave Trade and colonization.

So call me an optimist, but I have my reasons. And one of them is that knowledge experts in the West are more open than we've ever been when it comes to learning from and working alongside the wise people of the majority world.

Cultural Ecology

C ultural ecology examines the complex relations between people and their environment. Anthropologists using this approach distinguish themselves with their collective adherence to 2 big cultural ecology ideas. First, they argue that the natural environment sets certain possibilities from which people and cultures may choose. Second, naturally inspired possibilities and the element of human choice challenge the idea of environmental determinism, or the idea that cultural differences are environmentally determined. Rather than environmental determinism, cultural ecology asserts environmental "possibilism."

Julian Steward

Julian Steward finished his Anthropology Ph.D. in 1929 at the University of California, Berkeley, under the tutelage of Alfred Kroeber. Steward taught at a handful of universities, but for a decade in the midst of his career, he applied his talents to the Bureau of American Ethnology (BAE).

Established by Congress in 1879, the BAE's mission was to collect and curate all records and materials relating to Native Americans. The BAE had the task of bringing those records and programs to the Smithsonian Institution, which opened the National Museum of the American Indian 125 years later in 2004.

Steward was an excellent choice for the BAE because he had a rather unique approach to anthropology. Steward was not an empirical purist, but he did privilege environmental factors and the collection of material artifacts over the use of human informants.

He understood cultural evolution as multilineal: All cultures are on their own path. He explained that no singular force determines the course of cultural evolution. Rather, factors like the environment, economy, social organization, political system, technology, and ideologies influence cultural evolutionism.

One of Steward's most-replicated anthropological approaches was to empirically examine human societies by focusing on what he called its culture core. He advised would-be field researchers to investigate subsistence strategies within a culture. His idea was that as these dynamic strategies evolve over time, those changes would influence other cultural features like language and social organization.

Cultural Ecology Methods

In the 1950s, cultural ecology remained close to Steward's approach. It focused on investigating and documenting the intersection of culture, technology, environment, and human behavior.

But in the 1960s, there began to emerge an ecosystem approach, as illustrated by the work of Robert Netting. This new layer added concepts like caloric expenditure and the carrying capacity of human habitats.

Moving further into the century, Steward's student, Marvin Harris, took his teacher's methods to the extreme, presuming that all cultural elements stem from human adaptations to environmental pressures.

This extreme materialist explanation for cultural evolution elicited heated debates from the growing numbers of more interpretivist and postmodern anthropologists of the late 20th century. Nonetheless, Steward remains a seminal influence on the way we understand culture change and the dynamic relationship between culture and environment.

Robert Netting's Work

Inspired by Steward's work on Great Basin hunter-gatherer societies in North America, Robert Netting took off for some field research of his own to see if Steward's ecological approach would work for a small-scale agricultural society.

His project led to the 1968 ethnography *Hill Farmers of Nigeria*. In this terrific cultural ecology classic, Netting documented some brilliant farmers in northern Nigeria who cultivated terraced fields year-round using organic fertilizers like manure and compost.

Terrace fields

Netting reinforced core elements of cultural ecology. First and foremost, he opposed cultural and environmental determinism. Netting focused mainly on quantitative data. Rather than exploring the inner lives and folklore of his study community, he liked to hone in on the things he could observe and measure, like the farming calendar, tools, and diet.

Through a comparative historical approach, married with empiricism and cross-cultural perspectives, Netting emerged as a champion of the small-scale farmer.

Marvin Harris's Work

Marvin Harris doubled down on Steward's empiricism. In fact, he went as far as taking Steward's environmental possibilism into strict environmental determinism.

Harris argued that our sociocultural lives arise out of our subsistence practices. In essence, he said that all aspects of culture, from our mode of production to religion and ideology, originate from the way we eek out a living from the surrounding environment. His book, *Our Kind*, discusses scores of cultures from all over the world.

Harris was a brilliant raconteur who was a master at drawing cross-cultural connections. But his work was different from people like Julian Steward. Harris, by focusing exclusively on the material lives of his subjects, eschewed subjective accounts and human informants.

Instead, he stuck to the material world. In his famous piece on the sacred cattle of India, he laid bare his approach. If asked why so many Indians refuse to eat cows, many locals would say that cattle are sacred and must not be slaughtered for human feed.

But there's not a scientific measure for the sacredness of cows. As such, Harris rejects this explanation. He argued there are other, deeper, things afoot.

Rather than interviewing people, Harris analyzed preexisting quantitative data. He didn't even go to India. Instead, he crunched some numbers and revealed that the sacred cow may be more of an economic taboo rather than a religious one. In short, he showed how cattle are far more valuable alive than butchered.

Marvin Harris showed how cattle are far more valuable alive than butchered.

In the Indian economy, cattle are more valuable as producers of milk, offspring, and dung, rather than ground beef and steak cutlets. Cattle dung, for example, is used as organic fertilizer or fuel for cooking.

Harris wasn't arguing against religious explanations for this taboo. But he was demonstrating that religious explanations, and all other cultural elements, have deep roots in our material and productive lives.

Harris and his approach are connected to cultural ecology in the Julian Steward tradition, but his strict materialist perspective was a sharp diversion that eventually required its

own name. Harris branded his theory and methods cultural materialism.

Cultural Materialism

Cultural materialism explains how the political, economic, domestic, ideological, and symbolic dimensions of society are ultimately rooted in how we meet our basic biological needs. Here's the basic methodology:

1. Focus on the observable and quantifiable.
2. When researching, embrace the scientific method. Develop methods and analyze

empirical data in testable and correctable ways. Methods and analysis must be replicable.

3. Using empirical data, reduce cultural phenomena like religion or political organization into observable, measurable variables for cross-cultural comparisons.

Harris's pyramid of cultural materialism provides a look at this approach.

1. Humans need to satisfy their basic human needs. That is the base of the pyramid, the infrastructure. For example, humans on Baffin Island would be wise to work out how to build shelters, fish, and hunt. Regarding reproduction: With limited resources, this island isn't capable of hosting large urban-level populations. Instead, they'll keep their population small yet viable.

2. The next level of the cultural pyramid is the structure. The structure level is where humans take care of their domestic and political economies. The domestic economy includes how people

The history of development anthropology clearly suggests that globalization generally weakens traditional knowledge.

organize families, gender and age roles, and kinship. The political economy is where people work out things like class structure, political organization, and trade.

3. The apex of the pyramid is the superstructure. With their basic needs met and political and domestic economies in place, people can turn to cultural elements like art, literature, music, dance, rituals, religion, science, and more.

Ironically, Harris and his cultural materialism outline the cultural diversity on our planet, only to reveal underlying structures that link directly back to our relative environments. Despite the surface-level diversity of the human family, Harris's cultural materialism pyramid shows that structurally, we're all the same. We're not different; our environments are.

Freelisting

Freelisting is an anthropology research method that opens new ways to for us to investigate the relationship between culture and environment. It's a text-analysis method.

Marsha Quinlan publishes widely on her freelist research, and she provides a great example of this method that is attuned to the ecological knowledge of rural Dominica.

She investigated local herbal medicine, and she wanted to see if local populations were losing generational knowledge on local medicinal plants. The history of development anthropology clearly suggests that globalization generally weakens traditional knowledge.

Quinlan interviewed her host community and asked them, one by one, to list all the local medicines they knew. This is freelisting. By getting these medicine lists from each respondent, she got a comprehensive view of every medicine this community knew.

Quinlan discovered some surprising results.

- She noticed that women knew more medicinal plants than men.
- Formal education was negatively associated with the number of medicinal plants people know. It seems that those who don't get a formal

education know more about medicinal plants.

- In sum, Quinlan discovered that globalization has complex effects on local knowledge systems. In the case of medicinal plant knowledge in Dominica, freelisting opened a window into the minds of her study community.

Through this method, her community taught her that prevailing ideas about the impact of globalization on local knowledge systems isn't as straightforward as once thought.

How Freelisting Works

Here's an example of how freelisting works: If Americans were asked to list all the animals they know, they'd likely have cats and dogs high on the list, but animals like emus and aardvarks would be less likely to appear.

When anthropologists collect and analyze these lists, they aggregate all the terms and sort them into 2 piles. One pile contains the popular terms; the rest of the terms go in the other pile. These piles are invaluable to anthropologists.

For example, an anthropologist could have people from a community freelist from the prompt "crops my family likes to grow." When the anthropologist aggregates those lists, the result will be an efficient snapshot of cultural consensus. It will be a master list of where there's consensus on the most important crops in this community as a whole.

The final, weighted results rank each term in the aggregate list based on both the popularity of each term—or the number of times it was listed— and the average rank of each term— or how high up it is on each list.

Anthropologists can also use freelist results to identify experts and other outliers within a culture. They can measure the similarity of each respondent to all other respondents, then plot the results so that 2 people with very similar results will show up as 2 dots next to each other.

Conversely, people who mention words that don't make everyone else's lists will have dots way off to the side. These people are either experts, or a little off, or don't fit within the cultural consensus for some other reason. Anthropologists go to the outliers for more information.

Cultural consensus modeling helps anthropologists examine how others see and organize their worlds. That's what makes it an ideal method for anthropologists like Quinlan who build upon the cultural ecology tradition.

Suggested Reading

Crawford, *Moroccan Households in the World Economy*.

Harris, *Good to Eat*.

Netting, *Smallholders, Householders*.

Steward, *Theory of Culture Change*.

Questions to Consider

1. What is cultural ecology, and how does it explain the complex relationship between people and their physical environment?

2. What are cultural consensus modeling and freelisting, and how do anthropologists use them to understand the ways different communities understand and manage their local ecology?

Cultural Ecology

L ike anthropological pioneers Franz Boas and Bronislaw Malinowski, I vividly remember my first day in the village. Oh, that humbling day where, as a rookie field researcher, you viscerally feel the truth of cultural relativity. Boas looked amongst igloos, as I looked at thatched roof, adobe huts. And, so very far from home, we both knew, deep deep down, that upon arrival, we had become overgrown infants. Yet, we were tasked with learning everything, nearly from scratch. Language, social structure, kinship, worldviews, daily lives, food production, socio-economics, even local hopes and dreams.

I tell you, getting a crack at that rare opportunity, that chance to start anew, making sense of our lives and universe—that experience drives us field researchers. And I assure you, it's like reality Sudoku but on a whole new level. Every puzzle or community we encounter is somewhat familiar, but, each one is configured a little differently.

So why would a rational person embrace such a complex, multi-dimensional, living-puzzle? We do this because we're explorers of the human condition. We probe the nature of humankind over time and space. And by living both among and between multiple cultures, anthropologists find inspiration and answers to some of the great human questions such as, who are we, and where do we come from?

Boas was drawn to anthropology as a discipline because he was interested in the relationship between people and their environment. That's why he went to the Arctic. He picked not just a far off place, but he wanted a stark environment in which to probe the people-environment relationship. And when he did, he helped establish cultural relativity.

Like those Sudoku boxes, structurally, there's something very similar going on, but each puzzle has its own unique patterns. And in human Sudoku, Boas helped us see how the uniqueness of cultures can be explained by environmental adaptation, not a predetermined, biological structure. And that's what we're up to today. What's the relationship between culture and the environment?

When we think about all of the intricacies that make natural systems operate, we lump them all into a single term, ecology. And, as living and breathing beings, we're part of that ecology. We're part of the natural world. So over the years, anthropologists have worked out ways to articulate this very thought. We're uniquely human, and we're not just part of the animal kingdom, we're part of the natural world. And just like ospreys collecting sticks, rodents, and fish to raise their young, our actions and lives literally reshape the environment around us. And that brings us to cultural ecology. Anthropologists who honed our complex relationship with the natural world, they eventually established a new theoretical thread in 20th-century anthropology called cultural ecology.

Cultural ecology examines the complex relations between people and their environment. And they distinguish themselves from other anthropological approaches with their collective adherence to two big cultural ecology ideas.

First, they argue that the natural environment sets certain possibilities from which people and cultures may choose. And second, this blend of naturally inspired possibilities plus the element of human choice challenged the idea of environmental determinism. The idea that cultural differences are environmentally determined. So, rather than environmental determinism, Cultural Ecology asserts environmental possibilism.

Julian Steward finished his anthropology Ph.D. in 1929 at UC Berkeley under the tutelage of Alfred Kroeber. Steward, the first anthropology lecturer at the University of Michigan, taught at a handful of universities, but for a decade in his mid-career, he applied his talents to the Bureau of American Ethnology—otherwise known as the Bureau of Ethnology.

Established by Congress in 1879, the BAE mission was to collect and curate all records and materials relating to Native Americans. Up until then, Native American ethnology was the domain of the Interior Department. The BAE had the task of bringing those records and programs to the Smithsonian Institution, which opened the National Museum of the American Indian some 125 years later in 2004.

Steward was an excellent choice for the BAE because he had a rather unique approach to anthropology. Steward was not just an empirical purist, but he did privilege environmental factors and the collection of material artifacts over the use of human informants. He understood cultural evolution as multilineal, meaning, culturally, we're all on our own path. He explained that the no singular force determines the course of cultural evolution. Not biology and not environment. Rather, it's things like environment, economy, social organization, political systems, technology, and ideologies that influence cultural evolutionism.

One of Steward's most-replicated anthropological approaches was to empirically examine human societies by focusing on what he called its culture core. What does that mean? Well, he advised would-be field researchers to investigate subsistence strategies within a given culture. His idea was that as these dynamic strategies evolve over time, those changes would influence other cultural features like language and social organization.

So if we focus on what people do to make a living more than what they say or think, according to cultural ecologists, we'll produce more useful analyses. Analyses grounded in the material daily lives of our study communities. Steward and his cultural ecology profoundly influenced mid-20th-century anthropology. Like his academic grandfather Frans Boas, the graduate students Steward trained are a veritable who's who of 20th-century anthropology. To name a few, there's Sidney Mintz, Eric Wolf, Elman Service, and Marvin Harris—who we'll hear from in just a bit.

But, back to Steward, one of the more remarkable distinctions of his brand of anthropology research was his openness to new data and ideas. And by the end of his career, he basically advocated the idea of changing your mind as an anthropological method in and of itself. And I quote, "As I reexamine some of my own cross-cultural formulations, I note a long history of changing my mind." And he adds, "There are perhaps others, who should also change their minds from time to time." And indeed, over time, his students and others continued changing their minds. Following Steward, his widely influential thread of anthropology grew into new ideas and methods.

First, in the 1950s cultural ecology remained close to Steward's approach. Simply put, it focused on investigating and documenting the intersection of culture, technology, environment, and human behavior. But into the 1960's we begin to see an ecosystem approach, as illustrated by the work of Robert Netting. This new layer added concepts like caloric expenditure and the carrying capacity of human habitats.

Moving further into the century, Steward's prolific student Marvin Harris, took his teacher's methods to the extreme, presuming that all cultural elements stem from human adaptations to environmental pressures. This extreme materialist explanation for cultural evolution elicited heated debates from the growing numbers of more interpretivist and post-modern anthropologists of the late 20th century.

Nonetheless, Steward remains a seminal influence on the way we understand culture change and the dynamic relationship between culture and the environment. In 2003, the American Anthropological Association initiated the Julian Steward Award to celebrate new books that carry forth his cultural ecology legacy. In 2009, for example, anthropologist David Crawford was given this prestigious prize. A modern take on the cultural ecology approach, Crawford's *Moroccan Households in the World Economy* examined Berber social relations and inequality through the lens of local modes of production and reproduction.

So let's return to the golden age of cultural ecology to visit two more anthropologists who embraced Steward's approach. First, the graduate advisor of my graduate advisor, Robert Netting. Inspired by Steward's work on Great Basin hunter-gatherer societies in North America, Netting took off for some field research of his own to see if Steward's ecological approach would work for a small-scale agricultural society. His project led to the 1968 ethnography *The Hill Farmers of Nigeria*. In this terrific cultural ecology classic, Netting documented some brilliant farmers in Northern Nigeria who cultivated terraced fields all year-round using organic fertilizers like manure and compost.

Netting reinforced core elements of cultural ecology. First and foremost, he opposed cultural and environmental determinism. Like his teacher, Netting focused mainly on quantitative data. Rather than exploring the inner lives and folklore of his study community, he liked to hone in on the things he could observe and measure. Things like the farming calendar, work activities, diet, tools, technological know-how, and local production levels by household and by unit of land. There was more, but you get the picture. That said, he explained that socio-cultural factors, not just biology, and the material world, are essential pieces of any anthropological puzzle.

Time and time again, Netting returned to West Africa, where his enduring contribution was to systematically and quantitatively explore the social and ecological consequences of long-term resource use. Through a comparative historical approach, married with empiricism and cross-cultural perspectives, Netting emerged as a champion of the small-scale farmer. His work celebrates the small-scale family farmers, as champions who manage to feed families in the face of economic and ecological scarcity. He was awarded a Guggenheim and was elected by his scientific peers to the National Academy of Sciences. Netting, like his cultural ecology contemporaries—Roy Rappaport and John Bennett—were heavily influenced and inspired by the anthropologist whose name has become synonymous with cultural ecology—Julian Steward.

Another one of Steward's most influential students is Marvin Harris. Harris doubled down on Steward's empiricism. In fact, he went as far as taking Steward's environmental possibilism into strict environmental determinism. The short of it is that Harris argued that our sociocultural lives arise out of our subsistence practices. In essence, he was saying that all aspects of culture— from our mode of production to religion and ideology—they all originate from the way we eek out a living from the surrounding environment. According to

Harris, differences in cultural elements like musical styles and art, believe it or not, are rooted in differences in the way we make a living.

I first came across Marvin Harris in high school, via his popular press book, *Our Kind*. As I flipped through those pages, Harris exposed me to ways of life I had never heard of. And I remember being fascinated by the scores of cultures that he discussed from all over the world. Harris was a brilliant raconteur who was a master at drawing cross-cultural connections. But his work was a lot different from people like Julian Steward. Harris, by focusing exclusively on the material lives of his subjects, eschewed subjective accounts and human informants.

Instead, he stuck to the material world. In the famous piece on sacred cattle of India, he laid bare his approach. If you go to India and ask locals why so many Indians refuse to eat cows, even in the presence of hunger, the answer would be telling. They'd say that cattle are sacred, and must not be slaughtered for human feed. But that didn't cut it for Harris. Harris was a strict materialist trying to make sense of why people do the things they do. There's not a scientific measure for sacred or non-sacred bovines. So as such, Harris rejects this explanation. Sure people say that the taboo of eating cattle is based on religious practice, but Harris doesn't buy that. He argued there are other, deeper, things afoot in this Angus avoidance.

Rather than talking with and interviewing people, Harris analyzed pre-existing quantitative data. He didn't even go to India. Instead, he crunched some numbers and revealed that the sacred cow may be more of an economic taboo than a religious one. In short, he showed how cattle are far more valuable alive than butchered. You see, in the Indian economy, cattle are far more valuable as producers of milk, offspring, and dung, rather than ground beef and steak cutlets. Cattle dung, for example, is used as organic fertilizer or fuel for cooking.

And, yeah, I know, cooking on cow dung briquettes doesn't sound so appetizing, but Harris crunched the numbers and found that cattle provided more economic benefits when they're alive. He wasn't arguing against religious explanations for this taboo, that just wasn't his concern. But he was demonstrating that religious explanations, and all other cultural elements, have deep, deep roots in our material and productive lives. Most certainly, Harris and his approach are connected to cultural ecology in the Julian Steward tradition, but his strict materialist perspective was a sharp diversion that eventually required its own name. Harris branded his theory and methods cultural materialism.

Cultural materialism, he said, explains how our political, economic, domestic, and even the ideological and symbolic dimensions of society are ultimately rooted in how we meet our basic biological needs. Cultural materialism, as we'll see, reveals our common humanity, despite the vast, surface-level cultural diversity of the human family. The basic methodology is to one, focus on the observable and quantifiable. Period. Two, when you do research, embrace the scientific method. Develop methods and analyze your empirical data in testable and correctable ways. Your methods and analysis must be replicable. And last, using your empirical data, reduce cultural phenomena like religion or political organization into observable, measurable variables for cross-cultural comparisons.

Let's nail this idea down by taking a look at Harris's Pyramid of Cultural Materialism. Now always keep in mind, first and foremost, Harris starts with the environment, and then he builds cultural systems up from there. From square one, everything a society does is based on its environment. So by building Harris's pyramid, we can build a culture. Step one, we need to satisfy our basic human needs. That is the base of our pyramid, our infrastructure. So as a society, we need to look around us and figure out how to make a living. How are we going to fulfill our basic needs?

If we're born on Baffin Island, we'd be wise to work out how to build shelters, fish, and hunt because that'll be our mode of production. Now, all we need to finish our infrastructure is our mode of reproduction. We need to work out, based on our Arctic island, our population dynamics. With limited resources, this island isn't capable of hosting large urban-level populations. Instead, we'll keep our population small yet viable. And that is our infrastructure. Our modes of production and reproduction. Once we work out how to take care of those, we can then move on to build the next layer of our cultural pyramid, the structure.

The structure level is where we take care of our domestic and political economies. Our domestic economy includes how we organize our families, gender and age roles, and even kinship. This middle level, the structure, also contains our political economy, in which we work out things like class structure, political organization, and trade.

And then finally, when we work all that out, we get to the apex of the pyramid, the superstructure. With our basic needs met in the infrastructure, we built our domestic and political systems, so now we can turn to cultural elements like art, literature, music, dance, rituals, religion, science, and more.

And now that we have our cultural materialism pyramid, let's start at the top and reverse engineer two cultures. A cattle herder in West Africa is unlikely to become a world-class sailor. Why? It's because he comes from a society or culture that doesn't build sailing-friendly economies and families. Similarly, I've yet to hear a West Coast hip-hop artist rap about cows. Cattle herding is not at the base of West Coast Hip Hop artist's cultural pyramid. It's not how they keep their families fed.

Or think religion. That herder isn't from a region that can support large communities or sedentary living. As such, I doubt their religious life includes massive cathedrals packed with candles and stained glass alcoves. It just won't happen there, but that doesn't mean they don't have religious practices in their superstructure. They do. But they're different because the environment and economies in which they were built are different.

Ironically, Harris and his cultural materialism outline the cultural diversity on our planet, only to reveal underlying structures that link directly back to our relative environments. Despite the surface-level diversity of the human family, Harris's cultural materialism pyramid shows that structurally, we're all the same. We're not different. Our environments are.

Harris and his work proved influential, and he was eventually elected as the President of the American Anthropological Association. And although he passed in 2001, streams of cultural materialism and cultural ecology still flow through the methodologies of contemporary anthropology, and we're always looking for new ways to fulfill our duties as bridge builders in terms of the relationship between culture and the environment.

Now, to wrap up our ecological anthropology discussion let's take a quick trip to Central America because I want to show you one of my favorite anthropology research methods, and how it opens up new ways to for us to investigate the relationship between culture and environment.

The method I'm talking about is a text analysis method called freelisting. Marsha Quinlan publishes widely on her freelist research, and she provides a great example of this method that is attuned to the ecological knowledge of rural Dominica. She investigated local herbal medicine, and she wanted to see if local populations were losing generational knowledge on local medicinal plants. You see, the history of development anthropology clearly suggests that globalization generally weakens traditional knowledge.

So, in Quinlan's study population, medicinal plants are a normal part of life. Here, herbal medicines are typically the first response to illness. And so it wasn't a surprise that every adult Quinlan spoke with knew some medicinal plants. But she was curious. In an era of transnational aspirin, is this local knowledge fading? Are people losing their traditional knowledge of medicinal plants?

To figure this out, Quinlan interviewed her host community and asked them, one by one, to list all the local medicines they knew. This is a method called freelisting. And by getting these medicine lists from each respondent, she got a comprehensive view of every medicine this community knew. And I'll explain the mechanics behind the analysis in a moment, but for now, simply by analyzing these lists, Quinlan discovered some surprising results.

First, by comparing lists, she noticed that women knew more medicinal plants than men. So if you're in rural Dominica, you know who to go to for that headache medicine. Another remarkable finding was that formal education was negatively associated with the number of medicinal plants that people know. It seems that those who don't get a formal education actually know more about medicinal plants.

In sum, Quinlan discovered that globalization has complex and even surprising effects on local knowledge systems. In the case of medicinal plant knowledge in Dominica, freelisting opened a window into the minds of her study community. And through this method, they taught her that prevailing ideas about the impact of globalization on local knowledge systems isn't as straightforward as we thought. And, to be a genuine and useful project partner with populations like Quinlan's study community, we've got to get a clear understanding of the depths of their ecological knowledge and values. And freelisting is a terrific and efficient way to do exactly that.

Let me show you how this freelisting method actually works. We'll do a freelist right here, you and me. Now, I'll give you a prompt, and I want you to say every word that comes to mind. It's simple. It's free association. No self-censoring or restraint. Just list every word that comes to mind. You ready? Here's the prompt. List all the animal types you know. One, two, three, go.

Normally, I'd push you to list more and more animals until you were fully tapped. And I didn't give you much time, but I bet you said dog and cat. And maybe mouse? How about emu? Is aardvark on your list? Well, why not? You've heard of them, right? I mean, I love aardvarks. Why not start with them? Exactly, they're not really salient to your daily life. And that's why it's not salient as an animal type, at least for most of us. So when we collect and analyze these lists, we aggregate all the terms, and we sort them into two piles. The most popular terms on one side, and all the rest on the other. The popular terms are a gold mine for anthropologists.

You see, if I'm out trying to figure out what types of crops people prefer to grow, I could have them freelist on the theme, crops my family likes to grow. And just like you blurting out cat and dog, these farmers will start listing the crops in their fields. That's informative, but when I aggregate all those lists, I get something much bigger. When I take the most popular terms from the farmers' lists, I get an efficient snapshot of cultural consensus. I get a master list of where there's consensus on the most important crops in this community as a whole.

I'll spare you the specifics, but when we process the lists, we look at two measures. First, we look at how many times each term made one of the lists. And then we also look at where each term sits on those lists. So, our final, weighted results rank each term in the aggregate based on both the popularity of each term—that would the number of times it was listed—and the average rank of each term—and that's how high up it is on each list.

Another great way to hear a community's collective opinion through freelists is to look not at the terms themselves, but at how individual lists relate to each other. We can use freelist results to identify experts and other outliers within a culture. You see, we can measure the similarity of each respondent to all other respondents. And when we do, we plot the results so that two people with very similar lists will show up as two dots, right next to each other.

Conversely, if you mention words that don't make everyone else's lists, your dot on the scatter plot will be way off to the side. There'll be a cluster of dots of people with common terms, but you won't be with them. That's how I identify you as an outlier. And as such, you're either an expert or just a little off. Or maybe there's some other reason why you don't fit within the cultural consensus. As an anthropologist, I run to those outliers for more information.

I've used freelisting in my ecological investigations into local sorghum varieties, cropping choices, and so much more. I even use it in development contexts to assess community consensus when we're setting priorities for projects. Solar panels for the school, or new fencing for the women's garden?

And that's why freelisting and other cultural consensus tools are invaluable methods for contemporary cultural ecologists and other environmental anthropologists. Cultural consensus modeling helps us to see how others see and organize their worlds. That's what makes it an ideal method for anthropologists like Quinlan who build upon the cultural ecology tradition.

Cultural ecology and environmental anthropology are part of my own anthropological family tree. And I embrace this thread of anthropology because when I am out doing field research, I've found that my ecological questions, my people, and environment curiosities actually forge fast friendships with people who make a living off the land. And I'm certain that Steward, Netting, and others who asked family farmers similar questions about making a living off the land also found that this is a subject that can bridge people across cultures.

After all, in the industrial world, we're becoming more aware, of exactly how our human choices impact the natural world that we're a part of. And as such, in the 21st century, our anthropological mission has completely reversed course from the ideas that helped launch our discipline back in the late 1800s.

Early anthropologists were once deployed to bring modernization to the so-called primitive peoples of the world. But now, anthropologists like me work with these same populations in search of new answers and inspiration. I don't use my cultural ecology to teach Malian farmers how to manage their ecosystem and farming practices. I live and learn among this population because modern science and scientific agricultural research has yet to wipe out extreme hunger. I'm there to learn.

These farmers, despite extreme poverty, depleted soils, and unfavorable precipitation, they survive. And even though there's no Starbucks, my Malian host community is one of the most generous, collaborative, and content communities in which I've ever had the honor to live. We have something to learn there, but we also have something to share. Alone, neither scientists nor farmers hold the key for bolstering food security and eradicating extreme hunger in our world. Or, as my Mali partners say, one finger can't lift a stone.

This collaborative approach to cross-cultural knowledge production is not just the future of cultural ecology, it's fast becoming a defining feature of anthropology in the 21st century.

The Anthropology of Happiness

Throughout this course, we've seen how anthropology has emboldened our understanding of the complex layers that make us human. The course has looked at everything from our origins and biology to gender, food, and religion. But a final, extremely important topic remains: human happiness. This lecture takes a look at what the 4-field lens of anthropology—linguistics, biological anthropology, archaeology, and cultural anthropology—can teach us about happiness.

Linguistics

Linguistics can help us get a handle on the term *happiness*. If we go back and trace the origins of that word in every Indo-European language, we find a single answer. Even back to ancient Greek, the word for *happiness* is a cognate with the word for *luck*!

In Old English and Old Norse, the root *hap* means "luck" or "chance." The same goes in French and German.

As the centuries wore on, with the growth of the Abrahamic faiths in many places, happiness emerged as a religious sentiment describing the bliss of communion with God, either here on earth and/or in the afterlife.

But with the Enlightenment, we turned to the individual. Something happened that makes us think of happiness as something we cultivate. Now, when we're unhappy, we try to do things to get happy.

Many people in the world see happiness as being somewhere along this historical spectrum. There are plenty of cultures that are more in line with luck-based happiness versus contemporary Western constructions.

A cursory look into the linguistic evolution of the word *happiness* reveals that happiness is not simply a joyful feeling. It's the result of good living. Where Socrates would teach us that the path to happiness lies in the education of our desires, we can also see that happiness is about living.

Beyond our material and physical well-being on any given day, happiness comes from our relationships, our productivity, and our personal freedom. Happiness lines up with the Socratic view that we can make ourselves happy if we put in the effort (and, ironically, suffer quite a bit on the way).

The joy of raising a child, for example, is the ultimate source of happiness in many people's lives. But that happiness is built and grows over a lifetime, including painful bumps along the way.

Biological Anthropology

Biological anthropology enables us to examine happiness as a biological phenomenon. When we're happy and when we're laughing, our bodies produce brain molecules that are directly linked to euphoria or well-being. When people laugh, for example, they produce endorphins, which are natural opiates in terms of their chemical structure. They can help assuage physical and emotional pain.

People also produce dopamine, which is linked to pleasure seeking. In addition, they produce oxytocin, the so-called bonding and trust molecule. Other examples include substances like endocannabinoids, serotonin, and adrenaline.

In a 2012 study, scientists checked in on over 2000 participants from a wide variety of countries. They tracked down what made the participants feel happy. Specific activities produced higher happiness ratings. Making love was top on the list by quite a margin, but it wasn't the only "happy" activity. Listening to music, talking with a friend, and working out also appear to be happiness-producing activities.

Laughter can also help produce happiness. For example, the nonprofit organization Rx Laughter runs comedy therapy programs

Listening to music appears to be a happiness-producing activity.

for children with serious medical conditions. The organization also does collaborative research on happiness and healing with the UCLA Jonsson Comprehensive Cancer Center.

According to their research, patients in the Rx Laughter program experienced increased pain tolerance and reduced anxiety. They found that even kids who didn't audibly laugh out loud benefited from the program.

Archaeology

Archaeology focuses us on material culture, or artifacts. Using material remains, anthropologists can reconstruct the lives of people across the world and across the generations. In turn, anthropologists can watch for indicators that could be interpreted as happiness.

For example, Egyptologists might point to research on the popularity of senet, one of Ancient Egypt's favorite board games, as a window into the pursuit of happiness among the Ancient Egyptian masses.

Research has shown that our happiness is fed by our relationships. The archaeological record is unequivocal about why our happiness is so dependent upon the quality of close, face-to-face relationships.

Humans lived in small, hunter-gatherer bands for most of *Homo sapiens'* 200,000 years on earth. About 10,000 years ago, the rise of agriculture led to the eventual rise of urban centers, but most communities were rather small by today's standards.

That matters because *Homo sapiens* evolved as social animals. For thousands upon thousands of years, individual success was heavily dependent on communal cooperation. For our ancestors, the happiness experience would be rooted on the community side of the community-individual continuum.

Cultural Anthropology

It's tricky to identify and compare happiness across cultures. Unfortunately, cultural anthropologists haven't paid a lot of attention to happiness. But happiness research has taken place outside of anthropology.

As a whole, happiness research tends to focus on 2 main types of happiness: life satisfaction happiness and subjective happiness. Life satisfaction happiness is an assessment of a person's life in whole. Subjective happiness is more about the happy-to-sad spectrum and where a person feels they are on that scale from day to day.

Researchers use both of these categories to rank all the world's countries from the happiest to the most miserable.

One of the most comprehensive and data-driven ranking systems is the *World Happiness Report*, which has been publishing its findings since 2012. It measures both life satisfaction happiness and subjective happiness.

To navigate cultural and linguistic differences, the researchers boil subjective happiness down into a few revealing questions, like these:

- If you were in trouble, do you have relatives or friends you can count on to help you whenever you need them?
- Are you satisfied or dissatisfied with your freedom to choose what you do with your life?
- Have you donated to charity in the past month?

Life satisfaction happiness is more straightforward. The researchers simply ask people to rate their own happiness on a scale of 0 to 10, with 0 being the worst life imaginable and 10 being the best life imaginable.

For 2015, the mean happiness rating across all countries was 5.4. But massive differences are hidden in that statistic. For example, while North America weighs in at 7.1, sub-Saharan Africa is a meager 4.3.

Despite researchers' best efforts, identifying and measuring happiness is culturally complicated—and biased. These world rankings are based not only on people's responses to questions about their happiness, but also on life expectancy and GDP. Psychologists have proven that more money does not equate to more happiness, so it seems shortsighted to quantitatively discount the happiness of the majority world on the basis of GDP.

But even with unintentional cultural blinders and bias, the *World Happiness Report* is terrific. It produces more questions with each published edition, and the results have been fairly consistent over the years. And, just like anthropology, the more people critically test and correct their happiness theories and methods, the closer they'll get to a more inclusive approach.

In the 2016 report, a few interesting findings beg additional attention:

- There was a 4-point gap between top 10 and bottom 10 nations.

- Scandinavian countries dominate the top 5 happy slots, whereas Togo scored the lowest at 2.84.

- There was a net loss in happiness over the past decade for a handful of countries. These countries have been suffering greatly from enduring conflict or economic devastation: Yemen, Venezuela, Botswana, Saudi Arabia, Egypt, and Greece.

- After the Japan earthquake of 2011, generosity, trust, and solidarity during the crisis response led to increased happiness. This interesting

finding is certainly in line with the connection between happiness and community.

Happiness Examples: Bhutan and Okinawa

In the country of Bhutan, the people measure and actively work to increase what they call Gross National Happiness. The revered former king, Jigme Singye Wangchuck, abandoned the monarchy. But before he stepped out of his role, he helped lead the creation of a constitutional democracy.

The relatively new Bhutan government considers it the official responsibility of the government to create an environment for individuals to pursue happiness. They measure GDP too, but they place more value on the pursuit of happiness than the pursuit of sheer economic profit.

Another society with an interesting take on happiness can be found in Okinawa. This community, Ogimi-son, has the world's highest concentration of centenarians. These people live remarkably happy lives through a shared ethos of compassion through togetherness.

They take this ethos all the way to the grave. When people pass away in Ogimi-son, they are cremated, and their ashes are poured into a communal cemetery. Ogimi-son is a single community with a single destination that is clear from the start.

The Meaning

What does all this anthropology of happiness mean? Our reality is shaped by the language we use every day, and in particular the questions we ask. If our language is largely negative, it's no surprise that the world we see will seem negative too. Asking the wrong questions can mean we spend a lot of time and effort focusing our attention on all the wrong things.

Anthropology can be a great way to ask the "right" questions—the questions that reveal the unity and cultural diversity of humankind. They're the questions we've been answering all along in this course: Who are we? Where do we come from? How are we related to each other?

Perhaps the biggest human question is: What is the purpose of life? Everyone has to answer that for themselves, but the anthropology of happiness suggests that the more we search for an inclusive, universal expression of happiness, the more we'll see it as the fuel that sustains us.

Biologically, culturally, and emotionally, happiness keeps us healthy and motivates us to be more compassionate and connected to humankind. In a similar vein, the science of happiness shows us that there are definite connections between happiness, hope, gratitude, forgiveness, and altruism.

Suggested Reading

Collen, *10% Human*.

Hanh, *No Mud, No Lotus*.

Journal of Happiness Studies.

Ricard, *Happiness*.

Questions to Consider

1. How has humanity's definition of happiness changed since the years of our hunter-gatherer ancestors?

2. Is happiness a universal experience or does it differ from culture to culture?

3. How might anthropology help us increase happiness in our own lives?

The Anthropology of Happiness

Hello and welcome to the capstone lecture of our course on *Anthropology: The Study of Humanity*. It's always a challenge deciding where to end an epic exploration, but when you think about human nature—our desires, our motivations, and needs, and goals—one topic seems to stand out perhaps more than any other. And so I saved that all-embracing topic, that big kahuna, if you will, for the last. Welcome to the anthropology of happiness. It's not lost on us that our world today can be a rough one. More people are suffering from stress, anxiety, addiction and depression, hunger, and at all ages too. And all this despite rising economic wealth and improved health technologies.

Throughout this course, anthropology has emboldened our understanding of the complex layers that make us human, everything from our origins and biology to gender, food, religion, and so much more. Well today, let's see what the four-field lens of anthropology—linguistics, biological anthropology, archaeology, and cultural anthropology—can teach us about happiness.

Let's start with linguistics. Happiness is one of those sandy terms. Like dry ocean sand in your palm, the tighter you try to grasp it, the more it squeezes right out. The same thing goes for happiness, we're going to have to start off accepting the fact that the best we can do is to get a loose grip on this core, yet elusive phenomenon. And it's linguistics that can help us get a handle on the term. For example, studying the semantics and etymology quickly reveals that our modern idea of happiness, the idea that happiness is something we cultivate, is not what the origins of the word would suggest.

In previous lectures, we learned that linguistic anthropologists, among other things, trace the evolution of language and meaning And they are quite clear about happiness. If we go back and trace the origins of that word in every Indo-European language, we find a single answer. Even back to Ancient Greek, the word for happiness is a cognate with the word for luck. Fortune. In Old English and Old Norse, the root hap means luck or chance. Same thing goes in French and German it's all about luck. It appears that happiness was something that happened to you, kind of like the weather. As if some god or alien somewhere could press a button and snap you're happy and snap you're in tears.

Anyway, as centuries wore on, with the growth of the Abrahamic faiths in many places, happiness emerged as a religious sentiment describing the bliss of communion with God, either here on earth and/or in the afterlife. But with the

Enlightenment, we turn to the individual, to ourselves. Something happened that makes us think of happiness as something we cultivate, for ourselves and for our families and communities. Unlike the old days, now, when we're unhappy, we try to do things to get happy.

Another way to look at it through the diagnostic criteria for mental illness. In the official manual, there's depression, sure but not happiness. Depression is an illness and happiness is a goal. Maybe even our ultimate goal. Many people in the world see happiness as being somewhere along this historical spectrum. There are plenty of cultures out there that are more in line with luck-based happiness versus our contemporary, western constructions. But how could a modern human still think happiness is luck-based? Thomas Malthus hypothesized that cultures that evolved in harsh and scarce climates would be more likely to think of happiness as a function of luck. Ultimately, viewing happiness as luck-based could be conditioning from generations living in scarcity. Living successfully but in scarcity.

Contrary to the linguistic origins of happiness, classical Greek philosophers added a new layer to the idea of happiness, or what they called *eudemonia*. While happiness may sometimes be elusive, Socrates and the gang all agreed that happiness could be earned. The right actions could lead to *eudemonia*, which we might translate as well-being, or human flourishing.

This old idea was made new again in the Enlightenment when folks like John Locke reclaimed happiness as not only something that we can control but something that itself is the mission of humankind. "The business of man is to be happy," Locke wrote. And with this new perspective, we not only claim our individual control over our happiness but now it becomes our primary objective. A cursory look into the linguistic evolution of the word happiness reveals that happiness is not simply a joyful feeling. It's the result of good living. Where Socrates would teach us that the path to happiness lies in the education of our desires, we can also see that happiness is about living.

Beyond our material and physical well being on any given day, happiness with a capital H, that is, comes from our relationships, our productivity, and our personal freedom. And capital H happiness lines up with the Socratic view that we can make ourselves happy if we put in the effort, and ironically suffer quite a bit along the way. The joy of raising a child, for example, is for so many of us, the ultimate source of happiness in our lives. But that happiness is built and grows over a lifetime, including painful bumps all along the road.

Linguistics cast some light on the complexity of the word happiness, but biological anthropology can ground us a little. It enables us to examine happiness as a biological phenomenon. And when we do this, we see happiness as the foundation to a long and healthy life. Yeah, I know, you're laughing because we can all think of some seriously unhappy people who may also happen to be a little older but wait until you see the data on this.

First, let's put a hypothetical test subject into an MRI scanner. What happens in his brain when he's recounting a happy memory? Do certain spots light up? They most certainly do. When we're happy, the left side of our prefrontal cortex and our right posterior cingulate both fire up with activity. And that's just the architecture of the brain. What about our neurochemistry?

When we're happy and when we're laughing, our bodies produce brain molecules that are directly linked to euphoria or well-being. When you laugh, for example, you produce endorphins, which are natural opiates in terms of their chemical structure. They can help assuage physical and emotional pain. And that's not all we also produce dopamine, linked to pleasure seeking and oxytocin, the so-called bonding and trust molecule and there are others like endocannabinoids, serotonin, and adrenaline, and more. So to sum it up, our behavior alters our neurochemistry. If we find ways to produce these brain molecules through running, for example, we experience happiness. Happiness, indeed, has a biological dimension.

But how do we strike this neurochemical balance? What can we do to get these molecules running? Well, in a 2012 study, for example, scientists checked in randomly on over 2000 participants from a wide variety of countries, and they tracked down what makes us feel happy. And it turns out that specific activities produced higher happiness ratings. Making love was on top of the list, and by quite a margin, but it wasn't the only happy activity. There is listening to music, talking with a friend; working out also appear to be happiness-producing activities.

One thing that is definitely not correlated with happiness is a wandering mind. Beware. It may be really difficult to hear when we're down in the dumps, but the case is clear. If we're looking to trigger the happiness effect, the key is to do something. Focus on a mission and, if that's too much, well, there's always Billie Holiday, Lightning Hopkins and making love. Or maybe it's as simple as laughter?

The non-profit RX Laughter is walking the walk when it comes to healing with happiness. They run comedy therapy programs for kids with serious medical conditions from bone marrow transplants all the way to something like depression. And just as important, they do collaborative research on happiness and healing with the UCLA Jonsson Cancer Center. According to their research, patients in the RX Laughter program experienced increased pain tolerance and reduced anxiety. They found that even kids who didn't audibly laugh out loud, even they benefited from regular doses of Laughter Project medicine.

The physiology of happiness and laughter speak to the universality of this core human experience, and we don't have to get deep into biological anthropology to learn something that can prolong our lives. Multiple studies, including one conducted by University College London, have found that people who reported enjoying life age slower, experience a slower rate of decline, and that's especially when compared with their gloomy counterparts. It even proved true for smokers and drinkers. So if you can't get your friend to quit smoking, just be sure to keep them laughing.

OK, so far we've looked at happiness from the perspective of linguistics and biology. Now let's turn to a third subfield of anthropology—archaeology. With archaeology, remember, the anthropological mission of exploring humankind over time and space focuses us on material culture or artifacts. But do we have happiness artifacts? I mean, that happy face emoticon is quite a recent device, so we're unlikely to find one in the parietal art of the Chumash and Dogon ancestors. So how do archaeologists search for evidence of happiness?

The archaeology of happiness is challenged by the ideas we previously uncovered with linguistics. First, happiness evolves, so what is happy to you and me may not relate to the point of view of a 17th-century farmer. And second, even now, across contemporary cultures, people experience happiness in different ways. Nonetheless, using material remains, we can reconstruct the lives of people across the world and across the generations and when we do, we watch for indicators that could be interpreted as happiness.

For example, Egyptologists might point to research on the popularity of senet, one of Ancient Egypt's absolute favorite board games, as a window into the pursuit of happiness among the Ancient Egyptian masses. But more to the nitty gritty, research has shown that our happiness is fed by our relationships. And the archaeological record is unequivocal about why our happiness is so dependent upon the quality of close, face-to-face relationships.

Humans lived in small, hunter-gatherer bands for most of *Homo sapiens'* 200,000 years on earth. Then just 10,000 years ago, the rise of agriculture led to the eventual rise of urban centers, but still, most communities were rather small by today's standards. Why does that matter? Well, *Homo sapiens* evolved as social animals—for thousands upon thousands of years individual success was heavily dependent on communal cooperation. So for our ancestors, the happiness experience would be rooted on the community side of the community individual continuum. That's quite different from how most folks in the US would describe happiness today. So maybe we individualists should ask ourselves if we're forgetting something quite important. Keep that thought in mind as we shift to the fourth sub-field of American anthropology and use cultural anthropology to see how happiness may or may not differ across cultures.

As we've already noted, happiness is an elusive term that's difficult to pin down. And if it's ambiguous in English it gets even more tricky to identify and compare happiness across cultures. It's a great anthropology puzzle. If we want to explore happiness across cultures, we need to create an inclusive definition for happiness, as well as a happiness scale that includes English and non-English speakers, all of whom have their own unique ways of describing and being happy.

Unfortunately, cultural anthropologists haven't really paid a lot of attention to happiness. It's astounding. I mean, it was my amazing, non-anthropology parents who taught me that happiness was the ultimate goal, far more important than money. And it is. But, how have we cultural anthropologists overlooked happiness for so long?

One of the earliest anthropology publications, the *Popular Magazine of Anthropology*, actually argued that the central purpose of anthropology is to help all humanity achieve material prosperity and, wait for it, happiness. But that was 1866, and the anthropology of happiness never really took off as a specialization.

So, for our cultural anthropology mission today, we're going to double-down on the interdisciplinary ethos of anthropology, and we're going to need some help from some folks outside of our discipline. That's where you're going to find happiness studies. Looking through your library database, you'll find plenty of studies on the economics of happiness, and especially the psychology of happiness. As a whole, happiness research tends to focus on two main types of happiness. There's life satisfaction happiness, and there's also subjective happiness. This may not be perfect, but we need a common starting point or else we'd be comparing apples to oranges.

Life satisfaction happiness is an assessment of your life in whole. Whereas the second type, subjective happiness, is more about the happy to sad spectrum, and where you feel you are on that scale from day to day. Researchers use both of these categories to rank all the world's cultures from the happiest to, frankly, the most miserable.

Now there's quite a few of these world rankings, but one thing is for sure, no matter which one you consult the US doesn't win the gold medal for the happiness Olympics. One of the most comprehensive and data-driven ranking systems is the World Happiness Report, which has been publishing its findings since 2012.

How does it work? Well, let me put it this way. Malinowski would approve. One of the reasons we're focusing on this specific report is that, like Malinowski, this report, and the researchers, go out to collect data. With some help from Gallup, the World Happiness Report surveys thousands of respondents in over 150 countries. And it measures both life satisfaction happiness and subjective happiness. To navigate through a sea of cultural and linguistic differences, the researchers boil subjective happiness down into a few revealing questions. Think how you might reply.

If you were in trouble, do you have relatives or friends you can count on to help you whenever you need them? Are you satisfied or dissatisfied with your freedom to choose what you do with your life? Have you donated to charity in the past month? Is corruption widespread in government, business? Those questions help score subjective happiness.

Life satisfaction happiness, on the other hand, is a lot more straight-forward. The researchers simply ask people to rate their own happiness on a scale of zero to ten. Zero, worst possible life imaginable versus ten, the best possible life. Where do you think you'd rate your life satisfaction happiness? Maybe an eight? Well, if that's your score, you're officially happier than most everyone on this planet. For 2015, for instance, the mean happiness rating across all countries was 5.4. But, massive differences are hidden in that statistic. For example, while North America weighs in at 7.1, the place I love so much, sub-Saharan Africa, is a meager 4.3.

Of course, we have to be careful with these numbers, as we do with any statistic or finding. Despite our best efforts, identifying and measuring happiness is culturally complicated and, quite frankly, a little biased. Maybe our approach to calculating world happiness, despite our best attempts, is still influenced by scientific and western values. And if it is, that means that we may not be getting an accurate read. Let me give you an example from my work in Mali.

One of my long-time partners, Tamba, visited me in Bamako, Mali where I had reserved a special room for him to make him happy because he was coming to help me work on a documentary film project that I was involved in. At dinner, I told him the news. He gets his own deluxe room. But his spirits sunk the moment told him. Visibly distressed, yet politely trying to cover his true feelings, he diplomatically helped me see a different shade of happy. What makes me, a guy from a more individualistic society, happy namely my own hotel room and privacy, was the exact opposite for Tamba. His happiness is being with people Long story short, I slept on a cot and put him in my room. Happiness restored.

And it goes deeper than differences between individualistic versus more collaborative cultures. You see, these world rankings are based not only on people's responses to questions about their happiness, but they also incorporate life expectancy and Gross Domestic Product. I see the logic, but psychologists have proven that more money does not equate to more happiness. So it seems short-sighted to me to quantitatively discount the happiness of the majority world, a.k.a. the third world, on the basis of Gross Domestic Product.

But don't get me wrong, even with unintentional cultural blinders or bias, the World Happiness Report is terrific. It produces more questions with each published edition, and the results have been fairly consistent over the years. And, just like anthropology, the more we critically test and correct our happiness theories and methods, the closer we'll get to a more inclusive approach.

In the meantime, we're gaining valuable insight into happiness. In the 2016 report, for example, there were a few really interesting findings that beg additional attention. One, there was a four-point gap between top ten and bottom 10 nations. Scandinavian countries dominate the top five happy slots, whereas Togo scored the lowest at 2.84.

We also see a net loss in happiness over the past decade for only a handful of countries, but anyone who keeps up with world affairs will quickly perceive that all of these countries have been suffering greatly from enduring conflict or economic devastation: Yemen, Venezuela, Botswana, Saudi Arabia, Egypt, and Greece.

On the sunny side of the street, the 2016 report offered some surprising news. It seems like great catastrophes like a massive natural disaster may sometimes actually help increase our happiness. Great Japan Earthquake of 2011: the generosity, trust, and solidarity of the crisis response actually led to increased happiness.

This interesting finding is certainly in line with the connection between happiness and community that we noted earlier when we were talking about archaeology. And it underscores the question I raised about the relationship between GDP and happiness. In fact, I'd like to return to that question by taking a quick glance at two cultures where GDP seems to have very little bearing on the overall happiness of people.

First, let's go to Bhutan where they actually measure and actively work to increase what they call Gross National Happiness. The revered former king, Jigme Singye Wangchuck, abandoned the monarchy and has developed a strong affinity for biking. But before he stepped out of his role, he helped lead the creation of a constitutional democracy. And that constitution identifies basic human rights, including the pursuit of happiness. That's how the Declaration of Independence phrases it, and it's how they phrase it in Bhutan as well.

The key difference is that the relatively new Bhutan government considers it the official responsibility of the government to create an environment for individuals to pursue happiness. To create fertile soil for widespread happiness. Sure, they measure GDP too, but they're trying something new. They are placing more value on the pursuit of happiness than the pursuit of sheer economic profit. Thomas Jefferson would be proud. He actually wrote that "the care of human life and happiness, and not their destruction, is the only legitimate object of good government."

Like any anthropological observation, the cross-cultural perspective teaches us as much about ourselves as it does the people we study. And as is quite often the case, it can teach us different ways to think and do just about anything. Given the social, emotional, and physical benefits of happiness, I think we could learn something from Bhutan and their approach to happiness. In fact, they truly are ahead of the pack.

The other place I'd like to visit briefly is Okinawa. Unlike most of Japan, which is often fairly low on the world happiness rankings, a community in Okinawa seems to have discovered the secret to a long and happy life. In fact, this community, Ogimi-son, has the world's highest concentration of centenarians. These people live remarkably happy lives through a shared ethos of compassion through togetherness. And they take this ethos all the way to the grave. When people pass away in Ogimi-son, they are cremated, and their ashes are poured into a communal cemetery. They are mixed in with everybody else. A single community with a single destination that is clear from the start. Togetherness, from birth through death. Surely that's an insight we Westerners can learn from.

And it's insights like this that make me love what I do as an anthropologist. While in the field, anthropologists dedicate months and years to learning how to see the world through completely new eyes and ears. This can be an incredibly rewarding experience. In fact, Bronislaw Malinowski once wrote that "Realizing the substance of the happiness in our host communities is not only part of our mission, but it is the greatest reward which we can hope to obtain from the study of man."

Wow. Well, I would have said hu-man, but you get the point. Malinowski deems essential something most anthropologists never really talk about. Doing anthropology transforms your consciousness. And if you do your job right, one of the perks is that you get to experience new flavors of happiness. I'll tell you, this gig expands your emotional range, and it can get painfully nostalgic after you've been doing it for a while.

For one, I've seen the generation of elders who taught me, pass on. And I miss them dearly. But on the brighter side, infants that I once held in my arms now hold infants of their own. And it's all happiness. The built up memories. Over two decades of relationships. Happiness isn't about that fleeting moment of joy or the thing that makes us smile. Rather, it is the net sum of living a meaningful life, and that's something we all get to figure out for ourselves.

So great. What does all this anthropology of happiness mean? Well, our reality is shaped by the language we use every day, and in particular, the questions we ask. If our language is largely negative, it's no surprise that the world we see will seem negative too. Asking the wrong questions can mean we spend a lot of time and effort focusing our attention on all the wrong things. So let's think about that. What are the right questions?

Well, if there is one thing I hope we can all take home from this course, it's that anthropology can be a great way to ask these right questions. The big questions. Questions that reveal the unity and cultural diversity of humankind. They're the questions we've been answering all along in this course. Wo are we? Where do we come from? How are we related to each other? And so many more. And as we wrap up our anthropological exploration of happiness, we've inadvertently stumbled across an answer to one of those right questions, and perhaps the biggest human question yet: what is the purpose of life?

We've all got to answer that question for ourselves, but the anthropology of happiness suggests that the more we search for an inclusive, more universal expression of happiness, the more we'll see it as the fuel that sustains us. Biologically, culturally, and emotionally, happiness keeps us healthy and it motivates us to be more compassionate and connected to humankind. And in a similar vein, the science of happiness shows us that there are definite connections between happiness and hope, gratitude, forgiveness, and altruism. So if you'd like to put yourself on a path to greater happiness, those are qualities and actions that you should cultivate.

For what it's worth, our colleagues in cognitive psychology have demonstrated that regularly counting your blessings is one simple way, and proven way, to grow happiness. So, as you test these ideas out for yourself—as you seek to become happier by becoming more hopeful, grateful, forgiving, and generous, you'll find yourself asking the one more right question. How can I spread happiness to others? In the spirit of that question, and its surprising importance to the continued survival of our species, I'll share with you the parting message I share with my students on the final day of the semester.

I remind them of our anthropological journey and how, despite the incredible diversity of humankind, we are ultimately one human race. Cultural differences, however, can lead to confusion, misunderstanding, and at its worst, violence and war. So over the years, I've developed a mantra that helps me get clear when I'm at a loss for what to do in the face of these problems: Spread Love. It's as simple as that. Spread Love.

And with that, my dear friends, we've come to the end of our lecture and our course. Thank you, for coming along with me on this anthropological journey. It's been a joy talking with you about our astonishing and epic, 7 million year story as upright walking apes. But since we must part, let me leave you with my favorite Malian blessing: ka tile here caya which sounds just as transcendent in English: May the sun increase your peace.

Bibliography

Baker, Lee. *From Savage to Negro: Anthropology and the Construction of Race, 1896–1954*. Berkeley: University of California Press, 1998. Baker's essential study traces roles of early American anthropologists and the erroneous ideas that unscientifically divided our one human race, *Homo sapiens*, into multiple, socially constructed races.

Becker, Howard. *Art Worlds*. 25th anniversary ed. Berkeley: University of California Press, 2008. Sociologist Howard Becker's classic work helps visual anthropologists and others reconsider art not as objects produced by artists, but as collective action including artists, dealers, gallery managers, publishers, consumers, and just about anyone else engaged in the world of art.

Berwick, Robert and Noam Chomsky. *Why Only Us: Language and Evolution*. Cambridge, MA: MIT Press, 2015. Berwick, a computer scientist, and Chomsky, a celebrated linguist, join forces to explore humanity's unique biological foundations that render complex language possible.

Black, Edwin. *War Against the Weak: Eugenics and America's Campaign to Create a Master Race*. Westport, CT: Dialog Press, 2012. Edwin Black recounts the seldom-told story of the early days of genetics research and the direct connections between the American eugenics movement and the atrocities of Nazi Germany's concentration camps and holocaust.

Bolotin, Norman and Christine Laing. *The World's Columbian Exposition: The Chicago World's Fair of 1893*. Champaign: University of Illinois Press, 2002. Take a beautifully illustrated trip with Bolotin and Laing to the 1893 Chicago World's Fair where early anthropologists first introduced the American public to what was an emerging academic discipline.

Bostrom, Nick and Milan M. Cirkovic. *Global Catastrophic Risks*. Oxford: Oxford University Press, 2011. To assess the sobering existential threats facing humankind in the 21st century, Bostrom and Cirkovic turn to 25 leading experts to suss out the potential end of our species via nuclear war, biological weapons, terrorism, asteroids, social collapse, the rise of artificial intelligence, and more.

Brown, Peter J. and Svea Closser. *Understanding and Applying Medical Anthropology*. 3rd ed. New York: Routledge, 2016. Used in classrooms across the United States and world, this classic medical anthropology reader is one of the best and most efficient ways to survey the remarkable breadth and contributions of one of the most popular specializations in contemporary anthropology.

Burling, Robbins. *The Talking Ape: How Language Evolved*. Oxford: Oxford University Press, 2007. Thinking back on the origins of language, Burling shares some fascinating theories on how and why humans first started using words and grammar.

Clarke, Kamari Maxine. *Fictions of Justice: The International Criminal Court and the Challenge of Legal Pluralism in Sub-Saharan Africa*. Cambridge, U.K.: Cambridge University Press, 2009. A riveting example of legal anthropology at work in the 21st century, this compelling study of the International Criminal Court demonstrates that ideas like human rights and justice are culture bound, and thus, far more complicated (and political) than one might think.

Cleveland, David. *Balancing on a Planet: The Future of Food and Agriculture*. Berkeley: University of California Press, 2013. In this integrative, forward-thinking study, David Cleveland looks to the future of food production by embracing lessons and practices from scientific and small-scale farming systems.

Cohen, Mark Nathan and George Armelagos. *Paleopathology at the Origins of Agriculture*. Cambridge, MA: Academic Press, 1984. Considered a foundational text in bioarchaeology, this volume brings together authors from across the discipline to uncover fascinating discoveries about the impact of agriculture on the health of early farmers in Asia, the Middle East, Europe, and the Americas.

Collen, Alanna. *10% Human: How Your Body's Microbes Hold the Key to Health and Happiness*. Repr. ed. New York: Harper, 2016. From a biological perspective, biologist Alanna Collen explains that, remarkably, our happiness (and much more) has roots in the microbes our bodies carry.

Crawford, David. *Moroccan Households in the World Economy: Labor and Inequality in a Berber Village*. Baton Rouge: Louisiana State University, 2008. This thoughtfully crafted Moroccan ethnography provides an excellent example of how Steward's cultural ecology continues to influence 21st-century anthropology.

Darwin, Charles. *The Origin of Species*. 150th anniversary ed. New York: Signet, 2003. One of the foundational texts of modern biology and anthropology, Darwin's most celebrated title ushered in a new era in science, an era engrossed with deciphering the mysteries of biological evolution.

Dawkins, Richard and Yan Wong. *The Ancestor's Tale: A Pilgrimage to the Dawn of Evolution*. Rev. ed. New York: Mariner, 2016. Inspired by the form of Chaucer's *Canterbury Tales*, the authors of this meticulous account retell the story of life on Earth.

De Waal, Frans. *The Bonobo and the Atheist: In Search of Humanism Among the Primates*. New York: Norton, 2014. Known for his creative and insightful writing on primates, Frans de Waal unveils the biological and cultural roots of human morality in the observed behavior of the primate order and the wider animal kingdom.

Errington, Shelly. *The Death of Authentic Primitive Art and Other Tales of Progress*. Berkeley: University of California Press, 1998. Errington examines the category of primitive art to reveal it as a flawed and inadequate classification that tells more about the worldview of 20th-century anthropologists rather than the cultures and artists who produce what we once called primitive art.

Evans-Pritchard, Edward. *The Nuer: A Description of the Livelihood and Political Institutions of a Nilotic People*. New York: Oxford University Press, 1969 (orig. 1940). A revered model of early-20th-century ethnography, Evans-Pritchard's detailed account of the Nuer people of East Africa examines their religious practices, social structure, and much more.

Fagan, Brian. *Ancient Lives: An Introduction to Archaeology and Prehistory*. 4th ed. Upper Saddle River, NJ: Prentice Hall, 2009. World-renowned archaeologist Brian Fagan lays out the basic ideas and methods of archaeology as he surveys some of the biggest developments in human prehistory, including the spread of modern humanity and the emergence of agriculture.

Ferllini, Roxana. *Silent Witness: How Forensic Anthropology is Used to Solve the World's Toughest Crimes*. 2nd ed. Ontario: Firefly Books, 2012. This well-illustrated volume describes specific techniques and procedures used by forensic anthropologists through a discussion of 32 remarkable cases ranging from a train collision and airplane crash to genocide and serial killers.

Fouts, Roger and Stephen Tukel Mills. *Next of Kin: My Conversations with Chimpanzees*. New York: William Morrow, 1998. One of the first primatologists to work with apes using sign language, Roger Fouts, along with his coauthor, share terrific stories of friendship, humor, and discovery through interspecies communication and ape conservation.

Frazer, James. *The Golden Bough*. Abridged ed. Mineola, NY: Dover Press, 2002 (originally written in 1890). This early anthropological classic captured the imagination of many of the scholars who went on to establish anthropology as an academic discipline, and Frazer's compendium of the various beliefs and social institutions of world cultures was so popular and comprehensive that it eventually grew into a 12-volume epic.

Goodall, Jane. *In the Shadow of Man*. New York: Mariner, 2010. Starting with the story of a life-altering conversation with and suggestion from Dr. Louis Leakey, Goodall tells the inspirational story of how her life among the chimpanzees of Gombe unfolded, and how she came to know these chimps as complicated individuals and even toolmakers.

Graeber, David. *Debt: The First 5,000 Years*. New York: Melville House, 2014. Graeber's comprehensive history of debt draws on hard evidence from archaeology, cultural anthropology, and more to challenge and revise one of the foundational theories in classical economics, namely that human exchange started with bartering prior to the creation of debt and currency.

Graf, Kelly, Caroline Ketron, and Michael Waters, eds. *Paleoamerican Odyssey*. College Station: Texas A&M Press, 2014. By analyzing ancient skeletal remains as well as the genomes of living populations, the authors provide remarkable new insight on the peopling of the Americas.

Hanh, Thich Naht. *No Mud, No Lotus: The Art of Transforming Suffering*. Berkeley: Parallax Press, 2014. This world-renowned monk and former associate of Martin Luther King, Jr. explains an ironic secret to the art of happiness: acknowledging and then transforming our suffering.

Harari, Yuvai Noah. *Sapiens: A Brief History of Humankind*. New York: Harper, 2015. Harari boils down over a century of discoveries and evidence on the origins and spread of *Homo sapiens* in a compelling narrative that underscores the remarkable journey that created our modern humanity.

Harris, Marvin. *Good to Eat: Riddles of Food and Culture*. Long Grove, IL: Waveland, 1998. This engaging example of Harris's cultural materialism decodes food taboos and preferences from throughout the world, from the sacred cow in India to the troubles many parents have getting their children to eat spinach.

Hart, Keith. *The Memory Bank: Money in an Unequal World*. Cambridge, UK: Profile Books, 1999. Hart examined the impact of revolutionary technologies like the Internet, predicting that these changes will usher in truer forms of economic democracy across the globe.

Harvey, Joy Dorothy. *Almost a Man of Genius: Clémence Royer, Feminism, and Nineteenth-Century Science*. New Brunswick: Rutgers University Press, 1997. A little-known feminist and scientist hero, Clémence Royer once served as Darwin's French translator, but her contributions and critiques of the male-dominated scientific community were revolutionary and, arguably, over a century ahead of the mainstream.

Henley, Paul. *The Adventure of the Real: Jean Rouch and the Craft of Ethnographic Cinema*. Chicago: University of Chicago Press, 2010. This is an essential account of one of the most remarkable ethnographic filmmakers of the 20th century, whose methods and work did nothing short of launching the genre we know today as cinema verite.

Henrich, Joseph. *The Secret of Our Success: How Culture Is Driving Human Evolution, Domesticating Our Species, and Making Us Smarter*. New Brunswick, NJ: Princeton University Press, 2015. Our evolution as social beings produced a collective intelligence that, like cultural DNA, has fueled our success as a species that builds on the technological achievements of past generations.

Hrdy, Sarah Blaffer. *Mothers and Others: The Evolutionary Origins of Mutual Understanding*. Washington DC: Harvard University Press, 2009. Hrdy looks to primates and human history to reconstruct the development of our capacity to care for each other, and to examine our modern emotional state as human beings.

Humphrey, Caroline and Stephen Hugh-Jones. *Barter, Exchange and Value: An Anthropological Approach*. Cambridge, UK: Cambridge University Press, 1992. Humphrey and Hugh-Jones provide one of the most thorough and evidence-based discussions of the anthropology and history of barter, and they demonstrate that this human institution is far more complex than one may think.

Hurston, Zora Neale. *Mules and Men*. New York: Harper Perennial, 2008. Hurston returned to her hometown in southern Florida to document the rich oral traditions, wisdom, humor, and resilience of Black communities in the Deep South.

Jablonski, Nina. *Living Color: The Biological and Social Meaning of Skin Color*. Berkeley: University of California Press, 2012. Jablonski provides a definitive evolutionary account of the development of diverse skin colors across the human family, and she shows us how this basic adaptation to sunlight has impacted our sociocultural relations and health in profound ways that contradict popular constructions of race.

Jennings, Justin. *Killing Civilization: A Reassessment of Early Urbanism and Its Consequences*. Albuquerque: University of New Mexico Press, 2016. Jennings reviews excavation and survey data from Cahokia (Mississippi Valley), Jenne-jeno (Mali), Tiwanaku/Tiahuanaco (South America), and elsewhere to develop new perspectives on the nature and consequences of these celebrated early civilizations.

Journal of Happiness Studies: An Interdisciplinary Forum on Subjective Well-Being. Dordrecht, Netherlands: Springer, 2000 to present. This peer review journal features scientific research into understanding the happiness spectrum.

King, Barbara. *Evolving God: A Provocative View on the Origins of Religion.* New York: Doubleday, 2007. Noted primatologist Barbara King's remarkable study incorporates non-human ape behavior, archaeology, and even biology to give us a comprehensive and integrative window into the origins of the human religious experience.

Kolbert, Elizabeth. *The Sixth Extinction: An Unnatural History*. Repr. ed. London: Picador, 2015. Kolbert's widely popular and rather disquieting assessment of the impact of humankind on Earth warns us with urgency that previous mass extinctions foreshadow yet another massive die-off that, quite likely, could include *Home sapiens*.

Kroeber, Alfred and Clyde Kluckholn. *Culture: A Critical Review of Concepts and Definitions*. New Haven, CT: The Peabody Museum, 1952. In the twilight of his career, early anthropologist Alfred Kroeber worked with Cylde Kluckholn to document the evolution of anthropology through an investigation of the ever increasing ways that anthropologists conceive and define culture.

Kurzweil, Ray. *The Singularity Is Near: When Humans Transcend Biology*. London: Penguin Books, 2006. Technologist and futurist Ray Kurzweil offers his optimism that humankind could achieve an immortality of sorts through the exponential growth and application of genetics, nanotechnology, and robotics.

Lacy, Scott. "Nanotechnology and Food Security: What Scientists Can Learn from Malian Farmers" in *Can Emerging Technologies Make a Difference in Development?* Eds. E. Parker and R. Appelbaum, 86–98. New York: Routledge. In narrative form, this chapter recounts the author's work as a bridge builder, connecting knowledge experts across linguistic, cultural, and geographic barriers.

Leick, Gwendolyn. *Mesopotamia: The Invention of the City*. London: Penguin, 2003. Archaeologist Gwendolyn Leick takes us to Mesopotamia and the first urban revolution to help us understand what life was like in 10 ancient Mesopotamian villages.

Lévi-Strauss, Claude. *The Elementary Structures of Kinship*. Boston: Beacon, 1969 (orig. 1949). Reviewing an impressive array of ethnographic evidence and accounts, Levi-Strauss interprets a classic anthropological theme, kinship, as a form of exchange that reveals what he described as a universal basis for marriage prohibition (e.g. incest avoidance) and a formalization of male-female relationships.

Little, Walter and Timothy Smith. *Mayas in Postwar Guatemala: Harvest of Violence Revisited*. Tuscaloosa: University of Alabama Press, 2009. Little and Smith uncover the non-violent resilience of Mayan people and communities to improve political, social, and economic conditions in postwar Guatemala, despite the enduring violence and insecurity they grapple with daily.

Lowie, Robert H. *Primitive Religion*. New York: New York: Liveright Publishing Corporation, 1948 (originally published in 1924). This early classic on the anthropology of religion takes the reader on an epic journey across the world to learn about the diverse religious beliefs and practices of humankind.

Malinowski, Bronisław. *Argonauts of the Western Pacific*. Long Grove, IL: Waveland, 1984 (originally published in 1922). Considered the go-to guide for foundational principles in anthropology field research, Malinowski's pioneering and comprehensive study of the Trobriand Islanders cemented participant observation as the hallmark of cultural anthropology.

Manhein, Mary. *The Bone Lady: Life as a Forensic Anthropologist*. Baton Rouge: Louisiana State University Press, 1999. As someone who has worked on hundreds and hundreds of cases, forensic anthropology superstar Mary Manhein artfully tells the story of some of the most remarkable moments of her career—including the recovery and identification of the 7 astronauts who perished in the 2003 *Columbia* disaster.

Marion, Jonathan and Jerome Crowder. *Visual Research: A Concise Introduction to Thinking Visually*. London: Bloomsbury Academic, 2013. This comprehensive guidebook for visual anthropologists examines the potential opportunities and pitfalls of doing visual anthropology research in an age of globally ubiquitous cellphones and other image-making technologies.

Mintz, Sidney. *Sweetness and Power: The Place of Sugar in Modern History*. London: Penguin, 1986. This celebrated study by Mintz digs deep into the history of sugar and reveals that a comprehensive examination of a single, global commodity can reveal a compelling history of the exploitative nature of international relations and exploitation.

Nelson, Sarah Milledge. *Women in Antiquity: Theoretical Approaches to Gender and Archaeology*. Lanham, Maryland: AltaMira Press,

2007. Through archaeology we can see that contemporary ideas and constructions of gender and womenhood are anything but consistent.

Netting, Robert. *Smallholders, Householders: Farm Families and the Ecology of Intensive, Sustainable Agriculture*. Redwood City: Stanford University Press, 1993. Integrating case studies from across the globe, Netting's classic text makes a compelling and thorough case for the efficiency and sustainability of small-scale farmer over industrial agriculture.

Newkirk, Pamela. *Spectacle: The Astonishing Life of Ota Benga.* New York: Amistad, 2015. A surprisingly revealing and tragic account of the early days of modern science and regrettably unscientific constructions of race.

Nolan, Riall. *Development Anthropology*. Boulder, CO: Westview Press, 2001. Applied anthropologist Riall Nolan describes how international development projects work, including the roles and contributions of anthropologists.

Nordstrom, Carolyn. *A Different Kind of War Story*. Philadelphia: University of Pennsylvania Press, 1997. In this riveting ethnography of civil war in Mozambique, Nordstrom shows us how, in the face of extreme violence and danger, regular citizens emerge as remarkable heroes, healers, and peacemakers.

Packard, Randall M. *The Making of a Tropical Disease: A Short History of Malaria*. Baltimore: Johns Hopkins University Press, 2011. In this essential resource for anyone curious about medical anthropology, Packard explains the social and natural complexities that foster the spread of malaria, which still kills 1 to 3 million people a year.

Pauketat, Timothy. *Cahokia: Ancient America's Great City on the Mississippi*. Repr. ed. London: Penguin Books, 2010. Archaeologist Timothy Pauketat goes back 1000 years ago to unearth a great Mississippi Valley civilization with extensive trade relations, remarkable technological innovation, and a massive earthen pyramid.

Peletz, Michael. *Gender Pluralism: Southeast Asia Since Early Modern Times*. London: Routledge, 2009. In this historical and ethnographic study of gender pluralism in Southeast Asia, Peletz examines the inconsistent ways people construct human diversity in ways that lead us to reject some forms of human variation while embracing others.

Peterson, Dale and Richard Wrangham. *Demonic Males: Apes and the Origins of Human Violence*. New York: Mariner, 1997. Peterson and Wrangham turn to non-human apes to explore the potential biological and evolutionary roots of human violence.

Podolefsky, Aaron, Peter J. Brown, and Scott M. Lacy. *Applying Anthropology*. New York: McGraw-Hill, 2012. This popular reader contains compelling chapters written by a host of anthropologists who take you right into the field to see the myriad ways different anthropologists apply the 4 fields of their discipline to demonstrate anthropology's unique roles as a tool for social transformation and human understanding.

Price, David. *Weaponizing Anthropology: Social Science in Service of the Militarized State*. Petrolia, CA: CounterPunch Books, 2011. In this revealing and comprehensive history of the complex and often nefarious relationship between anthropologists and the CIA, FBI, and US military, David Price critically examines the ethical and practical dimensions of putting anthropology to use in war zones and counterinsurgency programs.

Redfield, Robert. *The Little Community and Peasant Society and Culture*. Repr. ed. Chicago: University of Chicago Press, 1989. This reprint edition includes 2 of Redfield's foundational and remarkably influential case studies on the nature of what he referred to as peasant communities.

Ricard, Mathieu. *Happiness: A Guide to Developing Life's Most Important Skill*. Boston: Little, Brown and Company, 2007. Drawing on his experience as both a molecular biologist and a Buddhist monk, Ricard (who has been described by scientists as the happiest man alive) reveals what it takes to achieve a lasting happiness.

Richards, Audrey. *Land, Labour, and Diet in Northern Rhodesia: An Economic Study of the Bemba Tribe*. London: Oxford University Press, 1939. Based on extensive participant observation fieldwork among agriculturalists in East Africa, Richards set a new anthropological standard by incorporating diet and other factors into her carefully documented study of the social and economic life of the Bemba.

Rodney, Walter. *How Europe Underdeveloped Africa*. Washington DC: Howard University Press, 1982. Used by social scientists and scholars of development studies, Rodney's classic text provides a view of international development from the perspective of the so-called third world.

Roughgarden, Joan. *Evolution's Rainbow: Diversity, Gender, and Sexuality in Nature and People*. 10th anniversary edition. Berkeley: University of California Press, 2013. Compiling a remarkable array of scientific studies on humans and the animal kingdom at large, Roughgarden's seminal and unparalleled study helps us better articulate the diversity of humankind with an exhaustive deconstruction and re-articulation of biological sex, sexuality, and gender.

Russell, Bernard. *Research Methods in Anthropology*. 4th ed. Lanham, Maryland: AltaMira, 2006. This definitive go-to guide for all anthropologists is a rich treasure trove that outlines the anthropological method.

Shanahan, Murray. *The Technological Singularity*. Cambridge, MA: MIT Press, 2015. Shannahan peers into the potential future of humankind, not to sensationally stir up fear, but to soberly consider a number of scenarios that could result from the exponential growth in artificial intelligence.

Shostak. Marjorie. *Nisa: The Life and Words of a !Kung Woman*. Cambridge, MA: Harvard University Press, 1981. Shostak lived amongst hunter-gatherers in the Kalahari and gained unparalleled insight into foraging lifestyles by documenting the surprisingly candid life history of a female hunter-gatherer named Nisa.

Steadman, Dawnie Wolf. *Hard Evidence: Case Studies in Forensic Anthropology*. 2nd ed. New York: Routledge, 2008. By exploring numerous cases, this collection of readings exposes the reader to the dynamics of forensics anthropology, casting some light on what it's like to work in this fascinating field.

Sterling, Eleanor, Nora Bynum, and Mary Blair. *Primate Ecology and Conservation: A Handbook of Techniques*. Oxford: Oxford University Press, 2013. From planning field research to observation, data collection, and lab analysis, this reference book surveys the amazing methodological range of primatologists concerned with primate ecology and conservation.

Steward, Julian. *Theory of Culture Change: The Methodology of Multilinear Evolution*. Repr. ed. Champaign: University of Illinois Press, 1990. Spanning 2 decades of work, this collection articulates how Steward unpacks and understands the intricacies of culture change.

Stocking, George. *Observers Observed: Essays on Ethnographic Fieldwork*. Madison: University of Wisconsin Press, 1983. Stocking's edited collection of essays portrays the inner lives of a number of anthropologists, including Franz Boas, and in so doing reveals the challenges and triumphs of ethnographic fieldwork.

Stocking, George W., Jr. *Victorian Anthropology*. New York: The Free Press, 1987. Stocking documents the troubling origins of anthropology in the Victorian Era, when armchair scholars typically relied on the extremely biased accounts of intrepid travelers and missionaries to understand the cultures of far-away people.

Stringer, Chris. *Lone Survivors: How We Came to Be the Only Humans on Earth*. New York: St. Martin's Griffin, 2013. Based on a stunning reevaluation of the fossil record, archaeological sites, and recent DNA research, paleoanthropologist Chris Stringer revises long-held theories on how *Homo sapiens* exclusively came to populate nearly every corner of the globe.

Tannen, Deborah. *That's Not What I Meant!: How Conversational Style Makes or Breaks Relationships*. New York: William Morrow, 2011. Linguist Deborah Tannen has some important lessons to help you maintain healthy relationships through communication, including ways to make sure the things you say to your loved ones are actually what they hear.

Tattersall, Ian. *Masters of the Planet: The Search for Our Human Origins.* New York: St. Martin's Griffin, 2013. Tattersal, curator emeritus of the American Museum of Natural History, peers into the extensive fossil record of humanity to reveal the truly remarkable and rapid rise of *Homo sapiens*.

Tylor, Edward B. *Primitive Culture.* Vols. 1 and 2. New York: Harper & Brothers, 1958 (originally published in 1873). In one of the original anthropological texts, Tylor spells out his enduring definition of culture, and tirelessly reviews a wide variety of sources on what he characterized as primitive cultures throughout the world.

Waters, Michael and Thomas Jennings. *The Hogeye Clovis Cache.* College Station: Texas A&M University Press, 2015. Waters and Jennings help us understand one of the original American techno-cultural trends, the Clovis point.

Wilkinson, Iain and Arthur Kleinmann. *A Passion for Society: How We Think about Human Suffering.* Berkeley: University of California Press, 2016. With a medical anthropology giant as one of its co-authors, this text makes the case for emboldening contemporary social science by returning to its roots as a discipline that is fundamentally committed to caring for others and critically fostering a more human world.

Image Credits

Page 249: © somethingway/iStock/Thinkstock.

Page 251: © FlairImages/iStock/Thinkstock.

Page 252: © filipefrazao/iStock/Thinkstock.

Page 254: © monkeybusinessimages/iStock/Thinkstock.

Page 266: © ValuaVitaly/iStock/Thinkstock.

Page 268: © fotohunter/iStock/Thinkstock.

Page 272: © oneinchpunch/iStock/Thinkstock.

Page 284: © anyaberkut/iStock/Thinkstock.

Page 287: © maked/iStock/Thinkstock.

Page 289: © gutaper/iStock/Thinkstock.

Page 291: © Shooter_Sinha_Images/iStock/Thinkstock.

Page 303: © Witthaya/iStock/Thinkstock.

Page 305: © tibu/iStock/Thinkstock.

Page 306: © denizbayram/iStock/Thinkstock.

Page 321: © StudioThreeDots/iStock/Thinkstock.

Page 323: © Photos.com/Thinkstock.

Page 325: © AbleStock.com/Thinkstock.

Page 327: © Stocktrek Images/Thinkstock.

Page 329: © Stocktrek Images/Thinkstock.

Page 341: © stevanovicigor/iStock/Thinkstock.

Page 343: © Stockbyte/Thinkstock.

Page 345: © AlexRaths/iStock/Thinkstock.

Page 347: © yankane/iStock/Thinkstock.

Page 359: © Kkolosov/iStock/Thinkstock.

Page 361: © auimeesri/iStock/Thinkstock.